Complexity, Chaos,
and Biological Evolution

NATO ASI Series

Advanced Science Institutes Series

A series presenting the results of activities sponsored by the NATO Science Committee, which aims at the dissemination of advanced scientific and technological knowledge, with a view to strengthening links between scientific communities.

The series is published by an international board of publishers in conjunction with the NATO Scientific Affairs Division

A	**Life Sciences**	Plenum Publishing Corporation
B	**Physics**	New York and London
C	**Mathematical and Physical Sciences**	Kluwer Academic Publishers
D	**Behavioral and Social Sciences**	Dordrecht, Boston, and London
E	**Applied Sciences**	
F	**Computer and Systems Sciences**	Springer-Verlag
G	**Ecological Sciences**	Berlin, Heidelberg, New York, London,
H	**Cell Biology**	Paris, Tokyo, Hong Kong, and Barcelona
I	**Global Environmental Change**	

Recent Volumes in this Series

Series B: Physics

Complexity, Chaos, and Biological Evolution

Edited by

Erik Mosekilde

Technical University of Denmark
Lyngby, Denmark

and

Lis Mosekilde

University of Aarhus
Aarhus, Denmark

Plenum Press
New York and London
Published in cooperation with NATO Scientific Affairs Division

Proceedings of a NATO Advanced Research Workshop
on Complex Dynamics and Biological Evolution,
held August 6–10, 1990,
in Hindsgavl, Denmark

Library of Congress Cataloging-in-Publication Data

NATO Advanced Research Workshop on Complex Dynamics and Biological
   Evolution (1990 : Hindsgavl, Middelfart, Denmark)
    Complexity, chaos, and biological evolution : [proceedings of a
   NATO Advanced Research Workshop on Complex Dynamics and Biological
   Evolution, held August 6-10, 1990, in Hindsgavl, Denmark] / edited
   by Erik Mosekilde and Lis Mosekilde.
        p.   cm. -- (NATO ASI series. Series B, Physics ; v. 270)
    "Published in cooperation with NATO Scientific Affairs Division."
    Includes bibliographical references and index.
    ISBN 0-306-44026-1
    1. Biophysics--Congresses. 2. Evolution (Biology)--Congresses.
   3. Biological systems--Congresses. 4. Chaotic behavior in systems-
   -Congresses. 5. Morphogenesis--Congresses. 6. Molecular biology-
   -Congresses. 7. Dynamics--Congresses.   I. Mosekilde, Erik.
   II. Mosekilde, Lis. III. Title. IV. Series.
   QH505.N333 1991
   574'.01'1--dc20   91-26491
                                                            CIP

ISBN 0-306-44026-1

© 1991 Plenum Press, New York
A Division of Plenum Publishing Corporation
233 Spring Street, New York, N.Y. 10013

Printed in the United States of America

SPECIAL PROGRAM ON CHAOS, ORDER, AND PATTERNS

This book contains the proceedings of a NATO Advanced Research Workshop held within the program of activities of the NATO Special Program on Chaos, Order, and Patterns.

Volume 208—MEASURES OF COMPLEXITY AND CHAOS
   edited by Neal B. Abraham, Alfonso M. Albano,
   Anthony Passamante, and Paul E. Rapp

Volume 225—NONLINEAR EVOLUTION OF SPATIO-TEMPORAL STRUCTURES
   IN DISSIPATIVE CONTINUOUS SYSTEMS
   edited by F. H. Busse and L. Kramer

Volume 235—DISORDER AND FRACTURE
   edited by J. C. Charmet, S. Roux, and E. Guyon

Volume 236—MICROSCOPIC SIMULATIONS OF COMPLEX FLOWS
   edited by Michel Mareschal

Volume 240—GLOBAL CLIMATE AND ECOSYSTEM CHANGE
   edited by Gordon J. MacDonald and Luigi Sertorio

Volume 243—DAVYDOV'S SOLITON REVISITED: Self-Trapping of Vibrational Energy
   in Protein
   edited by Peter L. Christiansen and Alwyn C. Scott

Volume 244—NONLINEAR WAVE PROCESSES IN EXCITABLE MEDIA
   edited by Arun V. Holden, Mario Markus, and Hans G. Othmer

Volume 245—DIFFERENTIAL GEOMETRIC METHODS IN THEORETICAL PHYSICS:
   Physics and Geometry
   edited by Ling-Lie Chau and Werner Nahm

Volume 256—INFORMATION DYNAMICS
   edited by Harald Atmanspacher and Herbert Scheingraber

Volume 260—SELF-ORGANIZATION, EMERGING PROPERTIES, AND LEARNING
   edited by Agnessa Babloyantz

Volume 263—BIOLOGICALLY INSPIRED PHYSICS
   edited by L. Peliti

Volume 264—MICROSCOPIC ASPECTS OF NONLINEARITY IN CONDENSED MATTER
   edited by A. R. Bishop, V. L. Pokrovsky, and V. Tognetti

Volume 268—THE GLOBAL GEOMETRY OF TURBULENCE: Impact of Nonlinear Dynamics
   edited by Javier Jiménez

Volume 270—COMPLEXITY, CHAOS, AND BIOLOGICAL EVOLUTION
   edited by Erik Mosekilde and Lis Mosekilde

PREFACE

From time to time, perhaps a few times each century, a revolution occurs that questions some of our basic beliefs and sweeps across otherwise well guarded disciplinary boundaries. These are the periods when science is fun, when new paradigms have to be formulated, and when young scientists can do serious work without first having to acquire all the knowledge of their teachers.

The emergence of nonlinear science appears to be one such revolution. In a surprising manner, this new science has disclosed a number of misconceptions in our traditional understanding of determinism. In particular, it has been shown that the notion of predictability, according to which the trajectory of a system can be precisely determined if one knows the equations of motion and the initial conditions, is related to textbook examples of simple, integrable systems. This predictability does not extend to nonlinear, conservative systems in general. Dissipative systems can also show unpredictability, provided that the motion is sustained by externally supplied energy and/or resources. These discoveries, and the associated discovery that even relatively simple nonlinear systems can show extremely complex behavior, have brought about an unprecedented feeling of common interest among scientists from many different disciplines.

During the last decade or two we have come to understand that there are universal routes to chaos, we have learned about stretching and folding, and we have discovered the beautiful fractal geometry underlying chaotic attractors. Hand in hand with this development we have seen a rapidly growing interest in the application of concepts from far-from-equilibrium thermodynamics. In analogy with the behavioral complexity that can arise in nonlinear dynamic systems, we observe the spontaneous unfolding of structure in spatially extended systems as the throughflow of energy and resources lift them further and further above the state of thermal equilibrium.

In the years to come, much of this research will be directed towards understanding the types of complexity that follow after chaos: hyperchaos, higher-order hyperchaos, spatio-temporal chaos and, perhaps, fully developed turbulence. This will lead to the integration of nonlinear dynamics and irreversible thermodynamics into a theory of complex physical systems. However, it is equally important to try to apply the ideas of nonlinear dynamics and irreversible thermodynamics to living systems. Such systems clearly depend on a continuous supply of energy and resources to maintain their functions and, frankly speaking, the concepts of conventional physics have never been of much help to the biological sciences. In many contexts, these concepts are inappropriate, if not directly meaningless, and it is hard to imagine that a greater contrast can exist than that which is found between the simple ideas of classical mechanics and equilibrium thermodynamics and the spontaneous morphogenesis, differentiation, and evolution that we observe in the living world.

With the concepts and ideas of complex systems theory it now appears that the biological sciences have acquired a set of tools which will allow us to describe the behavior, function and evolution of living systems in much more detail. As a step in this direction, small as it may be, this book reproduces a collection of papers which were delivered at the NATO Advanced Research Workshop on Complex Dynamics and Biological Evolution which took place at Hindsgavl Conference Center near Middelfart, Denmark, August 6-10, 1990.

Attended by some 60 participants, this workshop succeeded as one of the relatively unusual events where clinically oriented medical doctors, experimental biologists, chemists, physicists and mathematicians are able to find a common language and to communicate freely across conventional barriers. Surely, confronted with the real life observations of biologists and doctors, some of the physicists may have been struck by the extreme oversimplification of their approach to the complexity of the living world. Some of the doctors, on the other hand, may have felt a little intimidated by their lack of mathematical background.

However, mediated in part by the pleasant surroundings of the conference site, all such feelings rapidly disappeared to make way for a common experience of a worthwhile and interesting endeavor. We would like to extend our thanks to all the participants and invited speakers for contributing to the success of the meeting.

The workshop offered lectures by leading experts in the fields of theoretical biology, morphogenesis, evolution, artificial life, hormonal regulation, bone remodeling, population dynamics, and chaos theory. In addition to the presentations given by the invited speakers, many of those attending the workshop also reported on their recent results within these and related areas. A list of participants is given at the end of the proceedings.

As editors of the proceedings we would like to thank The NATO Science Committee, The Danish Natural Science Foundation, and The Technical University of Denmark for sponsoring the meeting. We would also like to thank Janet Sturis and Ellen Buchhave for their assistance in preparing the proceedings.

Erik and Lis Mosekilde

Copenhagen, February 1991

CONTENTS

Section I
An Introductory Overview

With the development of nonlinear dynamics we have acquired a completely new set of tools for the description of the complex dynamic phenomena we observe in the living world.

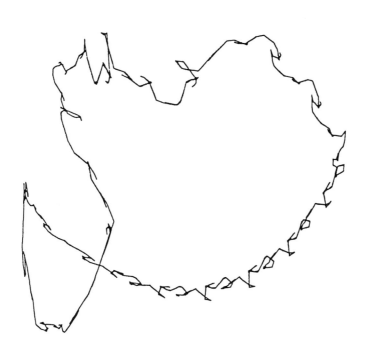

STRUCTURE, COMPLEXITY AND CHAOS IN LIVING SYSTEMS

Erik Mosekilde*, Ole Lund* and Lis Mosekilde¤

*System Dynamics Group, Physics Laboratory III
 Technical University of Denmark
 DK-2800 Lyngby, Denmark

¤Department of Connective Tissue
 Institute of Anatomy, University of Aarhus
 DK-8000 Aarhus C, Denmark

ABSTRACT

Although homeostasis of the internal environment is almost axiomatic in the physiological sciences, it is apparent that rhythms occur in the levels of a variety of bodily functions, ranging in time from a few cycles per millisecond for the firing of the central nervous system to the monthly period of the menstrual cycle. As an example of such a rhythm, this introductory overview describes how self-sustained oscillations arise in the pressure and flow regulation of individual nephrons as a result of a delayed action of the juxtaglomerular apparatus. The same problem is dealt with in considerably more detail in the contribution by Marsh et al., and one of the purposes of the present discussion is to speculate on the biological significance of such cycles.

A viral or bacterial infection is an example of an instability where a small initial population of foreign agents is allowed to multiply over several decades before the immune system or other reactions of the body finally establishes the defence required to cope with the infection. In a recent paper by Anderson and May, it was suggested that the response of the immune system to the simultaneous infection by HIV and another virus that activates the same T-cells can produce chaotic bursts of free HIV with intervals of the order of 20 weeks. We have performed a more detailed mathematical analysis of this model, including a construction of phase space trajectories as well as Poincaré

Complexity, Chaos, and Biological Evolution, Edited by E. Mosekilde and
L. Mosekilde, Plenum Press, New York, 1991

sections and return maps. At least one route to chaos in the model has been identified and found to proceed through a cascade of period-doubling bifurcations. Even though more recent observations seem to indicate that the Anderson and May model is incorrect in certain respects, it can still be expected that the immune system is capable of producing very complicated responses.

In contrast to the conventional picture of a relatively quiescent dynamics, it has always been acknowledged that the spatial structure of physiological systems is exceedingly complicated. Advances in experimental techniques over recent decades have made it possible to illuminate this complexity in even more detail, and the time has come when we must try to relate the complexity of the physiological structures with the dynamical processes involved in their formation and maintenance. As a final problem we thus discuss some of the difficulties involved in relating bone structure with bone remodeling processes. This problem will also be dealt with at greater length in a subsequent contribution by Lis Mosekilde.

INTRODUCTION

A variety of new experimental techniques in molecular biology, physiology and other fields of biological research constantly expand our knowledge and enable us to make increasingly more detailed functional and structural descriptions of living systems. Over the past decades, the amount and complexity of available information have grown manyfold, while at the same time our basic understanding of the nature of regulation, behavior, morphogenesis, and evolution in biological systems has made only modest progress. A key obstacle in this process is clearly the proper handling of the wealth of data. This requires a stronger emphasis on mathematical modeling through which the consistency of the adopted explanations can be checked and general principles may be extracted. As a much more serious problem, however, it appears that the proper concepts for the development of a theoretically oriented biology have not hitherto been available. Classical mechanics and equilibrium thermodynamics, for instance, are inappropriate and useless in some of the most essential biological contexts. Fortunately, there is now convincing evidence that the concepts and methods of the newly developed fields of nonlinear dynamics and complex systems theory will enable us to establish much more detailed descriptions of biological processes (Nicolis and Prigogine 1989).

Contrary to the conventional assumptions of homeostasis, many biological and biochemical control systems are unstable and operate in a pulsatory or oscillatory mode (Glass and Mackey 1988, Degn et al. 1987, Holden 1986). This is true, for instance, for the release of hormones such as growth hormone, luteinizing hormone, and insulin. The latter case is described in considerable detail in the contribution to this proceedings by Sturis et al. As discussed by Prank et al., the hormonal release process may also become more erratic, and the question arises whether the information associated with the temporal variation in hormone concentration has significance for the regulatory function. While this problem still remains at the speculative level, it is evident that disruption of certain rhythms can be associated with states of disease while, on the other hand, new types of oscillations may appear in connection with other diseases.

Many cells exhibit pulsatory variations in their membrane potential with extremely complicated patterns of slow and fast spikes. Heart cells, for instance, have been found to produce chaos and complicated forms of mode-locking when stimulated externally (Glass and Mackey 1988). Similarly, as discussed in the contribution by Colding-Jørgensen, the interaction between nerve cells can give rise to nonlinear dynamic phenomena with frequency locking and chaotic firing influencing the flow of information. Liebovitch and Czegledy, and Østergaard et al. present more detailed models of neural function while Babloyantz shows how one can characterize signals from the central nervous systems by means of fractal dimensions and other measures from nonlinear dynamics. In line with this research, Herzel et al. analyze examples of newborn infant cries and voiced sound to illustrate the occurrence of bifurcations and chaos.

Rhythmic signals also seem essential in intercellular communication (Goldbeter 1989). Besides neurons and muscle cells which communicate by trains of electrical impulses, examples range from the generation of cyclic AMP pulses in the slime mold *Dictyostelium discoideum* to the pulsatile release of hormones. While in these instances the oscillatory dynamics characterize the extracellular signal, recent observations indicate that signal transduction itself may be based on oscillations of intracellular messengers. In his contribution to this volume, Goldbeter assesses the efficiency of pulsatile signaling. It appears that periodic signals are more effective than constant, stochastic or chaotic stimuli in eliciting a sustained physiological response.

By virtue of the positive feedback associated with replication, and because of maturation and other delays, many problems in ecology, microbiology and population dynamics lead to complex dynamic phenomena involving different types of competition between erratic bursts and deterministic oscillations. Olsen et al. give a detailed analysis of this type of behavior in their contribution on epidemics of children's diseases. As shown in an example below, similar phenomena can arise in the response of the immune system of an AIDS patient to an opportunistic infection.

With recognition of the fractal geometry underlying chaotic attractors, and with the establishment of universal scaling laws for the transition to chaos, we can now describe behaviors which previously appeared to be hopelessly complex and which, for that reason, were usually neglected or ascribed to random exogenous processes (Devaney 1986, Holden 1986, Christiansen and Holden 1989). Similarly, the understanding of spontaneous structure formation in open thermodynamic systems under far-from-equilibrium conditions has provided us with a new paradigm for the description of fundamental biological processes such as morphogenesis, evolution and differentiation (Nicolis and Prigogine 1977, Haken 1978).

As demonstrated in the contribution by Meinhardt, biological pattern formation can be described in terms of instabilities and nonlinear dynamic phenomena in biochemical reaction-diffusion equations. A similar approach is adopted by Hunding in his study of early *Drosophila* embryogenesis. This model rests on the idea of Turing systems of the second kind in which a prepattern generates position dependent rate constants for a subsequent reaction-diffusion system. Maternal genes are assumed to be responsible for setting up gradients from the anterior and posterior ends as needed to stabilize the double period prepattern suggested as underlying the read out of the gap genes. The resulting double period pattern again stabilizes the following prepattern in a hierarchy of increasing structural complexity. Without such hierarchical stabilization, reaction-diffusion mechanisms yield highly patchy short wavelength patterns.

By combining reaction-diffusion equations with equations for the mechanical behavior of the cell membrane, Goodwin illustrates how similar ideas can be used to address the problem of biological morphogenesis for the single-celled alga *Acetabularia*. Because of its basic simplicity, this organism lends itself to experimental and theoretical studies of form formation. A model of the morphogenetic field and a finite element simulation of its behavior are presented which show that spatial patterns generically similar to those observed in the alga arise naturally, suggesting that normal morphogenesis can be described as an attractor of a moving boundary process. This approach can form the basis for a whole new discipline of theoretical study of developing organisms with the goal of identifying the generic properties which result in robust but highly modifiable structures.

At present this is mostly a vision, and there are enormous difficulties to overcome. One major difficulty is associated with experimental problems of identifying the morphogenes which appear in the various reaction-diffusion equations, not to mention the problems of measuring the nonlinear rate constants involved in their interactions. Another difficulty is associated with the strong preoccupation of many biologists with genetic control mechanisms. It is clear, however, that the genetic approach can never stand alone, as this approach does not help us understand how a gene codes for a specific form.

Parallel with the above developments we see attempts to describe evolutionary processes from a more formal point of view, including attempts to investigate life forms in alternative chemistries (Ebeling and Feistel 1982, Langton 1988, Stein 1989). This line of research is represented by the contributions on evolutionary theory and artificial life by Ebeling, by Blomberg and by Knudsen et al. In the latter field, one is primarily concerned with finding ways of formulating evolutionary dynamics as an open-ended problem, i.e., as a problem where the various roads that evolution can take have not already been laid down by the modeler. This has brought about a study of life forms in other media with the aim of elucidating the very definition of life. Emmeche, on the other hand, in his contribution on formalization of biological systems, challenges this approach with the claim that formal evolutionary models show general aspects and higher order behavior of living systems for which there is no existing experimental background.

Within the scope of this short introduction it has not been possible to give credit to all contributors to the present volume. There are, for instance, important contributions by Müller on vortex formation in excitable media, by Bohl on structural amplification in chemical networks, by Das et al. on boundary operator and distance measure for cell lineage of *Caenorhabditis elegans* and for the pattern in *Fusarium solani*, by Lloyd on timekeeping for intracellular dynamics, and by Rössler et al. on an optimality approach to ageing. In addition, Andresen and Rauch-Wojciechowski have contributed a couple of more mathematically oriented papers. Finally, we have included two contributions on higher-order chaos, even though these are not directly related to concrete biological problems. The idea has been to convey to the reader an impression of the richness of new ideas which arise from the interaction between the biological sciences and complex systems theory. We would also like to convey a little of the enthusiasm with which these novel developments are met. In the rest of this contribution we shall illustrate a few of the ideas in more detail by means of three examples associated with widely different biological problems.

BIFURCATIONS AND CHAOS IN KIDNEY PRESSURE REGULATION

To the surprise of most physiologists, experiments performed some 5 years ago by Leyssac and Holstein-Rathlou at the Institute of Experimental Medicine, University of Copenhagen revealed an oscillatory variation in the proximal tubular pressure in rat kidneys (Leyssac and Baumbach 1983, Leyssac and Holstein-Rathlou 1986). With typical periods of 0.3 - 0.5 min, these oscillations were far too slow to be caused by breathing or heart beat, and detailed analyses suggested that they arose from an instability in the regulation of glomerular filtration. While for normal rats the oscillations had the appearance of a regular cycle highly irregular oscillations were observed for spontaneously hypertensive rats (Holstein-Rathlou and Leyssac 1986).

Figure 1a shows a typical example of the experimental results obtained for normal rats. The intratubular pressure as measured with a thin micropipette is found to

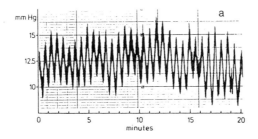

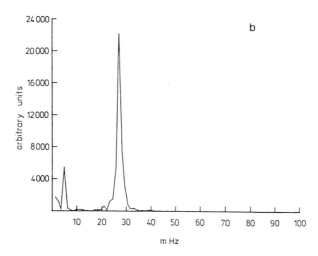

Figure 1. Experimental results for the variation in intratubular pressure for normal rats (a). Frequency distribution of the experimental results (b). The spectrum is dominated by a single sharp peak at 28 mHz.

oscillate with an amplitude of about 5 mmHg around a mean value of approximately 10 mmHg. Respiration and heart beat are reflected as superimposed, small amplitude ripples with much higher frequencies (1 Hz and 4 Hz, respectively). Figure 1b shows the corresponding spectral distribution. This spectrum is dominated by a single sharp peak at 28 mHz. The sharpness of this peak is a measure of the coherence of the oscillation which, considering the noisy environment, is quite astonishing and typical for a self-sustained oscillation.

Figure 2 shows similar results obtained for genetically hypertensive rats. The intratubular pressure here oscillates in a highly irregular fashion, and the frequency analysis reveals a broad-banded noisy spectrum with strong components in the low-frequency end. This is characteristic for a chaotic signal. Additional experimental results for a variety of different conditions can be found in the contribution to this volume by Marsh et al.

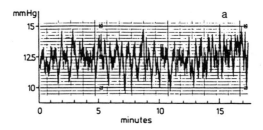

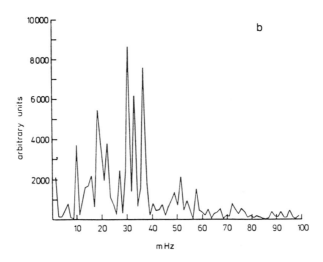

Figure 2. Experimental results for the variation in intratubular pressure for hypertensiv rats (a). Spectral distribution of experimental results (b). The intratubular pressure her oscillates in a highly irregular fashion, and the frequency analysis reveals a broad-bande noisy spectrum with strong components in the low-frequency end. This is characterist for a chaotic signal.

In collaboration with Holstein-Rathlou and Leyssac, Klaus Skovbo Jensen and Erik Mosekilde developed a dynamical model which could account for the observed behavior (Jensen et al. 1986, Holstein-Rathlou and Leyssac 1986). By means of this model we have shown how the nephron pressure and flow regulation becomes unstable because of a delayed action in the juxtaglomerular apparatus. In this apparatus, a signal from the distal tubule, presumably representing the sodium concentration of preurine leaving Henle's loop, is processed through a series of reactions to finally influence the nephron blood supply. With realistic parameter values, the model exhibits a limit cycle behavior. Variation of those parameters which are likely to differ between normal and hypertensive rats causes the model to develop chaotic behavior through a Feigenbaum cascade of period-doubling bifurcations.

Figure 3 shows a schematic drawing of a nephron, the functional unit of the kidney. There are approximately 1 mill. nephrons in a human kidney and 30,000 nephrons in a rat kidney. In the upper end, the nephron contains 20-40 capillary loops arranged in parallel inside a double layered capsule. This configuration, the glomerulus, is supplied with blood through the afferent arteriole, a short vessel which is capable of

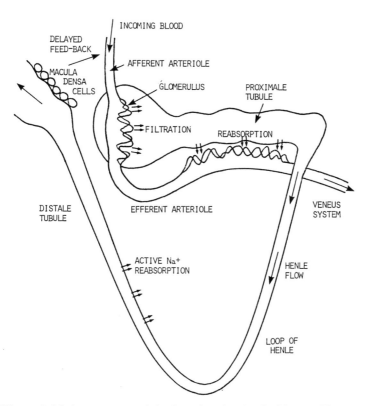

Figure 3. Main structure of the functional unit of a kidney: The nephron.

changing its diameter and thereby the resistance to the blood flow. The glomerulus and the inner layer of the capsule act as a filter through which blood constituents with molecular weight less than 68,000 can pass into the proximal tubule. In rats, typically 25-35% of the plasma is filtered out as the blood passes the capillary system. Blood cells and proteins are retained, and the blood viscosity therefore rises while at the same time the colloid osmotic pressure increases. The blood leaves the glomerulus through the efferent arteriole to pass through a new capillary system which embraces the tubule. Through selective absorption and secretion, the constituents of the filtrate are changed as it passes through the various segments of the tubule: The proximal tubule, Henle's loop, the distal tubule, and the collecting duct. In practice, more than 95% of the water and salts are reabsorbed while metabolic byproducts and foreign chemicals remain in the tubular fluid to be excreted as urine.

It has long been recognized that the kidneys are capable of compensating for changes in arterial blood pressure, and that this ability partly rests with feedback mechanisms in the individual nephron. By adjusting the flow resistance in the afferent arteriole, the pressure in the proximal tubule can thus be maintained at a relatively constant level. At the same time, the function of the kidneys plays an essential role for the regulation of the blood pressure, and it was precisely the wish to understand this interaction which motivated the detailed experimental studies performed at the University of Copenhagen.

The physiological mechanisms responsible for adjusting the flow resistance of the afferent arteriole are not known in detail. This problem will also be discussed in the contribution by Marsh et al. A widely accepted hypothesis is that the arteriolar wall is stimulated by a signal arising from specialized cells situated where the terminal part of the distal tubule passes the afferent arteriole. Some of these cells, the macula densa cells, are assumed to be sensitive to Na^+ and/or Cl⁻ ions. Due to the delicate balance between absorption and secretion processes in the distal tubule, the Na^+ and Cl⁻ concentrations in the exiting fluid depend upon the flow rate, and the activity of the macula densa cells therefore effectively represents the Henle flow.

If the flow rate is too high, the macula densa cells signal to the afferent arteriole to contract, and thereby reduce the incoming blood flow. However, due to a finite transition time through Henle's loop, the Na^+ concentration at the macula densa cells does not respond immediately to a change in the incoming blood flow. An additional delay is introduced in transmission of the signal from the macula densa cells to the arteriolar wall, presumably because this transmission involves a cascade of enzymatic processes. As a result of these delays, the negative feedback loop becomes unstable and self-sustained oscillations are produced.

The above model integrates already existing but hitherto separate physiological theories of glomerular filtration, proximal tubular dynamics, and tubuloglomerular feedback into a coherent description. With parameters which in most cases can be obtained from the literature or from independent measurements, the model quite accurately reproduces the characteristic limit cycle oscillations observed for normal rats, both with respect to the amplitude and the period of the oscillations.

Figure 4 reproduces a set of simulation results. In the base case, the transit time through Henle's loop and the signal processing delay in the juxtaglomerular apparatus are represented by a third order lag with a total delay time of 4.5 sec. This presumably corresponds to normal physiological conditions. Indirect evidence suggests that

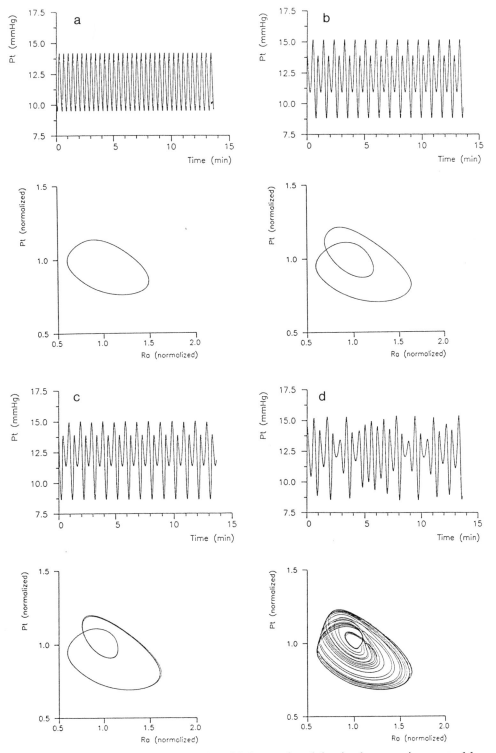

Figure 4. Series of simulation results with increasing delay in the negative control loop. The model passes through a cascade of period-doubling bifurcations.

spontaneously hypertensive rats have a reduced transport rate of Na^+ ions across the macula densa cells. We have therefore investigated changes in the model behavior as the delay time is increased from 4.5 sec to 9.0 sec. Figures 4a-c display some of the observed results. In each figure we have shown the proximal tubular pressure P_t as a function of time (top) together with a projection of the stationary trajectory into the (P_t, R_a)-plane (bottom). R_a here denotes the flow resistance in the afferent arteriole. It can be observed how the model develops through period-doubling bifurcations to reach a chaotic state before the delay time equals 9.0 sec. It is worth noticing that the model will pass through a similar cascade of period-doubling bifurcations if other parameters such as, for instance, the arterial pressure or the slope of the regulatory characteristic for the arteriolar resistance are changed. It is also interesting to note that period-2 behavior can be induced in normal rats by administering small doses of Furosemid, a drug that is known to influence kidney function. Figure 5 shows a somewhat different situation where period-2 behavior has been provoked by an accidental disturbance of the nephron. It can be observed how the nephron relaxes back towards its normal period-1 limit cycle during a period of the order of 5 min.

As discussed in the introduction, instabilities are increasingly being recognized as playing a significant role in the regulation of normal physiological systems. From a thermodynamic point of view, these instabilities are caused by the fact that physiological systems are open systems which are maintained in far-from-equilibrium conditions through dissipation of energy. In a nephron, for instance, the cells in the tubular wall depend on a steady supply of energy to perform their reabsorption of water and salts from the filtrate. In addition, the heart continuously pumps blood through the afferent-efferent arteriolar system and thereby maintains a hydrostatic pressure difference between capillary blood and intratubular fluid which permits filtration and passive reabsorption to take place.

Since biological rhythms have evolved under selective pressure, one can speculate about the advantages to the physiological system of an oscillatory behavior. In certain cases, the same structure can perform different functions during various phases of a cycle. This may be associated with a need to synchronize the various physiological processes by means of a biological clock. In other cases, efficiency may be improved through periodic shifts between an active and a dormant phase, or the oscillatory excursions may serve to guard the system against drift.

We cannot point to a particular advantage of the observed oscillatory behavior in the nephron pressure and flow regulation. Considering that the oscillations disappear if the arterial pressure is reduced, corresponding to a reduction of the average blood flow through the individual nephron, it is possible that the filtration capacity is increased by driving the nephron into an oscillatory state. Alternatively, the nephron could be stabilized by reducing the delay in the signal transmission from the macula densa cells to the arteriolar wall. Thus, one may speculate that "nature" has accepted the oscillatory regulation because it implies a higher capacity utilization and imposes lower demands on the speed of the feedback regulation than would stable operation.

As described, for instance, in the contributions by Prank et al., by Goldbeter, and by Sturis et al., oscillatory or pulsatory release is characteristic for many hormones. With this behavior, the supply can be regulated, at least partly, through variation of the inter-pulse period. This is similar to the on-off variation utilized in many technical control systems such as air-conditioners, refrigerators, freezers, oil burners, etc. The reasons for choosing this type of regulation are that it is extremely simple, and that it is often more stable towards long-term drift than is continuous-time regulation. Precisely the same reasons could apply in the biological realm.

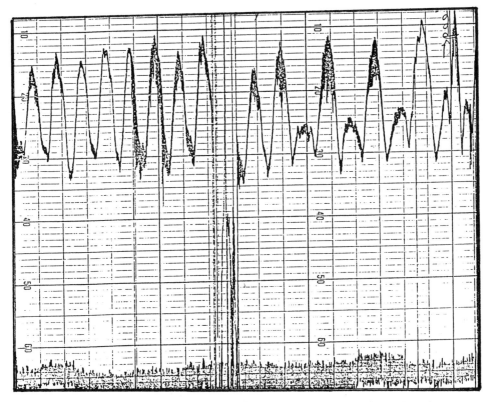

Figure 5. Proximal intratubular pressure for a normal rat showing relaxation from a period-2 to a period-1 cycle after a disturbance.

However, if two such systems interact with one another they will tend to lock their periods into a rational ratio (Jensen, Bak and Bohr 1984), leading to a situation, perhaps, where the various hormones are released in synchronism. Hereby, the regulatory ability of the individual system might be significantly reduced with growth hormones being released together with insulin in response to high blood glucose concentrations, etc. One way to avoid this situation and restore the regulatory ability of the individual hormonal control would be for the systems to operate in a chaotic mode without a well-defined periodicity. Thus chaotic behavior can be advantageous in control systems and, in fact, this type of dynamics seem to be quite characteristic of living systems.

THE INTERMEDIARY ORGANIZATION OF THE SKELETON - A DESCRIPTION OF FUNCTIONAL UNITS AND THEIR WORK PATTERN

A hundred years ago, the cell was recognized by most physiologists as the basis of health and disease in the kidneys, bones and other physiological structures. What they didn't perceive was the fundamental role played by the intermediary organization of the cells (Frost 1983). Micropuncture studies of intact, living nephrons ended this renal "atomism" by revealing the control roles of nephrons in disease and in pharmacologic, biochemical and endocrine responses. It required this special technique to understand nephron functions because studies of intact kidneys and of cells in vitro had not been able to expose them.

Similarly, dynamic bone histomorphometry as introduced by Frost (1969) provided a new, powerful methodology for studying the intermediary organization of the skeleton. This technique revealed a variety of new phenomena that had remained hidden in studies of isolated bone cells or intact subjects. Dynamic bone histomorphometry involves the use of a tissue time marker (Tetracycline) which is deposited at places where new bone is being actively formed. This marker has made it possible to identify sites with bone formation and also - when dosed more than once - to obtain information on the rate of bone formation (Melsen et al. 1978).

The different functions associated with the intermediary organization of the skeleton were thereby revealed: (i) growth which determines size, (ii) modeling which determines shape, and (iii) remodeling which maintains subsequent functional competence.

In the adult skeleton of man (and other large mammals) bone remodeling is the basic process in relation to the intermediary organization. The bone remodeling units (BMU) thereby play the same functional role for the skeleton as the nephrons for the kidney. Bone remodeling is the summation of the contributions of thousands of individual units - just as muscle contraction is a summation of the twitches of separate motor units, and urine production is the summation of the function of the separate nephrons (Frost 1983).

During the last 20 years, dynamic bone histomorphometry has provided researchers with a vast amount of information on bone remodeling in disease and during therapeutic regimes - but even now little is known about the natural regulation of bone remodeling: the role of hormones, of applied stresses and strains, and of the interaction between hormones and biomechanical forces (Bailey and McCulloch 1990).

Bone histomorphometry has the disadvantage of providing information on the remodeling process in only 2 dimensions. By introducing scanning electron microscopy (at high magnification) it has recently become possible to describe not only the architecture of loadbearing and non-loadbearing bone, but also the bone remodeling units and their work pattern in 3 dimensions (Mosekilde, Lis 1990). Figure 6 shows an example of such a scanning electron microscopy photograph.

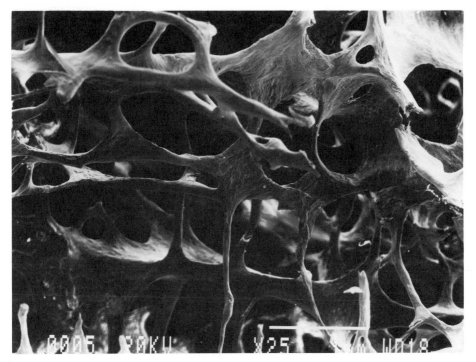

Figure 6. Scanning electron microscopy photograph taken inside the vertebral body of a 90 year old man. The picture shows the rod-strut type of trabecular architecture typical for loadbearing bone. Big remodeling sites can be identified on several of the surfaces. Also, several perforations are visible (autopsy specimen).

The architecture of loadbearing bone, such as the central part of the vertebral bodies, typically has thick vertical plates or columns supported by thinner horizontal struts (trabeculae). This special arrangement provides maximum strength in the vertical direction, which is the normal loading direction. The biomechanical competence of such a structure can be assessed in a materials testing machine, and by obtaining bone specimens with different orientation a "map" of characteristic biomechanical properties can be drawn. In younger individuals this structure is normally slightly anisotropic, with strength-values in the vertical direction twice as high as in the horizontal direction. The same is true for other biomechanical parameters: stiffness and energy absorption capacity (Fig. 7) (Mosekilde, Lis et al. 1987).

With age, there is a loss of bone mass because the balance during the remodeling process is negative. For unknown reasons the remodeling process seems to affect the horizontal supporting struts most, causing a thinning of these and, later, fortuitous perforations. Such age-related changes in bone architecture can be followed by the use of 400 μm thick sections of plastic embedded specimens investigated in polarized light with a λ-filter (Mosekilde, Lis 1988).

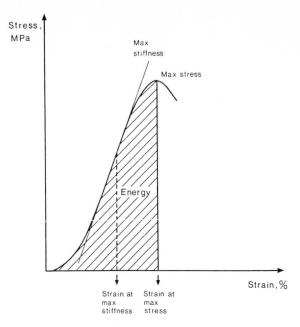

Figure 7. Typical stress-strain curve for vertebral trabecular bone compressed in the vertical direction. Maximal stress is defined as the top point of the curve, maximal stiffness as the steepest slope of the curve, and energy absorption capacity as the integral under the curve below maximal stress.

The biomechanical consequences of these age-related changes in bone mass and structure are massive: there is primarily an increase in anisotropy followed by a pronounced decline in bone strength, stiffness, and energy absorption capacity in both the horizontal and the vertical directions (Mosekilde, Lis et al. 1987).

A model of this vertebral trabecular bone architecture and its biomechanical properties has recently been developed by Jensen et al. (1990). The architecture described by this model is based on bone samples taken from the central part of vertebral bodies from normal individuals aged 30 to 90 years. The purpose of the study was to put into perspective the experimental data on trabecular architecture, bone mass, and mechanical behavior, and to test current hypotheses on the relationship between bone mass and trabecular bone biomechanics.

For the analyses, an idealized, semi-analytic mathematical model was used. The trabecular architecture was modeled as a three-dimensional lattice with a unit cell geometry. In order to make the model realistic without destroying its generality, a measure of relative lattice disorder α was defined. In all calculations, it was assumed that the solid bone material was linear elastic. Also, the overall dimensions of the lattice in the model were the same as in the tested bone specimens.

The model, expressed in terms of a set of lattice equations (Bernoulli - Euler beam theory), was analyzed on a computer using a structural design program that enabled calculations to be made of displacement, rotation, and reaction of any joint in the lattice.

The conclusion from the study was that the model accounted reasonably well for the age-related changes in vertical and horizontal stiffness as seen in experimental data. By the introduced measure for the randomness of lattice joint positions in the modeled trabecular network, it could be demonstrated that the apparent stiffness varies by a factor of 5-10 from a perfect cubic lattice to a network of maximal irregularity, even though bone mass remains almost constant. Also, when the bone material was slightly redistributed between vertical and horizontal trabeculae, a considerable change in mechanical behavior was seen.

Combined with data derived from dynamic bone histomorphometry and with observations from scanning electron microscopy concerning location and size of the remodeling sites in the three-dimensional lattice, the described model provides important information concerning the influence of stress and strain on bone dynamics.

But neither dynamic bone histomorphometry nor scanning electron microscopy has enabled description or identification of the feedback mechanisms between mechanical usage of bone and its architecture or the biological mechanisms necessary for maintaining this architecture. In particular, the transducer mechanism is not known. One theory is that the basic mechanical transducer is interstitial fluid flow - another theory that it is piezoelectric fields. But in both cases the signal is caused by the tissue strains which, in turn, are caused by the loads on the tissue.

As the transducer mechanism is not known, it has been difficult to define the stresses or strains (magnitude, frequency, and period) necessary for the maintenance of bone biomechanical competence. This is not only a theoretical problem - it has become increasingly relevant in connection with prevention of the most common bone disease in the western world - osteoporosis.

During the last 20-25 years, the incidence of osteoporotic fractures has increased dramatically, and this increase is much higher than the increase in the population of elderly people. Also, the rate of increase is now more pronounced for men than for women. These new data suggest that factors other than hormonal status or ageing as such are important for the general decline in bone strength over the last generation (Obrant et al. 1989). The new data lead investigations to focus directly on mechanical usage of bones, i. e., physical activity.

To proceed with these crucially important investigations, an interaction between biology, physics, computer science, and medicine is essential - as are a comprehensive approach to the problem and a common terminology.

COMPLEX DYNAMICS IN THE RESPONSE OF AN HIV-INFECTED IMMUNE SYSTEM

Although the initial rapid spread of AIDS is slowing down as the disease propagates from the high risk groups of male homosexuals and intra-venous drug users to groups of less pronounced risk behavior (Colgate et al. 1989), the human immunodeficiency virus (HIV), continues to represent a severe threat to mankind. As far as we know today, infection with this virus is nearly 100% fatal, even though the average life expectancy of a person from infection to death may be as long as 8-10 years. On the other hand, present data doesn't seem sufficient to assert the degree to which the epidemic will penetrate the various risk groups (May and Anderson 1988). A major determinant will be the changes that the epidemic induces in our sexual behavior. Various forms of vaccine are under development and can be expected to be available for large scale campaigns within this decade.

To delineate the prospects of this critical race between the spread of AIDS and the advance of biotechnologies required to fight the virus one must, of course, understand the dynamics involved in the propagation of the epidemic through various populations, in the interaction between HIV and the immune system, and in the modification of the virus produced by its relatively high rate of mutation. All of these processes involve nonlinear dynamic and evolutionary phenomena.

The human immunodeficiency virus belongs to a particular family of RNA viruses known as retroviruses. These viruses follow an unusual reproductive pathway in which the genomic RNA serves as template for synthesis of DNA. This process, which is catalyzed by an enzyme denoted transverse transcriptase, is an exception to the central dogma of biology and molecular dynamics according to which genetic information flows from DNA to RNA. The synthesized DNA is thereafter inserted into the genetic material of the host cell where it may remain silent for long periods of time. Activation may occur in connection with a secondary infection. In addition, HIV has developed a number of mechanisms through which it can control the processing of the inserted DNA, and at a certain stage this DNA may give rise to the production of a large number of new viral genomes.

The first step in the viral reproduction process is thus the reverse transcription of DNA from RNA. This process is not subject to an effective proofreading, and as a result the produced DNA often contains errors. This provides the retroviruses with a relatively high rate of mutation, and one can already distinguish between more than 20 different HIV strains. In the individual patient one typically finds of the order of 5 different variants 2-3 years after onset of the infection, rising to 10-15 variants near the end of the disease (J.E.S. Hansen, Hvidovre Hospital, private communication). Apparently this proliferation is accompanied by a shift in the character of the virus from "slow-low" viruses, which reproduce slowly and have low rates of infection, to "rapid-high" viruses which infect and kill T4 cells in a couple of days. One may therefore speculate that more aggressive variants could develop as the epidemic spreads to a larger fraction of the population.

Within the retroviruses, HIV belongs to a subfamily known as lentiviruses because they slowly destroy the infected biological system. The strategy adopted by these viruses is to combine a low infectiveness with a slow rate of reproduction and thus an extended infective period. The main targets for HIV are the various cells of the immune system such as the T4 lymphocytes (also known as helper cells), the macrophages, the blood monocytes and the stem cells in the bone marrow. To a lesser extent other variants of the white blood cells such as the T8 lymphocytes (killer cells) and the B lymphocytes may also be attacked (Nara 1989). The ability of HIV to attack precisely those cells which are supposed to defend the body against invading pathogens is a unique property of these viruses. In addition, HIV has developed a variety of different strategies by which it can evade the immune defence.

To better understand the dynamics of the interaction between HIV and the immune system, Anderson and May (1989) have developed a mathematical model which includes the influence of an opportunistic infection on the production of free HIV. It has long been speculated that bursts of free HIV might occur at more or less regular intervals after the initial strong antigenemia (Weber et al. 1987). As we have just described, such bursts could be caused by opportunistic infections which activate the HIV-infected T4 cells. Alternatively, the bursts of free HIV could occur whenever a new HIV variant was produced by mutation. In the beginning, as long as the immune system had not yet recognized the new variant, this would cause a renewed viraemia.

The occurrence of such bursts of free virus could help to explain the significant variations in the infectiveness of the disease, i.e., in the probability that an infected individual transfers the virus to a partner in a particular type of risk behavior. As deduced from the rate at which the epidemic spreads compared with the reported activity of the infected persons, this infectiveness is, in general, very low. However, strong interpersonal variations seem to occur, and it is also likely that the infectiveness varies significantly over time for the individual person. On the other hand, it must be admitted that it is not certain whether the alleged bursts are indeed associated with free virus or whether they represent variations in free HIV antigenes.

Another problem with the Anderson and May model is that it neglects the role that other parts of the immune system, particularly the macrophages, play in HIV reproduction and neutralization. Recent observations suggest that there is a background production of HIV by the macrophages of about the same magnitude as the production by the T4 lymphocytes (J.E.S. Hansen, private communication). One reason why this contribution has been overlooked for a number of years is that, in contrast to the T4 cells, the macrophages are not destroyed in the virus replication process.

Figure 8 shows a flow diagram for the model developed by Anderson and May. In this diagram, box symbols represent populations of cells, viruses and other agents, valve symbols represent the rates at which these populations change through recruitment, reproduction, activation, infection, etc., and the dotted arrows represent the causal relations assumed to control these rates.

The model describes the change in state and number of the T4 lymphocytes as they interact with HIV and with a second infectious agent. The population of T4 cells is subdivided into populations of immature, non-activated cells P, of activated but uninfected cells X, and of infected cells Y. Immature lymphocytes are assumed to be produced at a constant rate Λ. They are removed either by "natural" death at a per capita rate μ or by activation through contact with HIV. The rate of activation is taken to be γPV with V denoting the population of free HIV. Besides by activation of immature cells in the thymus, the population of activated T4 cells also grows through proliferation in the activated state. In the absence of a second infectious agent, the rate constant for this process is r. It is assumed, however, that the presence of the opportunistic infection adds a term kXI to the rate of proliferation of activated T4 cells. Here, I denotes the population of infectious agents.

The population of activated lymphocytes is reduced at a rate βXV as a result of infection by HIV. It is also assumed that there is a nonlinear restraining term $-dX^2$ in the rate of growth of activated T4 cells, representing the effects of suppressor cell regulation of the activated lymphocyte population. The population of infected T4 cells increases as activated cells become infected and decreases as the infected cells lyse at the release of free virus. The per capita death rate of infected T4 cells is α, and the burst size, i.e., the

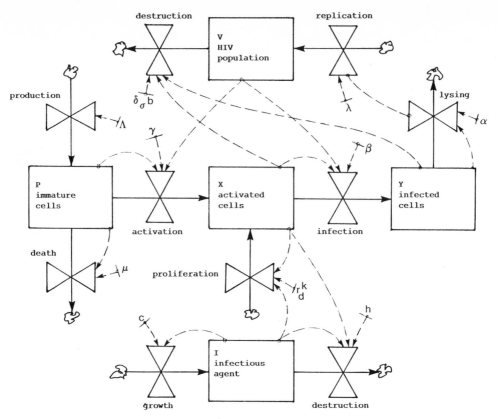

Figure 8. Flow diagram for the Anderson and May model of the interaction between a HIV infected immune system and a second infectious agent.

number of free HIV produced in average per T4 cell is denoted λ.

Following the changes in the population V of free HIV, the diagram shows how this population increases through replication in the infected T4 helper cells and decreases by normal mortality (loss of coat proteins) at a per capita rate b. In addition, the population of free HIV decreases by absorption by activated and infected T4 cells at the rate $\delta(X + Y)V$ and by cell mediated viral destruction at the rate σXV. Finally, the population of infectious agents I is assumed to grow through reproduction at a per capita rate c and to decrease through cell mediated destruction at a rate hXI, controlled by the population of activated T4 cells.

The flow diagram corresponds to a set of five coupled nonlinear differential equations, one for each type of population (state variable). Generally speaking, such a model can show a great variety of different behaviors, depending on the values of the various parameters. The dissipative, stabilizing terms are primarily connected with the exponential death processes for the various agents. The shorter the life times, the more stable the system tends to be. On the other hand, destabilising influences arise from the proliferation and replication processes. Except for the infection probability β, which we have used as a bifurcation parameter, we have adopted similar values for the various parameters as used by Anderson and May, i.e., $\Lambda = r = b = c = 1.0$, $\mu = \sigma = 0.1$, $\gamma = \delta = k = h = 0.01$, $d = 0.001$, $\alpha = 2$, and $\lambda = 5$. The time unit is one week. These parameters are such that in the absence of any HIV infection, the activated T4 cells always eliminate the opportunistic infection. In the presence of HIV, however, the second infectious agent introduces a destabilizing mechanism: With more infectious agents, the activated T4 cells proliferate faster and more T4 cells will be available for infection by HIV. Thus more T4 cells will be infected, and more free HIV will be produced by these cells as they lyse. This appears to represent a realistic phenomenon,

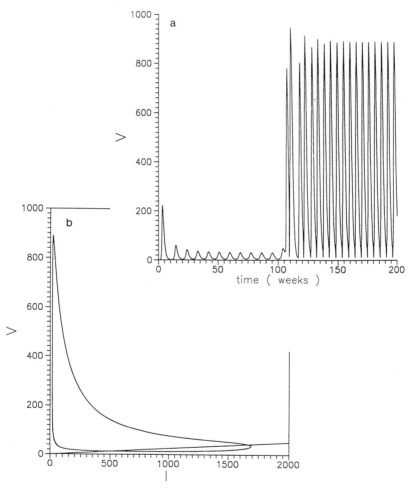

Figure 9. Temporal variation in the abundance of free HIV (a). A secondary infection occurs at time 100 weeks. Also shown is a projection of the stationary solution onto the (V,I) plane (b).

namely that a sound reaction of the immune system to the secondary infection worsens the situation with respect to the HIV infection.

Figure 9 shows the base case result obtained with the model. Here, we have taken $\beta = 0.1$. 9a shows the temporal variation in the abundance of free HIV. After an initial viraemia during which the immune system becomes activated a balance between the HIV infection and the immune system is established in which self-sustained oscillations with relatively small excursions in the various concentrations occur. At time t = 100 weeks, the secondary infection is introduced and as a result of the destabilizing influence of this infection the system switches into a new oscillatory state with much higher amplitudes. 9b shows a projection of the stationary phase space trajectory into the (V, I) plane. This illustrates the characteristic limit cycle behavior with alternating high abundances of free HIV and of the opportunistic agent I.

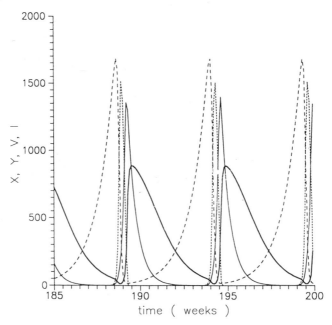

Figure 10. Illustration of the temporal variation in more detail. Peaks in abundances occur in the following order: secondary infectious agent, activated lymphocytes, infected lymphocytes, and free HIV.

To clarify the phase relationships between the various variables, figure 10 shows the temporal variation in more detail. Here, the leading dotted curve represents the abundance of secondary infectious agents, the second dotted curve represents the abundance of activated T4 cells, the thin line the infected T4 cells, and the heavy line the abundance of free HIV. It is possible to go through the model in detail and understand how this sequence of maxima occurs.

If the probability β of infecting a lymphocyte with HIV is increased, the model is forced further into the nonlinear region. In analogy with the results obtained for the kidney model (figure 4), this gives rise to a period-doubling bifurcation such that the system now oscillates with two alternating maxima. Figure 11a shows the obtained temporal variation. Note that it takes some time before the transient response to the secondary infection dies out. Figure 11b shows the corresponding phase plot. We clearly see how the system completes two full cycles before it repeats itself.

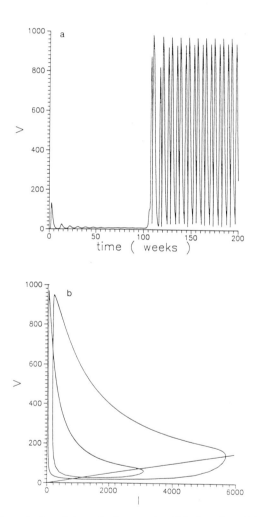

Figure 11. Temporal variation (a) and phase plot (b) for $\beta = 0.17$. The system now completes two full cycles before it repeats itself.

Also shown in figure 11b is a straight line through origo. This line represents a plane that we have used to obtain a Poincaré section, i.e., to determine the points of intersection with the phase space trajectory. For $\beta = 0.17$ the points of intersection in the upward direction correspond to $V \simeq 70$ and $V \simeq 140$, respectively. As β is increased, the system is further destabilized and passes through a Feigenbaum cascade of period-doubling bifurcations. This is illustrated in figure 12 where we have plotted the V-coordinates of the intersection points between the phase space trajectory and the Poincaré plane as a function of β. For $\beta \simeq 0.45$ four intersection points arise, from $\beta \simeq 0.69$ we have 8 intersection points, and for $\beta > 0.80$ we can find chaotic behavior. A phase plot of the chaotic attractor obtained for $\beta = 1.00$ is shown in figure 13. It is worth noticing that a similar bifurcation cascade can be expected if other parameters of the model are changed.

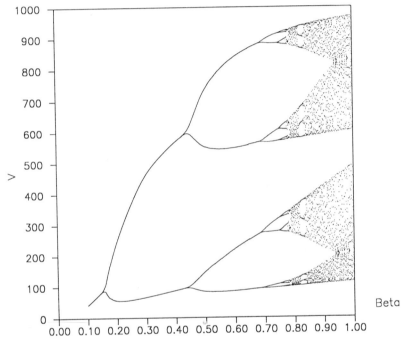

Figure 12. Bifurcation diagram for the response of the HIV infected immune system to a secondary infection. The bifurcation parameter is the probability of infecting a T lymphocyte with HIV.

One of the most useful techniques in nonlinear dynamics is the application of return maps. A return map can be constructed from the Poincaré section by plotting the value of the HIV abundance for one intersection point V_{n+1} as a function of the HIV abundance for the preceding intersection point V_n. For a periodic attractor this will produce a number of isolated points corresponding to the periodicity of the attractor, and to obtain a more complete picture of the underlying map we must also make use of the transient behavior. For a quasi-periodic or chaotic attractor, the stationary solution suffices to construct the return map.

Figure 14 shows a set of return maps for the HIV model with β as a parameter. Closed circles are for $\beta = 0.1$, open circles for $\beta = 0.17$, closed triangles for $\beta = 0.25$, open triangles for $\beta = 0.30$, closed squares for $\beta = 0.45$, open squares for $\beta = 0.78$, and stars for $\beta = 1.00$. All the maps are characterized by a steep upwards slope for small V_n. This represents the instability (or the stretching) in the model. The maps then pass through a maximum and bend downwards for higher values of V_n. This represents the nonlinear folding in the system by which high values of HIV abundance in one cycle give rise to very low values in the subsequent cycle. We clearly see how the slope of the return map at the point of intersection with the diagonal $V_{n+1} = V_n$ increases with β.

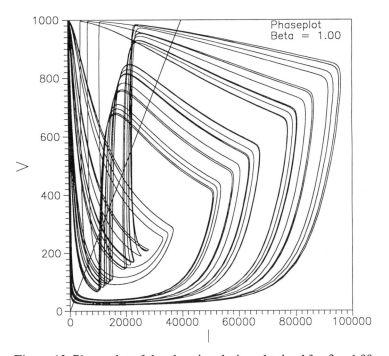

Figure 13. Phase plot of the chaotic solution obtained for $\beta = 1.00$.

CONCLUSION

Although the Anderson and May model only represents a first tentative step towards modeling the interaction between HIV and the immune system, it is of interest because it shows how oscillatory and chaotic behavior in the abundances of T4 cells and free HIV can result from a set of very simple descriptive equations. The causes of

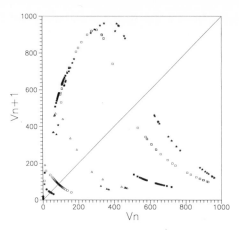

Figure 14. Return maps for the HIV-model with the probability of infection β as a parameter. The return maps determine the abundance of free HIV in one cycle as a function of the abundance in the previous cycle.

instability are the positive feedback mechanisms associated with proliferation of activated T lymphocytes and with replication of HIV in the infected cells. Similar phenomena may be found in many other population models.

In the same way, the nephron model is only a first step towards understanding the complex phenomena controlling nephron pressure and flow oscillations. This model should be extended to give a more comprehensive description of the processes that occur in the juxtaglomerular apparatus and to include reabsorption. We have tried to illustrate how seemingly disparate phenomena on different scales and within different biological specialities reveal a common complexity in dynamics. Combined with the rapid development of experimental techniques and with an increasing emphasis on mathematical modeling and computer simulation it seems that the insight provided by nonlinear dynamics and far-from-equilibrium thermodynamics bring us into a new era for the biological sciences, an era where general principles can be formulated on a much sounder theoretical basis.

Background Literature

Christiansen, P.L. and R.D. Parmentier (eds.): "Structure, Coherence and Chaos in Dynamical Systems", Proceedings in Nonlinear Science, Manchester University Press, Manchester (1989)

Degn, H., A.V. Holden and L.F. Olsen (eds.): "Chaos in Biological Systems", Plenum Publishing Corporation (1987)

Devaney, R.L.: "An Introduction to Chaotic Dynamical Systems", Benjamin, Menlo Park (1986)

Ebeling, W. and R. Feistel: "Physik der Selbstorganisation und Evolution", Academie Verlag, Berlin (1982)

Glass, L. and M.C. Mackey: "From Clocks to Chaos, The Rhythms of Life", Princeton University Press, Princeton (1988)

Goldbeter, A. (ed.): "Cell to Cell Signalling: From Experiments to Theoretical Models", Academic Press, London (1989)

Haken, H.: "Synergetics, An Introduction", Springer-Verlag, Berlin (1978)

Holden A.V. (ed.): "Chaos", Nonlinear Science, Theory and Applications, Manchester University Press (1986)

Langton C.G. (ed.): "Artificial Life", Santa Fe Institute Studies in the Sciences of Complexity VI, Addison-Wesley, Redwood City (1990)

Nicolis, G. and I. Prigogine: Self-Organization in Nonequilibrium Systems", Wiley and Sons, New York (1977)

Nicolis, G. and I. Prigogine: "Exploring Complexity", Freeman and Company, New York (1989)

Stein. D.L. (ed.): "Lectures in the Sciences of Complexity", Santa Fe Institute Studies in the Sciences of Complexity, Addison-Wesley, Redwood City (1988)

References

Anderson, R.M. and R.M. May: "Complex Dynamical Behavior in the Interaction between HIV and the Immune System", in "Cell to Cell Signalling", ed. A. Goldbeter, Academic Press (1989)

Bailey, D.A. and R.G. McCulloch: "Bone Tissue and Physical Activity." Can. J. Spt. Sci. 15,4, 229-239 (1990)

Colgate, A., E.A. Stanley, J.M. Hyman, C.R. Qualls, and S.P. Layne: "AIDS and a Risk-Based Model", Los Alamos Science 18, 2-39 (1989)

Frost, H.M.: "Editorial: Tetracycline Based Histological Analysis of Bone Remodeling", Calif. Tissue Res. 3, 211-237 (1969)

Frost, H.M.: "The Skeletal Intermediary Organization", Metab. Bone Dis. Relat. Res. 4, 281-290 (1983)

Holstein-Rathlou, N.-H. and P.P. Leyssac: "TGF- Mediated Oscillations in the Proximal Intratubular Pressure: Differences between Spontaneously Hypertensive Rats and Wistar-Kyoto Rats", Acta Physiol. Scand. 126, 333-339 (1986)

Holstein-Rathlou, N.-H. and P.P. Leyssac: "Oscillations in the Proximal Intratubular Pressure: A Mathematical Model", Am. J. Physiol. 252, F560-572 (1986)

Jensen, K.S., E. Mosekilde and N.-H. Holstein-Rathlou: "Self-Sustained Oscillations and Chaotic Behavior in Kidney Pressure Regulation", Proc. Solvay Institutes Discoveries 1985 Symposium, Laws of Nature and Human Conduct, I. Prigogine and M. Sanglier (eds.), G.O.R.D.E.S., Bruxelles (1986)

Jensen, K.S., N.-H. Holstein-Rathlou, P.P. Leyssac, E. Mosekilde, and D.R. Rasmussen: "Chaos in a System of Interacting Nephrons", Proc. NATO Advanced Research Workshop on Chaos in Biological Systems, H. Degn, A.V. Holden and L.F. Olsen (eds.), Plenum Press (1987)

Jensen, K.S., Li. Mosekilde, and Le. Mosekilde: "A Model of Vertebral Trabecular Bone Architecture and its Mechanical Properties", Bone 11, 417-423 (1990)

Jensen, M.H., P. Bak and T. Bohr: "Transition to Chaos by Interaction of Resonances in Dissipative Systems. I Circle Maps", Phys. Rev. A 30, 1960-1969 (1984)

Leyssac, P.P. and L. Baumback: "An Oscillating Intratubular Response to Alterations in Henle Loop Flow in the Rat Kidney", Acta Physiol. Scand. 117, 415-419 (1983)

Leyssac, P.P. and N.-H. Holstein-Rathlou: "Effects of Various Transport Inhibitors on Oscillating TGF Pressure Response in the Rat", Pfluegers Arch. 407, 285-291 (1986)

May, R.M. and R.M. Anderson: "The Transmission Dynamics of Human Immunodeficiency Virus (HIV)", Phil. Trans. Royal Soc. B 321, 565-607 (1988)

Melsen, F., B. Melsen and Le. Mosekilde: "An Evaluation of the Quantitative Parameters Applied in Bone Histology", Acta Path. Microbiol. Scand. Sect. A 86, 63-69, (1978)

Mosekilde, Li.: "Age Related Changes in Vertebral Trabecular Bone Architecture Assessed by a New Method", Bone 9 (1988) 247-250

Mosekilde, Li.: "Consequences of the Remodeling Process for Vertebral Trabecular Bone Structure - A Scanning Electron Microscopy Study (Uncoupling of Unloaded Structures)", Bone and Mineral 10, 13-35 (1990)

Mosekilde, Li., Le. Mosekilde and C.C. Danielsen: "Biomechanical Competence of Vertebral Trabecular Bone in Relation to Ash Density and Age in Normal Individuals", Bone 8, 79-85 (1987)

Obrant, K.I., U. Bengnér, O. Johnell, B.E. Nilsson, and I. Sernbo: Editorial: "Increasing Age-Adjusted Risk of Fragility Fractures: A Sign of Increasing Osteoporosis in Successive Generations?, Calsif. Tissue Int. 44, 157-167 (1989)

Nara, P.: "Aids Viruses of Animals and Man: Nonliving Parasites of the Immune System", Los Alamos Science 18, 54-89 (1989)

Weber, J.N., P.R. Clapham, R.A. Weis et al.: "Human Immunodeficiency Virus Infection in Two Cohorts of Homosexual Men: Neutralizing Sera and Association of Anti-gag Antibody with Prognosis", Lancet 119-121 (1987)

PROBING DYNAMICS OF THE CEREBRAL CORTEX

A. Babloyantz

Université Libre de Bruxelles
CP 231 - Campus de la Plaine
Boulevard du Triomphe
B-1050 Bruxelles, Belgium

1 Introduction

The central nervous system is certainly one of the most complex biological tissues, as it comprises some 10^{10} interconnected neurons with thousands of connections per each unit. The flow of information in this intricate network involves a great number of transmitters which modulate the electrical activity characteristic of brain tissue.

The existing large amount of biochemical information at cellular and synaptic level, although extremely important says very little about the mechanisms of information processing of cerebral cortex. This ability is the property of a large ensemble of neurons. Moreover the dynamical properties of such an ensemble are the relevant parameters for probing cortical activity.

The global activity of a neuronal mass may be assessed non-invasively with the help of electroencephalographic recording (EEG). Although EEG is related to the neuronal activity, the exact relationship between the cellular and global level of electrical activity is not yet completely elucidated. Inspite these shortcomings nevertheless the EEG reflects behavioral states of the brain and was a clinical tool for the diagnostics of functional as well as structural deficiencies of human brain before the advent of the more advanced devices of today. EEG is still a clinical tool for probing sleep disorders as it reflects behavioral states of the brain.

It is admitted that the dynamical states of the cortical tissue governs the behavioral states of the brain. The EEG of an aroused and active brain is characterized by waves of high frequency and low amplitude (beta activity). As the eyes are closed and the subject is relaxing, spindle-like waves, called alpha rhythm, appear. As the subject drifts towards sleep, the brain enters into a succession of stages (stage 1 to stage 4) where the EEG amplitude gradually increases as the mean frequency decreases. In a normal brain, the deep sleep (stage 4) is characterized by the highest EEG amplitude, while the mean frequency of oscillations is the lowest. With the onset of REM sleep (Rapid Eye Movement), where most of dreams occur, the EEG reverts back again to beta-like activity.

In animal experiments where the measurements of field potentials from the cortical

Complexity, Chaos, and Biological Evolution, Edited by E. Mosekilde and
L. Mosekilde, Plenum Press, New York, 1991

surface or depth are feasible, the global electrical activity is a valuable research tool and may enravel, some aspects of information processing capabilities of the brain. For example, the recent experiments of Gray and Singer [1] and Eckhon et al. [2] have demonstrated the relationship between the visual input and the onset of coherent oscillations in the visual cortex. They suggest this property as a mechanism for feature linking in the visual system.

Traditionally the dynamical study of the brain required model construction in which the evolution of several variables, or a network of neuronal elements were followed in time. Usually one tried to describe those brain states which exhibited more or less regular behavior, therefore could be related to the solutions of deterministic differential equations. An example of such a modeling is the description of epileptic seizures as a periodic solution of a set of nonlinear coupled differential equations [3]. Another example was the description of alpha waves, again in terms of a stochastic dynamical system [4]. The EEG of other behavioral states such as for example deep sleep seemed too irregular to be related to a deterministic dynamics.

However with the advent of deterministic chaotic dynamics, new alleys became open in brain research. Indeed in the theory of nonlinear dynamics it was shown that a set of three variable coupled nonlinear differential equations may exhibit solutions which show the irregular behavior in time reminiscent of the brain waves of normal individuals. Therefore the modeling of the brain in various behavioral states became feasible [5]. On the other hand recent progress in nonlinear time series analysis provides new methods for the assessment of systems dynamics from the measurement of a single parameter in time. From these time series several dynamical parameters such as dimensions, Lyapounov exponents and Kolmogorov entropies can be evaluated. These parameters provide quantitative information about some of the dynamical aspects of the system under consideration.

In 1985 Babloyantz, Nicolis and Salazar [6] applied these techniques to the study of human EEG. They could show that several key behavioral state of the brain followed deterministic chaotic dynamics with a rather low values of correlation dimensions D_2. This paper extends the analysis and the evaluation of D_2 to other physiological data such as cardiac rhythm [7], alpha waves [8], petit-mal epilepsy [9] and a severe disease of the brain namely Creutzfeld-Jakob coma [10].

Correlation dimensions are evaluated following four different embedding techniques. The results are compared with the same quantities obtained this time from the evaluation of the spectrum of Lyapounov exponent following Kaplan York and Mori conjectures [11, 12]. With the use of recurrence plots it is also shown how to assess the stationarity of time series. The latter technique may reveal the presence of hidden periodicities of longer wave length superimposed on otherwise chaotic dynamics.

2 Stationarity of time series

Recently Eckman and colleagues [13] proposed a novel, elegant graphical tool, called a "recurrence plot" for the diagnoses of the presence of drift and hidden periodicities in the time evolution of dynamical systems which are unnoticeable otherwise. The first step in the construction of the recurrence plot and the evaluation of all the dynamical properties is the digitalization of the EEG at regular time intervals. For a given "time series" m vectors are constructed from a single lead V(t) by introducing a time lag τ leading to the V(t), V(t + τ), and V(t + 2τ)... V(t+(m-1)τ) variables. The space spanned by these vectors is called a "phase space" and a point in this space defines completely the instantaneous state of the system. As time evolves, the point in the phase space travels on a trajectory which gives rise to the phase portrait. It has been

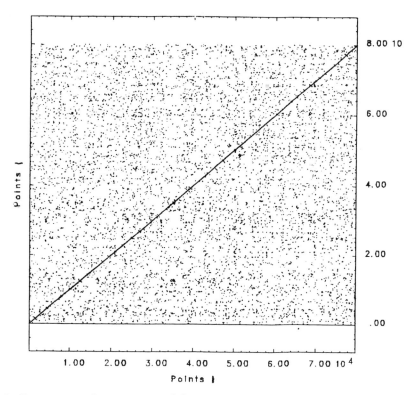

Figure 1. Recurrence plot constructed from 320 sec. of EEG recorded from a normal individual during deep-sleep. Sampling rate 250 Hz, $\tau = 6$ and $m = 6$. The uniform distribution of points indicates the absence of drift or slow periodicity.

shown that the topological properties of phase portraits constructed by the above procedure are equivalent to the original phase portraits constructed from known variables of the system [14, 15, 16].

Once the phase portrait is constructed, one chooses a point $x(i)$ on a trajectory and considers a ball of radius $r(i)$ centered on this point . $r(i)$ is chosen such as it contains a reasonable number of other points $x(j)$ on the orbit. The recurrence plot is constructed as an array of dots in a NxN square, where a dot is drawn at (i,j) whenever $x(j)$ is sufficiently close to $x(i)$ (see Fig. 1). Recurrence plots tend to be fairly symmetric with respect to the diagonal $i - j$. If $x(i)$ is close to $x(j)$ then $x(j)$ is also close to $x(i)$. However there is no complete symmetry as one does not require $r(i) = r(j)$ but rather a fixed number of points in the ball $r(i)$. Points i and j represent time, therefore the recurrence plot embodies natural and subtle time correlation information which locally shows a texture and in the same time exhibits a global topology.

If all characteristic times (the usual periodicities of EEG) are short compared with the total recording time, then the global aspect of the recurrence plot is homogeneous. Figure 1 shows a recurrence plot obtained form a 320 second recording of an episode of deep sleep. It is seen that the data set is stationary and does not contain a slowly varying drift. For a time series with drift the recurrence plot is much paler as the distance from the diagonal increases (see Fig. 2, ref. [13]).

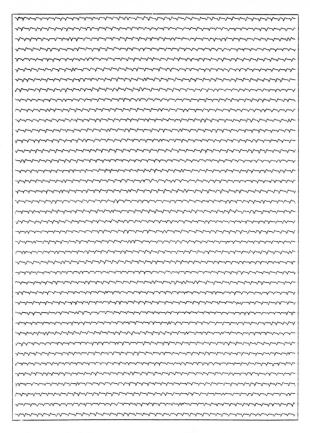

Figure 2. The EEG recorded from a patient suffering from Creutzfeldt-Jakob disease. The visual examination of twenty minutes of the time series does not show any obvious drift or periodicities. 30 sec. per line, 12 bit sampling at 250 Hz, 120 Hz filter.

Figure 2 shows few seconds of EEG recorded from a patient suffering from Creutzfeldt-Jakob disease. In this degenerative and rare pathology of the brain, which result from viral attack, the EEG shows synchronous oscillations of a pseudoperiod of the order of one sec.

At the first sight no drift or superimposed periodicity is seen in the data set of Fig. 2. However the recurrence plot constructed from this time series shows a checkerboard structure of a periodicity of the order of 60 sec. This fact indicates the presence of slow oscillations superimposed on a much more rapid chaotic motion of periodicity of the order of one second (see Fig. 3a). To make these observations more quantitative, the density of points is plotted as a function of $i - j$ as seen in Fig. 3b. The recurrence plot was redrawn after filtring the data and eliminating activities below 0,05 HZ. The structure of the plot remained the same.

The recurrence plots of other physiological data such as cardiac rhythms, and alpha waves also show that a sufficiently long stretches of stationary time series could be obtained. Long and stationary time series are a prerequisite to further dynamical study of experimental data. EEG of the deep sleep excepted, all the other data showed some kind of superimposed activity of very long duration on the otherwise much higher pseudo-frequency data sets. In all cases the contribution of the long wave activity to the dynamical parameter of the attractors is marginal and could be neglected.

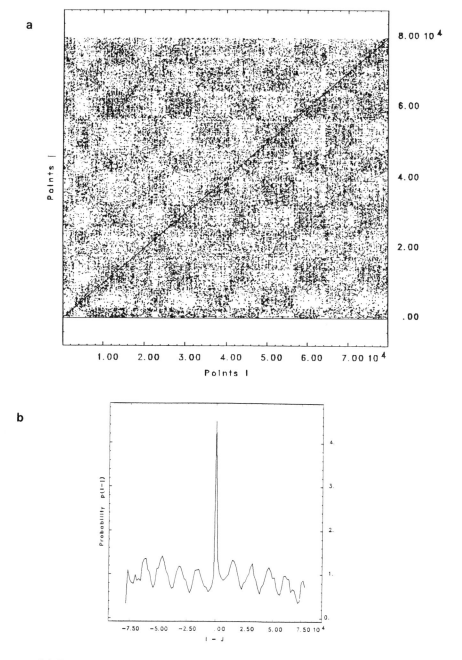

Figure 3. (a) Recurrence plot constructed from 320 sec. of EEG recorded from a patient suffering from Creutzfeldt-Jakob disease. $\tau = 6$ and $m = 6$. The regular checkerboard structure shows the presence of slow periodicity of about 53 sec. (b) A quantitative visualization of the same observation. The probability density of points $P(i - j)$ as a function of $(i - j)$.

3 Attractor dimensions

From experimental data recorded from a single or closely positioned multichannel electrodes, phase portraits could be constructed following four different procedures [17].

1. *Laging method [14, 15, 16]* In this procedure, one introduces a time lag τ, in principle arbitrary, in a single time series, thus creating an embedding space containing the attractor. When dealing with a finite set of data, not every value of the lag τ gives an acceptable embedding. Usually, only a narrow range of τ, which has to be found by comparative trials, provides phase portraits which can be used for dimension analysis.

2. *Multi-channel trajectories [14]* Eckmann and Ruelle conjectured that the phase space spanned by simultaneous measurements of the same variable in different sites also constitutes an embedding. This construction has the advantage of covering the system spatially and thus incorporates better statistics. Although in this procedure one avoids the difficulties inherent to the laging procedure, another subjective quantity, namely the inter-site distance, appears in the problem. Moreover the latter has a lower bound fixed by the experimental recording technique.

3. *Singular value decomposition [18]* This procedure, used by Broomhead and King provides a set of vectors for the construction of phase portraits. If $\{v_1, v_2, ..., v_N\}$ are a succession of measurements of a given variable v, then the method of delay is used to construct n-dimensional vectors $\{x_1, x_2, ..., x_{N-n+1}\}$ where $x_j^T = (v_j, v_{j+1}, ..., v_{j+n-1})$. $\{x_j^T\}$ are the rows of the rectangular trajectory matrix $\mathbf{X}$. The product $\mathbf{X}^T\mathbf{X}$ constitutes the covariance $n \times n$ matrix of the system. The singular values $\sigma_1 ... \sigma_n$ are the square roots of the eigenvalues of $\mathbf{X}^T\mathbf{X}$ and the corresponding eigenvectors constitute an orthonormal basis for the embedding space. If the system is projected onto the more important axes, this procedure is a noise reducing technique (the finer twists of trajectories which are comparable in magnitude to the noise probably are also suppressed) and provides smoother phase portraits which however are very sensitive to the sampling interval of the data. Also an important step in the analysis is the choice of the window $(n-1)$. Again the phase portrait is very sensitive to the choice of the window. As we shall see later, this choice is determinant for the evaluation of the topological dimension. Broomhead and King suggest that the window must be chosen such as the power spectrum contains no frequencies with significant magnitude greater than a cutoff frequency known as the band limit. In the case of complex and relatively high dimensional attractors, we find this procedure inapplicable. One is again led to finding the appropriate window by trial and error.

4. *Singular decomposition of multi-channel data* which combines the two above cited procedures.

The majority of algorithms in non-linear time series analyses require uniform attractors. The attractors obtained from alpha rhythm and deep sleep are more or less uniform. Therefore the algorithms could be applied with some confidence. However he phase portrait constructed from cardiac rhythm and Creutzfeld-Jakob deseases shows obvious non-uniformity. Nevertheless the global properties of these attractors could still be evaluated. Recently by applying techniques more appropriate to the inhomogeneous attractors, local attractor dimensions for Creutzfeld-Jacob and cardiac rhythms were obtained [19]. The average value of these partial dimensions gave a comparable

Table 1. Correlation dimensions evaluated from different phase space constructions. The standard Grassberger-Procaccia algorithm is used and 1000 equidistant origins are taken for each calculation. The data length N, the sampling frequency and the range of τ are respectively: (a) 60000, 250Hz, 120 to 240 ms., (b) 60000, 250Hz, 40 to 80 ms., (c1) 60000, 250 Hz, 12 to 40 ms., (c2) 18000, 1200Hz, 25 to 46 ms., (d) 6000, 100Hz, 100 to 200 ms., (e) 6000, 1200Hz, 14 to 21 ms., (f) 20000, 50 Hz (integration step h=0.02), 80 to 160 ms, (g) 40000. Parameters for the Lorenz attractor: r=28, Pr=10, b=8/3.

System	Single channel correlation dimension		Multi-channel correlation dimension	
	standard construction	singular vectors	standard construction	singular vectors
(a) cardiac rhythm (ECG)	3.6±0.1	3.43±0.01	2.9±0.1	2.8±0.1
(b) CJ EEG	3.7–5.4	3.78±0.05	3.8±0.1	3.8±0.1
(c) alpha EEG	(1) 6.1±0.5	6.2±0.1	–	–
	(2) 6.3–7.4	5.95±0.05	6.06±0.05	6.00±0.02
(d) deep sleep EEG	4.4±0.1	4.5±0.1	–	–
(e) epileptic EEG	2.05±0.09	2.03±0.05	–	–
(f) Lorenz model	2.05–2.1	2.05±0.01	–	–
(g) white noise (for $n=10$)	9.4	–	9.4	–

value to the ones obtained by applying the algorithms developed for uniform attractors, and applied to the same data.

The correlation dimensions for alpha waves, deep sleep, CJ coma and cardiac rhythms were evaluated using the well known Grassberger and Procaccia algorithm [20]. The attractors were constructed according all four procedures described above. The results are shown in table 1. It is seen that several key behavioral states of the human brain exhibit chaotic attractors of rather low dimension. Moreover as the cognitive power of the brain diminishes during the sleep cycle so does the numerical value of correlation dimension.

In healthy subjects the lowest value appears during the deep sleep. However in severe pathologies such as Creutzfeld-Jacob desease or the petit mal epilepsy the attractor dimension decreases further to a low value of $D_2 = 2.05$ [9]. Thus indicating a relationship between coherent behavior and loss of cognitive power in cortical tissue.

The dynamical properties of the attractors such as the spectrum of Lyapounov exponents may also be evaluated. The presence of positive Lyapounov exponents indicates the existence of chaotic dynamics. Moreover once the Lyapounov exponents are determined, with the help of Kaplan-Yorke and also Mori conjectures correlation dimensions could be evaluated [21].

Such a procedure was applied to the alpha waves, deep sleep and Creutzfeld-Jacob coma and the attractors were constructed by the usual laging method. Table 2 shows the number of positive Lyapounov exponent as well as the correlation dimensions computed from the two above cited conjunctures. These values compare favorably with the values seen in table 1.

4 Conclusions

There have been some discussions and doubts about the applicability of nonlinear time series analysis to the physological data. Characteristically in most cases the criticisms came from those who were not directly envolved neither in physiology nor in physiological time series analysis.

Table 2. Fractal dimensions computed by static and dynamic methods. Correlation dimensions (D_2), Kaplan-Yorke conjecture (D_{KY}) and Mori conjecture (D_{Mo}).

Stage	CJ Coma	Deep sleep	Alpha rhythm
D2	3.8 ± 0.2	4.4 ± 0.1	6.1 ±0.5
		4.05-4.37	6.3-7.4
dKY	4.49	5.28	6.97
	5.22	6.12	7.63
dMo	4.0	4.44	5.71
	4.1	4.56	6.19

The main concern has been the non stationary nature of the time series. Here we showed that with the help of the recurrence plots quite long and stationary strethes of EEG or ECG could be selected.

On the other hand in the case of the attractors obtained from differential equations, it was argued that in order to obtain accurate results, exhorbitant number of data points are necessary [22]. Such high accuracies are not necessary and as a matter of fact are useless when dealing with physiological time series. Here the relative values of dynamical parameters are only relevant. Therefore some degree of error is permissible in the evaluation of D_2. On the other hand if physiological attractors exist, there is no guarantee that they remain exactly the same in time. Thus the numerical values of the paramters give only an indicative and comparative value.

In our opinion the most serious shortcomings of the dynamical methods is that they do not give good results for high dimensional attractors. Therefore great care must be taken in choosing the data sets and one most avoid high dimensional systems.

On the other hand in our research we have analyzed the same data by several entirely different approaches. We have compared attractor dimensions computed from static properties with the ones evaluated from dynamical properties. They all give more or less the same results. Thus it seems reasonable to think that physiological attractors do exist and, provided they are not of very high dimension, they can be studied in the framework of non-linear data analysis.

Acknowledgements

I am indebted to A. Destexhe for his help in the preparation of the manuscript.

References

[1] Gray, C.M. and Singer, W. Stimulus-specific neuronal oscillations in orientation columns of cat visual cortex. *Proc. Natl. Acad. Sc. USA* **86**: 1698-1702 , 1989.

[2] Eckhorn, R., Bauer, R., Jordan, W., Brosch, M., Kruse, W., Munk, M. and Reitboek, H.J. Coherent oscillations: a mechanism of feature linking in the visual cortex ? *Biol. Cybernetics* **60**: 121-130, 1988.

[3] Kaczmarek, L.K. and Babloyantz, A. Spatiotemporal patterns in epileptic seizures. *Biol. Cybernetics* **26**: 199-208, 1977.

[4] Lopes da Silva, F.H., Hoeks, A., Smits, H. and Zetterberg, L.H. Model of brain rhythmic activity. The alpha rhythm of the thalamus. *Kybernetik* **15**: 27-37, 1974.

[5] Destexhe, A. and Babloyantz, A. Pacemaker-induced coherence in cortical networks. submitted to *Neural Computation*, 1990; see also Destexhe, A. and Babloyantz, A. in: *Self-Organization, Emerging Properties and Learning*, Ed. A. Babloyantz, Plenum press, in press, 1990.

[6] Babloyantz, A., Nicolis, C. and Salazar, M., Evidence for chaotic dynamics of brain activity during the sleep cycle. *Phys. Lett. A* **111**: 152-156 , 1985.

[7] Babloyantz, A. and Destexhe, A. Is the Normal Heart a Periodic Oscillator? *Biological Cybernetics* **58** 203, 1988.

[8] Babloyantz, A. and Destexhe, A., Strange attractors in the cerebral cortex. in: *Temporal Disorder in Human Oscillatory Systems*. Edited by Rensing, L., an der Heiden, U. and Mackey, M.C., *Springer Series in Synergetics*, **36**. Berlin: Springer, 1987, pp. 48-56.

[9] Babloyantz, A. and Destexhe, A., Low dimensional chaos in an instance of epileptic seizure. *Proc. Natl. Acad. Sc. USA* **83**: 3513-3517 , 1986.

[10] Babloyantz, A. and Destexhe, A., The Creutzfeldt-Jakob disease in the hierarchy of chaotic attractors. in: *From Chemical to Biological Organization*. Edited by Markus, M., Muller, S. and Nicolis, G., *Springer Series in Synergetics*, **39**. Berlin: Springer, 1988, pp. 307-316.

[11] Kaplan, J. and Yorke, J. Chaotic behavior of multi-dimensional difference equations. in: *Functional Differential Equations and Approximations of Fixed Points*, Eds. Peitgen, H.O. & Walther, H.O. (Springer, Berlin), *Lectures Notes in Mathematics* **Vol. 330**, 1979, pp. 228-236.

[12] Mori, H. Fractal dimensions of chaotic flows of autonomous dissipative systems. *Prog. Theor. Phys.* **63**: 1044-1047, 1980.

[13] Eckmann, J.P., Kamphorst, S.O. and Ruelle, D. Recurrence plots of dynamical systems. *Europhys. Lett.* **4**: 973-977, 1987.

[14] Eckmann, J.P. and Ruelle, D. Ergodic theory of chaos and strange attractors. *Rev. Mod. Phys.* **57**: 617-656, 1985.

[15] Packard, N.H., Crutchfield, J.P., Farmer, J.D. and Shaw, R.S. Geometry from a time series. *Phys. Rev. Lett.* **45**, 712 (1980) .

[16] Takens, F. Detecting strange attractors in turbulence. in: *Dynamical Systems and Turbulence*. Edited by Rand, D.A. and Young, L.S., *Lecture Notes in Mathematics* **898**. Berlin: Springer, 1981, pp. 366-381.

[17] Destexhe, A., Sepulchre, J.A. and Babloyantz, A. A comparative study of the experimental quantification of deterministic Chaos. *Phys. Lett. A* **132**: 101-106 , 1988.

[18] Broomhead, D.S. and King, G.P. Exctracting qualitative dynamics from experimental data. *Physica* **20 D**: 217, 1986.

[19] Gallez, D., Destexhe, A. and Babloyantz, A. Static and dynamical properties of nonhomogeneous physiological attractors, to appear, 1990.

[20] Grassberger, P. and Procaccia, I. Characterization of strange attractors. *Phys. Rev. Lett.* **50**: 346-349, 1983.

[21] Gallez, D. and Babloyantz, A. Predictability of human EEG: a dynamical approach. Biol. Cybernetics, in press, 1990.

[22] Smith, L.A. Intrinsic limits on dimension calculations. *Phys. Lett. A.* **133**: 283-288, 1988.

CHAOS AND BIFURCATIONS DURING VOICED SPEECH

Hanspeter Herzel, Ina Steinecke, Werner Mende and Kathleen Wermke

Humboldt University Berlin
Invalidenstr. 42
1040 Berlin
Federal Republic of Germany

ABSTRACT

Paralysis, laryngitis, cancer and other pathological conditions often lead to irregularities in the sounds of speech. Many of these irregularities seem related to intrinsic nonlinearities in the vibrations of the vocal cords and may thus be understood in terms of the concepts of nonlinear dynamics. In this paper, we analyze examples of newborn infant cries and voiced sound to illustrate the occurrence of bifurcations and chaos. A dynamic model of speech production is presented. Simulation with the model reveals a variety of complex dynamic behaviors, including subharmonic generation and chaos.

INTRODUCTION

The sounds of speech are among the most complex acoustic signals that exist in nature. The theory of speech production is of interest in many fields, including voice pathology, and speech synthesis and recognition (Helmholtz 1870, Fant 1960, Sorokin 1985).

The driving force of sound generation is the air flow, typically from the lungs via the bronchi and trachea. In order to discuss the basic physical mechanisms of speech production, we distinguish between voiced and voiceless sound. Voiceless signals as, e.g., fricatives are produced if the Reynolds number at a constriction of the vocal apparatus exceeds a critical value of about 1700 (Catford 1977). The generated

Complexity, Chaos, and Biological Evolution, Edited by E. Mosekilde and
L. Mosekilde, Plenum Press, New York, 1991

turbulent noise is quite irregular, reflecting the high dimensionality of fully developed turbulence.

By contrast, voiced sound is much more regular. According to the accepted myoelastic theory of voice production, the vocal folds are set in vibration by the combined effects of a subglottal pressure, the elastic properties of the folds, and the Bernoulli effect. The air passing into the vocal tract is usually in the form of discrete puffs (see Fig. 4b). The effective length, mass and tension of the vocal folds are determined by muscle action, and in this way the fundamental frequency ("pitch") and the glottal waveform can be controlled. The vocal tract acts as a filter which transforms the generated primary signals into meaningful voiced speech.

Normal voiced sound appears to be nearly periodic, although small perturbations are always present ("jitter" and "shimmer"). This variability is relatively small (in the order of percent) but is nevertheless of importance for the naturalness of speech and for the characterization of voice quality (see, e.g., Arends et al. 1990). The variability may even provide diagnostic information for the recognition of brain functions (Mende et al. 1990b).

Under certain circumstances much larger irregularities of voiced signals are observed (Dolansky and Tjerlund 1968, Howard 1989). Nowadays, quite a lot of terms are used to describe such irregularities: creaky voice, vocal fry, harshness, roughness, raucous voice (Laver 1980, Eskenazi et al. 1990).

On one hand, irregular trains of glottal pulses are observed if extreme values of pitch frequencies are reached (mainly at the lower end of the pitch range) (Laver 1980). On the other hand, many pathologies such as paralysis, laryngitis, cancer and nodules lead to irregularities (see, e.g., Hecker and Kreul 1971, Childers et al. 1984). Thus, an understanding of irregular vocal fold vibrations is of considerable interest. Some of the noisy components of voiced signals (as, e.g., breathiness) are related to an incomplete closure of the vocal cord orifice, which leads to turbulent noise, as in the case of fricatives (Catford 1977). However, in general the precise mechanisms of irregular vocal fold motion are still unknown (Laver 1980).

Many of the observed irregularities seem to be related to intrinsic nonlinearities of the vocal cord vibrations. Thus, the concept of nonlinear dynamics can be applied, and the observed phenomena can be understood in the framework of bifurcation theory and deterministic chaos. The excitation of glottal pulses is a highly nonlinear process and, therefore, the appearance of subharmonics and chaos is not unexpected.

In the next section, experimental observations of subharmonics and chaos are presented. The analyzed data are specific voiced signals, namely newborn infant cries. The third section is devoted to a two-mass model of vocal cord vibration (Ishizaka and Flanagan 1972). Integrating this model, besides periodic oscillations we have found beating-like behavior, subharmonic regimes and deterministic chaos. These simulations are qualitatively similar to observations during voiced speech and newborn cries. Thus we conclude that low-dimensional chaos plays an essential role in speech production.

Analyzing sustained vowels from patients with tumors, polyps and other complications, we have found a compendium of interesting dynamic types: period-doubling and -tripling, intermittent behavior, chaos and various types of toroidal oscillations with beat frequencies between 20 and 50 Hz.

Implications of nonlinear dynamics for the characterization of pathological voices will be discussed in a forthcoming paper.

NEWBORN INFANT CRIES

The acoustical analysis of infant vocalization is a helpful noninvasive tool to explore brain function at very early stages of child development (Lind 1965, Mende et al. 1990b). Newborn phonation is based on an already well-developed laryngeal coordination and, therefore, one can hope to detect disturbances of brain functions in cry signals.

Up to now, pain cries of 70 newborn infants between the ages of 1-5 days have been analyzed. About one-third of them had peri- and postnatal complications (Mende et al. 1990b). The signals were recorded on tape, antialiasing filtered and digitized with a sampling rate of 25 or 50 kHz and an amplitude resolution of 12 or 15 bit. Cries are more or less unstationary processes. Transitions between several oscillatory modes occur very suddenly (in a few milliseconds). However, in many cases time segments of 5-20 ms can be considered as approximately stationary episodes and, therefore, time series analysis makes sense.

The unstationarity of cries has the advantage that many changes ("bifurcations") become visible during a single cry. In order to visualize the evolution of the spectral density during a cry, so-called spectrograms (or sonagrams) are appropriate. They are based on many subsequent short-time power spectra of overlapping segments. We have chosen a shift of 7 ms, and the spectral amplitude is encoded in terms of 4 bit grey tone scale. In this way, spectrograms as shown in Fig. 1 are obtained.

At the beginning of the spectrogram, only the fundamental frequency of about 500 Hz with its harmonics is present. Then, in connection with increasing energy and pitch, remarkably sharp transitions to more complicated states are visible. These "sub-harmonic bifurcations" and especially the noisy segments are discussed in detail by Mende et al. (1990a). With the aid of Poincaré sections and dimension analysis, it is shown in that paper that the noise-like behavior can be identified as low-dimensional chaos.

Fig. 2 shows a cry segment of 128 ms which exhibits various interesting regimes. The signal changes from a periodic oscillation with a fundamental frequency of more than 400 Hz via a subharmonic bifurcation to a regime with exactly half the frequency. Then a backward octave jump to fast oscillations occurs. Comparable frequency breaks are observed also during human speech (Bowler 1964). In a recent paper, not only

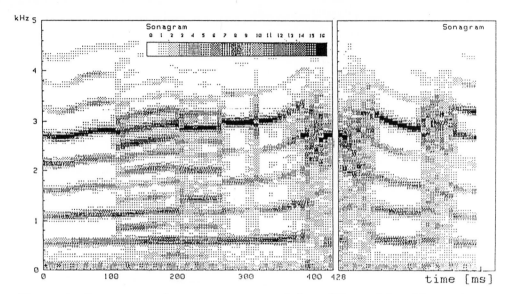

Figure 1. Spectrogram from a premature infant (28 weeks) with respiratory complications. Subharmonic bifurcations and chaotic episodes are clearly visible.

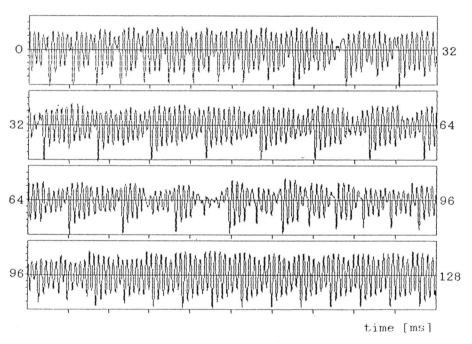

Figure 2. Cry segment with a sudden transition from a "basic cry" (pitch about 400 Hz) to a subharmonic regime with nearly the double pitch period.

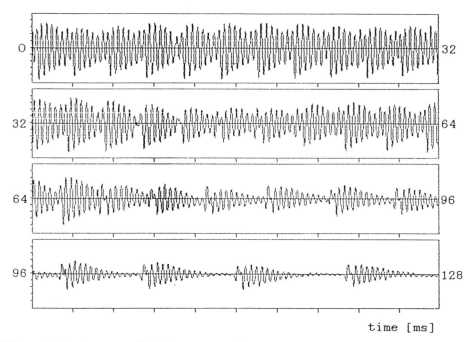

Figure 3. Newborn cry displaying a transition to very low fundamental frequencies.

octave jumps but also changes of the pitch by a factor of 3 are reported (Howard 1989).

Another common phenomenon in speech is very low-pitched regimes (Dolansky and Tjerlund 1968, Catford 1977). Such a beating-like behavior was found during newborn cries as well (see Fig. 3).

Subharmonics and chaos have been found in cries of healthy infants as well as in infants with complications. Perhaps the frequency of occurrence and the duration of chaotic episodes could have diagnostic relevance. It is worth mentioning that already 25 years ago, noise-like segments and sudden shifts of the fundamental frequency were recognized during newborn cries (Lind 1965).

MODELING VOICED SPEECH PRODUCTION

There have been several attempts to model vocal cord excitation and the vocal tract. The vocal cords are often treated with mass-spring models (see e.g. Ishizaka and Flanagan 1972) or as elastic materials capable of propagating compressional, shear and surface waves (Titze 1976). The vocal tract can be modeled with good accuracy as a series of tubes which are characterized by their acoustic inductances, capacitances and dissipative losses (Flanagan 1972).

45

In this section, we discuss simulations of the two-mass model developed by Ishizaka and Flanagan (1972). A vocal cord is modeled by two mechanical oscillators representing the lower and upper part of a cord.

$$m_i \frac{d^2 x_i}{dt^2} + r_i \frac{dx_i}{dt} + s_i(x_i) + k_c(x_i - x_j) = F_i \tag{1}$$

$$(i,j = 1,2)$$

$$s_i(x_i) = k_i(x_i + \eta_i x_i^3) + \theta(\Delta x_i) h_i(\Delta x_i + \mu_i \Delta x_i^3)) \tag{2}$$

$$\text{where} \qquad \theta(\Delta x) = \begin{cases} 0 & \text{for} \quad \Delta x \leq 0 \\ 1 & \text{otherwise.} \end{cases} \tag{3}$$

Here, m_i ($i=1,2$) denote the masses of the lower and upper parts, and x_i their respective elongations. Δx_i are the excess elongations which lead to contact forces when the cords collide, and $\theta(\Delta x_i)$ is Heaviside's step function. h_i and k_i are the elastic force constants with and without contact between the two cords, respectively, and η_i and μ_i are the corresponding nonlinear elastic parameters. r_i are the damping constants (viscous resistances), and $s_i(x_i)$ the restoring forces. F_i are forces that depend on the airflow and the elongations.

The cords are treated as bilaterally symmetric. If they collide (i.e. if $\Delta x_i > 0$), additional strong restoring forces are taken into account. The exciting forces F_i depend in a complicated manner on the elongations x_i (see Ishizaka and Flanagan 1972 for details). The terms F_i include the effects of subglottal pressure and contain phenomenological constants describing the pressure drop at constrictions. Bernoulli forces are also accounted for.

In addition to the above model, equations for the volume flow through the glottis and the tract have to be solved. These equations are coupled to the cord oscillations, since the Bernoulli forces depend on the volume flow. In our simulations we have chosen a two-tube approximation of the vocal tract.

A modified version of the described two-mass ansatz was used more recently by Childers et al. (1986) to connect vocal fold motion to electroglottographic waveform and to simulate vocal fry and a nodule or polyp on one fold.

Integrating the above model using physiologically relevant parameters, self-sustained oscillations are found which exhibit many features of human voiced speech. A characteristic example of regular oscillations is shown in Fig. 4. Entrainment of the two oscillators (see Fig. 4a) occurs over a wide range of parameter values. When varying the fundamental frequency, the mechanical properties of the cords and the coupling between the masses, regimes with more complicated behavior are observed.

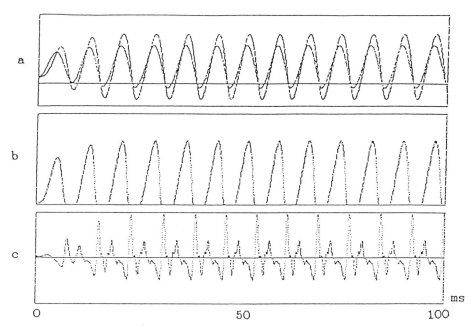

Figure 4. Simulation of voiced sound.

     (a) Elongations x_1 and x_2 of the two masses
     (b) Volume velocity at the glottis
     (c) Mouth sound pressure

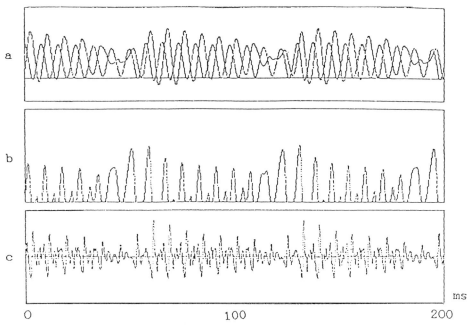

Figure 5. Toroidal oscillations leading to low pitched sound.

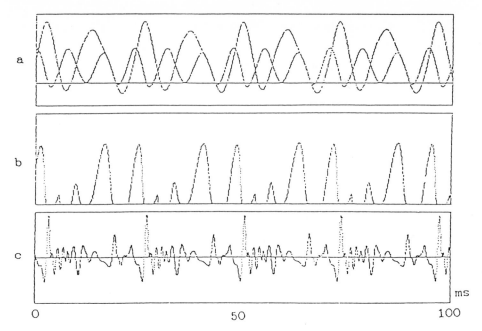

Figure 6. Entrainment of the two oscillators with a frequency ratio of 2:3. This resonance leads to an acoustic signal with a multiple period of normal voice as in Fig. 4. Consequently, subharmonics appear in the spectrum.

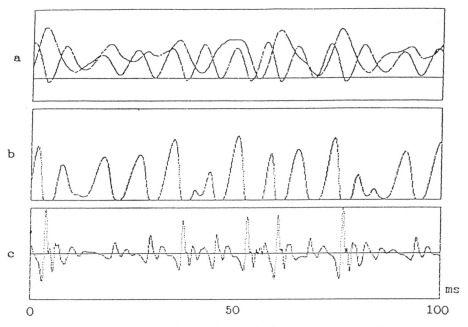

Figure 7. Chaotic oscillations in the two-mass model.

48

In particular for low pitch and weak coupling, we have found the rich bifurcation structure of coupled oscillators which is discussed for example in Ebeling et al. (1990). Characteristic simulations of toroidal oscillations, 2:3 entrainment and chaotic motion are given in Figs. 5-7. Thus, the two-mass model can reproduce several phenomena of experimentally observed phonatory signals: Low frequency oscillations, subharmonic regimes and irregular vocal fold motion.

In the two-mass model, complicated regimes result from the interaction of the lower and upper part of the vocal cords. However, in real speech the coupling of other oscillating modes can lead to subharmonics and chaos as well. For example, the asynchronous motion of the left and right cord, surface waves, or the effect of the ventricular folds ("false vocal cords") might play a certain role. It is interesting to note that chaos was found also in a more sophisticated biomechanical model (Titze, private communication).

CONCLUSION

Summarizing our experimental results, we observe that there is a rich dynamics during infant cries. Cries exhibit periodic oscillations, subharmonic regimes, beating-like behavior and chaotic episodes. Perhaps the frequency of occurrence and the duration of chaotic episodes could have diagnostic relevance. The qualitative agreement between observed irregularities and the simulations suggests that low-dimensional nonlinear dynamics plays a key role in understanding certain aspects of human speech. We are convinced that normal speech corresponds to a synchronized motion, whereas frequency breaks and irregularities can be interpreted as effects at the borderline of the 1:1 resonance zone.

REFERENCES

Arends, N., Povel, D.-J., van Os, E., and Speth, L., 1990, Predicting voice quality of deaf speakers on the basis of glottal characteristics, J. Speech Hearing Res. 33:116.

Bowler, N.W., 1964, A fundamental frequency analysis of harsh voice quality, Speech Monogr. 31:128.

Catford, J.C., 1977, "Fundamental Problems in Phonetics," Edinburgh University Press.

Childers, D.G., Alsaka, Y.A., Hicks, D.M., and Moore, G.P., 1986, Vocal fold vibrations in dysphonia: model vs. measurement, J. of Phonetics 14:42.

Childers, D.G., Smith, A.M., and Moore, G.P., 1984, Relationship between electroglottograph, speech, and vocal cord contact, Folia phoniat. 36:105.

Dolansky, L. and Tjerlund, P., 1968, On certain irregularities of voiced-speech waveforms, IEEE Trans. AU-16, 1:51.

Ebeling, W., Engel, H., and Herzel, H., 1990, "Selbstorganisation in der Zeit," Akademie-Verlag, Berlin.

Eskenazi, L., Childers, D.G., and Hicks, D.M., 1990, Acoustic correlates of vocal quality, J. Speech and Hearing Res. 33:298.

Fant, G., 1960, "Acoustic Theory of Speech Production" s'Gravenhage, Mouton.

Flanagan, J.L., 1972, "Speech Analysis, Synthesis and Perception," Springer, New York.

Hecker, M.H.L., and Kreul, E.J., 1971, Descriptions of the speech of patients with cancer of the vocal fold, J. Acoust. Soc. Am. 49:1275.

Helmholtz, H., 1870, "Die Lehre von den Tonempfindungen als physiologische Grundlage für die Theorie der Musik," Vieweg und Sohn, Braunschweig.

Howard, D.M., 1989, Peak-picking fundamental period estimation for hearing prostheses, J. Acoust. Soc. Am. 86:902.

Ishizaka, K. and Flanagan, J.L., 1972, Synthesis of voiced sounds from a two-mass model of the vocal cords, Bell. Syst. Techn. J. 50:1233.

Laver, J., 1980, "The Phonetic Description of Voice Quality," Cambridge Univ. Press.

Lind, J., 1965, "Newborn Infant Cry," Almquist and Wiksells Boktrycken, Uppsala.

Mende, W., Herzel, H. and Wermke, K., 1990a, Bifurcations and chaos in newborn infant cries, Phys. Lett. A 145:418.

Mende, W., Wermke, K., Schindler, S., Wilzopolski, K., and Höck, S., 1990b, Variability of the cry melody and the melody spectrum as indicators for certain CNS disorder, Early Child Development, in press.

Sorokin, W.N., 1985, "Theory of Speech Production," Moscow.

Titze, I.R., 1976, On the mechanics of vocal-fold vibration, J. Acoust. Soc. Am. 60:1366.

THE ULTRADIAN CLOCK: TIMEKEEPING FOR INTRACELLULAR DYNAMICS

David Lloyd[*]

Institute of Biochemistry
Odense University
Campusvej 55, DK-5230   Odense M

Evgenii I. Volkov

Department of Theoretical Biophysics
P. N. Lebedev Institute
Leninskii 53, Moscow, USSR

INTRODUCTION

"One good experiment is worth a thousand models" (BÜNNING); but one good model can make a thousand experiments unnecessary.

The living organism is an ensemble of systems with relaxation times which span a wide range of time-scales. The potential usefulness of periodic behaviour (Table 1) has not been overlooked by the selection processes of evolution, and examples are given for each class of functions currently recognized in biology.

Table 1. Some functions of Oscillations
in Living Systems

Process	Reference
1. Temporal organization	
a. Coordination of processes in separate subcellular compartments	3
b. Separation of incompatible processes	14
c. Phase entrainment	4
d. Frequency entrainment	5
e. Prediction (e.g. of dawn or dusk)	6
2. Spatial organization	
a. Positional information specification	7
3. Signaling	
a. Increased range	8
b. Frequency encoding	9
4. Increased energy efficiency	10

---

[*]Present address: Microbiology Group (PABIO), University of Wales College of Cardiff, P.O. Box 915, Cardiff CF1 3TL Wales, U.K.

Complexity, Chaos, and Biological Evolution, Edited by E. Mosekilde and
L. Mosekilde, Plenum Press, New York, 1991

Despite the fact that techniques now available provide access to many of these time domains at the molecular level in vitro[1], accessibility in vivo by non-invasive means is more limited, especially for the individual functional unit (the single cell or the intact organism). In order to discern the time structure (Fig. 1), it is necessary to work with single cells or with highly synchronized populations. Most of the biochemical information currently available has been obtained by the use of samples taken from heterogeneous populations of cells or organisms, and much of the "steady state" behaviour described represents a time-average[11].

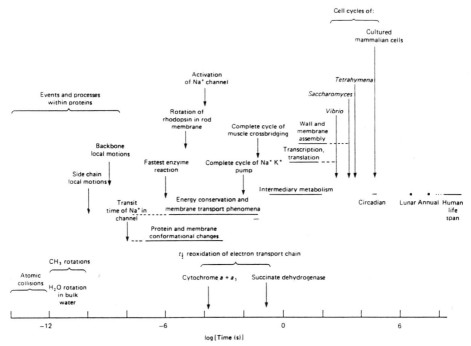

Fig. 1. The time domains of living systems (reproduced with permission from Biochem. J. **242**, 314 (1987).

METHODOLOGY

The experimental results that fit the model described in this paper were obtained with populations of lower eukaryotes growing and dividing synchronously[12]. To obtain these suspensions, organisms (e.g. the soil amoeba, Acanthamoeba castellanii) were sedimented by low-speed (10 g) centrifugation for a short time (2 min) while still in their growth media. The most slowly-sedimenting subpopulations (recently-divided organisms) were decanted and cultured separately. Control (asynchronous) cultures were also simultaneous studied: these were obtained by mixing sedimented and non-sedimented organisms after centrifugation. The asynchronous cultures enable assessment of possible perturbative influences of the experimental procedures. Perfect synchrony is never obtained in practice and high resolution of temporal organization requires observation of single organisms (e.g. Paramecium tetraurelia).

EXPERIMENTAL RESULTS

In synchronous cultures of lower eukaryotes we have observed oscillations of many different biochemical parameters; periods are of the order of one hour and are a characteristic[13] of the species studied. The most easily monitored is cellular respiration, as measured by oxygen consumption[12]. Other observables include total cellular protein[14], activities of various enzymes[15], spectrophotometrically or immunologically-detectable enzyme protein (e.g. for cytochrome c oxidase[15] or catalase[17] respectively), and a natural inhibitor of mitochondrial ATPase[18]. These parameters show no oscillatory behaviour in asynchronous controls, and the oscillations are therefore not the consequences of external perturbation: They are rather the expression of an endogenous rhythm. Originally termed "epigenetic oscillations "[19], the preferred term is now "ultradian rhythms"[20]. This is because they are outputs of an ultradian clock, and are different from all the other high frequency oscillatory phenomena hitherto described, in that they exhibit a temperature-compensated period. This indicates a timing function, and it was predicted that an important example would be that of cell division[21]. As the interdivision time of organisms is strongly temperature-dependent we suggested that control by a temperature-independent timer would give rise to an increase in cell cycle times by discrete quantal increments: hints of this process had been noted for A. castellanii. This model could explain previously observed quantal differences in cell cycle times in cultured mammalian cell lines[22].

Work with single cells of the large ciliate protozoon, Paramecium tetraurelia further confirmed the operation of an ultradian clock and added a behavioural rhythm, that of motility, to the list of coupled outputs[23]. Furthermore, measurement of the generation times of more than 600 individuals at different growth temperatures indicated clustering rather than random division times. The clusters are separated by a time interval which corresponds to the period of the motility rhythm. Again temperature compensation occurs, and values were 70.0 min at 33°C, 69.2 min at 27°C and 71.8 min at 21°C.

It has been repeatedly suggested that the period of the circadian clock might be generated by frequency demultiplication (i.e. counting every nth output) of a higher frequency oscillator. Recent evidence indicated the simultaneous operation of ultradian oscillations and the circadian clock (e.g. for chlorophyll accumulation in Chlamydomonas reinhardii[24] and motility in Euglena gracilis)[25].

MATHEMATICAL MODEL

The generation of quantized cell cycle times comes about when an ultradian clock output interacts with the cell cycle (mitotic) oscillator[26]. We have based the present model on that of Chernavskii[27] in which one slow (τ_L, of the order of hours) and one fast (τ_R, of the order of minutes) variables are described by the following system of equations:

$$\tau_L \frac{dL}{dt} = \eta - 2LR - DL$$

$$\tau_R \frac{dR}{dt} = \mathcal{H} + LR - R^2 - \frac{\gamma R}{(R + \delta)} \qquad (1)$$

where L is the concentration of oxidizable lipids, R is the concentration of membrane lipid free radicals, τ_L is the characteristic time for membrane lipid oxidation, τ_R is the characteristic time for membrane-lipid free radical formation, and η, D, $\mathcal{H}$, γ and δ are velocity constants for lipid insertion and decay, radical formation, and anti-oxidant insertion and decay respectively. Both the slow and the fast components oscillate with the same period, but with very different waveforms (Fig. 2a). The slow variable has a non-symmetrical time dependence and its flat part (where dL/dt ≃ 0) becomes more prolonged as the <u>system</u> approaches bifurcation (the non-proliferative state). The limit cycle behaviour stems from three different sources of non-linearity in the model based on membrane lipids (free radical chain reactions, radical recombination, and radical-anti-oxidant interaction) but the anti-oxidant component can be represented algebraically because of quasi-stationarity.

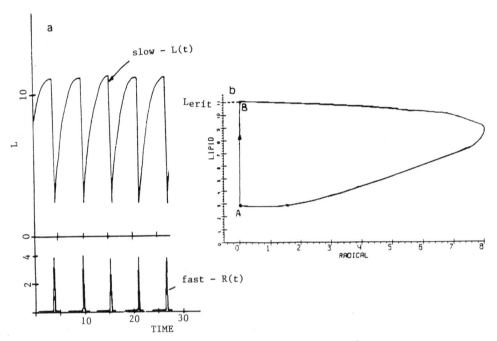

Fig. 2. The solution of the model: η = 5.7, D = 0.4, $\mathcal{H}$ = 0.15, γ = 1.5, δ = 0.15, τ_L = 1, and τ_R = 0.1. (a) Time dependence of slow and fast variables; (b) Phase portrait for oscillating regime. AB is slow part of limit cycle. Reproduced with permission from BioSystems **23**, 305 (1990).

Modulation by an output of the ultradian clock (period T_{UR}) requires introduction of a harmonic term into the slow equation:

$$\tau_L \frac{dL}{dt} = \eta - 2LR - DL + C \sin \Omega t \qquad (2)$$

where

$$T_{UR} = \frac{2\pi}{\Omega} \ll T_{\text{cell cycle}}$$

Figure 2a shows the auto-oscillating solution of the model when C = 0; the phase portrait of the system is also shown (Fig. 2b).S is triggered when L(t) reaches L_{crit} and the rapid part of the cycle initiates. The value of the rate constant η above the bifurcation value has a marked effect on generation time ($T_{cell\ cycle}$).

Real systems always show dispersion of generation times, and therefore, in order to add realistic fluctuations, a noise term ($\Delta\eta \cdot \zeta_i$) is included[28].

$$\tau_L \frac{dL}{dt} = \eta - 2LR - DL + \Delta\eta \cdot \xi_i + C\sin\Omega t$$

When ($\eta - \eta_{BIF}$) <<1, $T_{cell\ cycle}$ has either a bimodal solution or a polymodal one, depending on the value of η (Fig. 3).

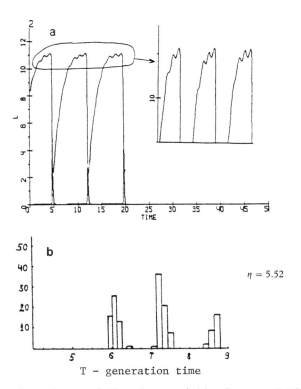

Fig. 3. (a) Time dependence of the slow variable for η = 5.52;
(b) Generation time distributions obtained for the system incorporating noise when τ_L = 1; τ_R = 0.1;$\mathcal{H}$ = 0.15; D = 0.4; γ = 1.5; δ = 0.15; Δη = 0.3; ξ = random numbers, ε[0.1]; C = 0.45; T_{UR} = 1.25 and for η = 5.52. Reproduced with permission from BioSystems **23**, 305 (1990).

Where biomodality disappears it may be restored by changing the value of C. Figure 3 also shows how the quantization is generated; the deterministic time dependence L(t) is shown for $\eta = 5.52$. The ultradian rhythm modulates the trajectory and makes some parts of it approach L_{crit}, and in the presence of noise each protrusion increases the probability of actually reaching the threshold for cell division. The number of peaks depends on the length of the flat part of L(t) when $(L_{crit} - L(t)) \ll 1$.

We conclude that quantization of cell cycle times requires : (i) operation in the neighbourhood of the bifurcation (where $\eta - \eta_{BIF} \ll 1$), (ii) that the period of the cell cycle depends strongly on η, and (iii) that the cell cycle oscillator be of the relaxation type rather than the saw-toothed type (with constant derivative DL/dt).

DISCUSSION

There have been many models of the mitotic oscillator[29-33] control of cell division and several suggestions for the identity of the putative oscillator. Experiments with cell cycle mutants of various lower eukaryotes (yeast, C. reinhardii, Physarum) in the late 1970's led to the conclusion that two major types of control were both essential[1]. These were, (i) a sizer (organisms only divide when they become large enough, and (ii) a timer (a certain time-span ensures completion of all necessary process). The past decade has seen the identification of cyclin, a molecular oscillator common to yeast (cdc-13 gene product), starfishes, frogs and mice, and of the protein kinase (cdc-2 gene product in yeast) that catalyses its phosphorylation[34]. This MPF (maturing promoting factor) system has been termed a cellular clock; indeed the cell cycle itself has often been called a clock[19]. But an oscillator is only one component of a clock.

The present model resolves the paradox. It has both sizer (threshold) and timer (ultradian clock) elements, and it explains how a temperature-dependent oscillator (the mitotic oscillator) is timed. Although conceived for a lipid-free radical mechanism it is really a general concentration-state dependent model, in which not only does a critical concentration (threshold) have to be attained, but a set value of the phase angle of the clock has also to be satisfied. The slow variable could be cyclin and the fast one activated (dephosphorylated) cdc-2 protein. The novel feature of the present model is the participation of the ultradian clock as a forcing function in the differential equation for the slow variable. Interaction of the ultradian clock pulses with the cell cycle oscillator generates quantized cell cycle times similar to those observed experimentally.

The output of the ultradian clock has many functions[19,20]. As well as those already mentioned, evidence from diverse systems suggest that the circadian clock is composed from ultradian rhythms (Fig. 4). Thus quantized departures from the wild-type period have been observed in clock mutants of Neurospora crassa[35]. The so-called arhythmic per[o] mutant of Drosophila melanogaster shows ultradian rhythmicity when held in constant low-level red illumination[36]; further work suggest that the function of the per[+] gene products is to mediate the coupling of multiple ultradian oscillators to produce wild-type circadian rhythms[37,38]. Disruption of normal rhythmicity of Calliphora vicina by constant bright light[39], or of rats and hamsters by exposure to low temperatures[40] or by mutation[41], reveals the composite ultradian rhythms. The damping of circadian rhythms at high unvarying levels of illumination[6] may be due to desynchronization of coupled oscillators. Finally, the hourly ultradian clock may not be the sole timekeeping instrument for intracellular processes, as there is also a very-high frequency clock that controls Drosophila courtship song[42].

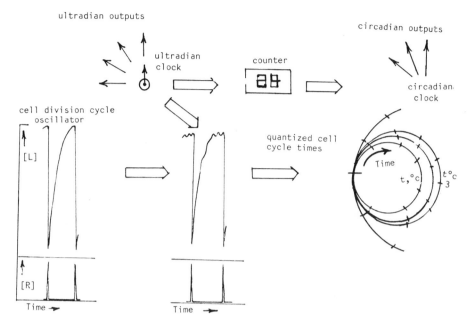

Fig. 4. The ultradian clock has multiple outputs (rhythms of respiration, adenine nucleotides, accumulating protein, enzyme activities and amounts): it also interacts with the cell division cycle oscillator to give quantal increments in cell cycle time as temperature is decreased. Ultradian clock pulsesmay also be counted to give circadian timekeeping of diverse cell functions, including cell division time.

ACKNOWLEDGEMENT

The authors wish to acknowledge support from the British Council and the Royal Society.

REFERENCES

1. D. Lloyd, R. K. Poole, and S. W. Edwards, "The Cell Division Cycle: Temporal Organization and Control of Cellular Growth and Reproduction", Academic Press, London (1982).
2. D. Lloyd, The cell division cycle, Biochem. J. 242:313 (1987)
3. G. Turner and D. Lloyd, Effects of chloramphenicol on growth and mitochondrial function of the ciliate protozoon Tetrahymena pyriformis strain ST, J. Gen. Microbiol. 67:175 (1971).
4. L. N. Edmunds Jr., Physiology of circadian rhythms in micro-organisms, Adv. Microb. Physiol. 25:61 (1984).
5. D. Noble, "The Initiation of the Heartbeat" Clarendon Press, Oxford (1975).
6. E. Bünning, "The Physiological Clock, Circadian Rhythms and Biological Chronometry, 3rd Edition, English Universities Press, London (1973).

7. B. C. Goodwin and M. H. Cohen, A phase-shift model for spatial and temporal organization in developing systems. J. Theor. Biol. 25:59 (1969).

8. V. Nanjundiah, Chemotaxis, signal relaying and aggregation morphology, J. Theor. Biol. 42:63 (1973).

9. P. E. Rapp, A. I. Mees, and C. T. Sparrow, Frequency dependent biochemical regulation is more accurate than amplitude dependent control, J. Theor. Biol. 90:531 (1981).

10. P. H. Richter and J. Ross, Concentration oscillations and efficiency: Glycolysis, Science 211:715 (1981).

11. D. Lloyd and S. W. Edwards, Temperature-compensated ultradian rhythms in lower eukaryotes: periodic turnover coupled to a timer for cell division, J. Interdiscipl. Cycle Res. 77:321 (1986).

12. D. Lloyd and S. W. Edwards, Oscillations of respiration and adenine nucleotides in synchronous cultures of Acanthamoeba castellanii: mitochondrial respiratory control in vivo, J. Gen. Microbiol. 108:197 (1978).

13. D. Lloyd and S. W. Edwards, Epigenetic oscillations in synchronous cultures of lower eukaryotes, in "Chronobiology and Chronomedicine: Basic Research and Applications", G. Hildebrandt, R. Moog and F. Rashke, eds., Peter Lang, Frankfurt am Main (1987).

14. S. W. Edwards and D. Lloyd, Oscillations in protein and RNA content during synchronous growth of Acanthamoeba castellanii: evidence for periodic turnover of macromolecules during the cell cycle, FEBS Lett. 109:21 (1980).

15. V. Michel and R. Hardeland, On the chronobiology of Tetrahymena III. Temperature compensation and temperature dependence in the ultradian oscillator of tyrosine aminotransferase, J. Interdiscipl. Cycle Res. 16:17 (1985).

16. D. Lloyd, S. W. Edwards, J. L. Williams, and J. B. Evans, Mitochondrial cytochromes of Acanthamoeba castellanii: oscillating accumulation of haemoproteins, immunological determinants and activity during the cell cycle, FEMS Lett. 16:307 (1983).

17. S. W. Edwards, J. B. Evans, and D. Lloyd, Oscillatory accumulation of catalase during the cell cycle of Acanthamoeba castellanii, J. Gen. Microbiol. 125:459 (1981).

18. S. W. Edwards, J. B. Evans, J. L. Williams, and D. Lloyd, Mitochondrial ATPase of Acanthamoeba castellanii: oscillating accumulation of enzyme activity, enzyme protein and F_1 - inhibitor during the cell cycle, Biochem. J. 202:453 (1982).

19. D. Lloyd and S. W. Edwards, Epigenetic oscillations during the cell cycles of lower eukaryotes are coupled t o a clock: Life's slow dance to the music of time, in "Cell Cycle Clocks, L. N. Edmunds, ed., Marcel Dekker, New York (1984).

20. D. Lloyd and S. W. Edwards, Temperature-compensated ultradian rhythms in lower eukaryotes: timers for cell cycle and circadian events?, in "Advances in Chronobiology, Part A", J. E. Pauly and L. E. Scheving, eds., Alan Liss, New York (1987).

21. D. Lloyd, S. W. Edwards, and J. C. Fry, Temperature-compensated oscillations in respiration and cellular protein content in synochronous cultures of Acanthamoeba castellanii, Proc. Natn. Acad. Sci. U.S.A. 79:3785 (1982).

22. R. R. Klevecz, Quantized generation times in mammalian cells as an expression of the cellular clock, Proc. Natn. Acad. Sci. U.S.A. 73:4012 (1976).

23. D. Lloyd and F. Kippert, A temperature-compensated ultradian clock explains temperature-dependent quantal cell cycle times, in "Temperature and Animal Cells", K. Bowler and B. J. Fuller, eds., Society Experimental Biologists, Cambridge University Press (1987).

24. H. Jenkins, A. J. Griffiths, and D. Lloyd, Simultaneous operation of ultradian and circadian rhythms in light-dark synchronized cultures of Chlamydomonas reinhardii, J. Interdiscip. Cycle Res. 21:75 (1990).

25. K. J. Adams, Circadian clock control of an ultradian rhythm in Euglena gracilis, in "Chronobiology and Chronomedicine: Basic Research and Applications", E. Morgan, ed., Peter Land, Frankfurt am Main (1989).

26. D. Lloyd and E. I. Volkov, Quantized cell cycle times: interaction between a relaxation oscillator and ultradian clock pulses, BioSystems 23:305 (1990).

27. D. S. Chernavskii, E. K. Palamarchuk, A. A. Polezhaev, G. I. Solyanik, and E. B. Burlakova, Mathematical model of periodic processes in membranes with application to cell-cycle regulation, BioSystems 9:187 (1977).

28. A. T. Mustafin and E. I. Volkov, On the distribution of cell cycle generation times, BioSystems 15:111 (1982).

29. E. E. Sel'kov, Two alternative self-oscillating stationary states in thiol metabolism - two alternative types of cell division normal and malignant ones, Biophysika 15:1065 (1970).

30. E. Zeuthen, Induced reversal of order of cell division and DNA replication in Tetrahymena, Expl. Cell Res. 116:39 (1978).

31. D. A. Gilbert, The nature of the cell cycle and the control of cell replication, BioSystems 5:197 (1974).

32. D. A. Gilbert, The cell cycle 1981. One or more limit cycle cycle oscillations? S. Afr. J. Sci. 77:541 (1981).

33. S. A. Kauffman and J. J. Wille, The mitotic oscillator in Physarum polycephalum, J. Theor. Biol. 55:47 (1975).

34. S. Pelech, When cells divide, The Sciences 1:23 (1990).

35. G. F. Gardner and J. F. Feldman, The frq locus in Neurospora crassa: a key element in circadian clock organization, Genetics 96:877 (1980).

36. C. Helfrich, Untersuchungen über das circadiane System von Fliegen, Dissertation, University of Tübingen, Tübingen (1985).

37. H. B. Dowse and J. M. Ringo, Further evidence that the circadian clock in Drosophila is a population of coupled ultradian oscillators, J. Biol. Rhythm 2:65 (1987).

38. H. B. Dowse, J. C. Hall, and J. M. Ringo, Circadian and ultradian rhythms in period mutants of Drosophila melanogaster, Behav. Genet. 17:19 (1987).

39. F. P. Gibbs, Temperature dependence of the hamster circadian pacemaker, Amer. J. Physiol. 244:R607 (1983).

40. W.-R. von Grosse, Zur endogenen Grundlage der circadianen Aktivität bei Calliphora vicina R. D., Zool. Jb. Physiol. 89:49 (1985).

41. D. Buttner and F. Wollnick, Strain-differentiated circadian and ultradian rhythms in locomotory activity of laboratory rat, Behav. Genet. 14:137 (1984).

42. C. P. Kyriacou and J. C. Hall, Circadian rhythm mutations in Drosophila melanogaster affect a short-term fluctuation in the male courtship song, Proc. Natl. Acad. Sci. U.S.A. 77:6729 (1980).

Section II
Complex Dynamics in Physiological Control Systems

Contrary to the conventional assumption of homeostasis, many physiological and biochemical control systems are unstable and operate in a pulsatory or oscillatory mode.

COMPLEX DYNAMICS IN THE KIDNEY MICROCIRCULATION

Donald J. Marsh, N.-H. Holstein-Rathlou, K.-P. Yip,
and Paul P. Leyssac

Department of Physiology and Biophysics,
University of Southern California, Los Angeles, CA, and
Institute for Experimental Medicine, Panum Institut
Copenhagen University, Copenhagen

INTRODUCTION

Maintaining the volume and composition of the body fluids within narrow bounds is one of the chief functions the kidneys perform. The successful achievement of this goal provides other tissues and organs the stable environment needed for their own particular functions. One of these other organs is the heart, whose action leads to perfusion with blood of organs like the kidney. The proper function of the kidneys depends on adequate blood perfusion, and the kidneys play an especially important role in blood pressure regulation by maintaining the volume of the extracellular fluid,and therefore of the blood. Normal blood volume is important for the heart to achieve stable blood pressure. The relationships between renal function and blood pressure regulation form an essential duality that is crucial to normal function, and that fails in a number of disease states. The interaction between kidney function and blood pressure regulation occurs at a number of points, and analysis of the dynamics invariably provides an informative point of departure.

BLOOD PRESSURE DYNAMICS

Although it is commonplace for physiologists to strive to achieve rigid control of experimental conditions, mammals normally live in an environment that changes continually. Animals must respond to these environmental stimuli, and the responses form a dynamic pattern that can be analyzed with the standard methods of dynamics. Animals can also generate their own dynamics. Figure 1 shows a blood pressure record made over 3 days in a conscious dog, allowed to move about a normal enclosure. The record reveals a pattern of variability, with no particular periodic signal visible.

Figure 2 shows the power spectrum calculated from this record. There is a linear regression between the logarithm of the power density and the logarithm of the frequency; the slope is -1.35. The spectrum is of the type usually referred to as $1/f$. Such spectra are widespread in nature, and the systems that generate them are the results of processes with many overlapping time scales (2). The blood pressure is the resultant of very many processes occurring in different vascular beds throughout the body, each of which has regulation of a local process as a design goal, so that the generation of a $1/f$ spectrum is quite consistent with the known physiology of blood pressure regulation. An implication of this description is that there is no

Complexity, Chaos, and Biological Evolution, Edited by E. Mosekilde and
L. Mosekilde, Plenum Press, New York, 1991

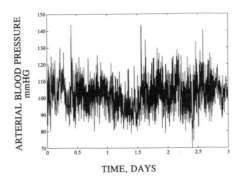

Figure 1. Continuous blood pressure tracing made from a conscious dog. The recording was low pass filtered to remove the heartbeat signal. From (1).

single dominant mode in which the system regulating blood pressure operates. The presence of some dominant frequency would signal such a mode, but none is present. The records that generate *1/f* spectra are fractal curves (3). The average fractal dimension estimated from recordings made from several animals was 1.82. There is no complete description of the system regulating blood pressure, so that the estimation of the fractal dimension is not yet more informative as a description of the system than the regression coefficient calculated from records like Figure 2.

The pressure shown in Figure 1 was recorded in the aorta, which distributes blood to the various organs of the body. As an organism goes about its daily life the needs for perfusion of different organs change, and the blood pressure must change accordingly. But a blood pressure change required to satisfy the perfusion requirements of one organ may be inappropriate for the functional needs of others. For example, there is a positive correlation between blood pressure power and activity of anti-gravity skeletal muscles in the frequency band 1 cycle/8hrs to 1 cycle/60 min (4). The variation in blood pressure is driven from the brain stem (5), which stimulates the heart to increase cardiac output so as to provide increased perfusion of skeletal muscle when muscular contraction is required. This variation in blood pressure could, if unopposed, cause a variation in kidney blood flow. Much of the activity of the kidney depends on the local blood flow rate. Should the kidney blood flow vary according to a pattern that is inappropriate for the needs for regulation of extracellular fluid volume, blood pressure regulation will fail and mammals would be forced to function differently than they do now. The kidney solves this problem by adjusting the admittance of arterioles as blood pressure varies, reducing the amplitude of fluctuations in flow that variations in blood pressure might cause. The process responsible for adjusting the admittance is known as autoregulation; it occurs in many different organs, but is especially effective in the kidney. Autoregulation in most organs, including the kidney, depends on sensors and effectors located within the organ.

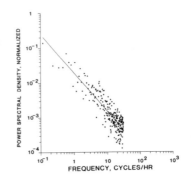

Figure 2. Power spectrum from record shown in Figure 1.

RENAL BLOOD FLOW DYNAMICS

Figure 3 indicates how effective autoregulation is in opposing the natural occurring fluctuations in blood pressure, and the frequency band over which it is active. The figure combines results from two different studies. Data covering the lower frequencies were obtained from measurements in conscious dogs, using the spontaneously occurring fluctuation in blood pressure as the input, and measured renal blood flow as the output (1). Data for the higher frequencies were taken from a study in anesthetized rats, using imposed fluctuations in blood pressure as the input (6). Because of the small amplitude of the inputs, linear methods were used in both studies to estimate the transfer function. The figure shows that fluctuations in blood pressure are reduced approximately 7 db - about 55% - by autoregulation, over a frequency band beginning at about 4 /day and extending to 150 mHz. When account is taken of phase shifts, the efficacy of autoregulation is considerably higher. Because the magnitude of the transfer function is relatively constant over the frequency band in which autoregulation is active, it is probably justified to assume that the mechanisms providing the adjustments of blood vessel admittance operate in a single mode over this band. A major issue of concern to physiologists has been to identify the mechanisms responsible for autoregulation. A strategy dictated by our interest in the dynamics is to concentrate attention on the narrower frequency bands that contain the transitions from high to low admittance. Virtually nothing is known of the mechanisms determining the lower frequency transition, and we will therefore confine our remarks to the higher frequency transition.

The mammalian kidney consists of a large number of similar tubular structures lined by a single layer of epithelial cells that originate in an enclosed cluster of capillaries called the glomerulus. The tubules, often called nephrons, are filled with fluid that is to become the final urine. The process of urine formation begins with the formation in the glomerulus of a filtrate of the blood plasma. This fluid is propelled through the tubule by a gradient of hydrostatic pressure and is modified by the tubular epithelial cells. Modifications to the original filtrate include selective reabsorption of many materials and secretion of others. The reabsorbed materials are returned to the blood while materials either not reabsorbed or secreted are eliminated from the body in the final urine. Formation of the glomerular filtrate depends in part on the hydrostatic pressure of blood in the glomerular capillaries, and the rate of excretion of materials in the final urine is dependent on the glomerular filtration rate. The fluctuations in arterial blood pressure could easily cause the rate of glomerular filtration to vary over a range so large that regulation of excretion would become impossible, and assuring some stability of glomerular capillary pressure is therefore essential for successful completion of the kidney's tasks. The primary responsibility for limiting the variation in glomerular capillary pressure rests with a negative feedback system that senses flow rate-dependent changes in the composition of the tubular fluid and transmits information to an arteriole

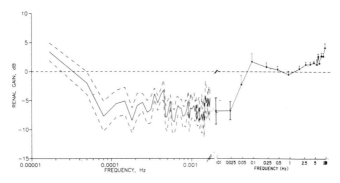

Figure 3. Magnitude of the admittance gain, with arterial pressure as input, and renal blood flow as output. The figure is a composite, using data from conscious dogs (1) for the lower frequencies, and from anesthetized rats(6), for the higher ones.

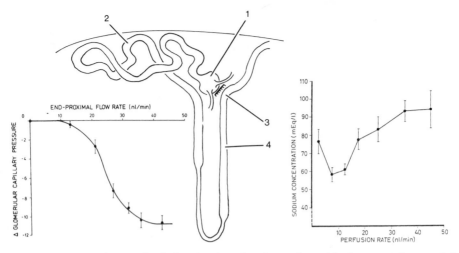

Figure 4. Diagram of a renal tubule, showing the glomerulus with the arterioles supplying and removing blood (1), the proximal tubule (2), the macula densa (3), and the loop of Henle (4). The panels show the effect of changing tubular perfusion rate on chloride concentration at the macula densa sensing site (right), and of changing tubular perfusion rate on the hydrostatic pressure available to cause glomerular filtration, in the open loop nephron (left). The perfusion rate in these experiments was varied at the end of the proximal tubule, where the normal operating point is 15 nl/min.

located between the main renal artery and the glomerulus. The diameter of the arteriole is adjusted in response to changes in the composition of the tubular fluid. The structures comprising the feedback system are shown in Figure 4. The glomerulus receives its blood supply from an arteriole which is also the effector site for the feedback system regulating blood flow. The filtrate is formed in the glomerulus and flows into and along the tubule, which modifies the composition and the volume of the tubular fluid. The tubule has several segments, each with a different function. The tubule forms a loop, and makes contact with the glomerulus again at a point about 1.5 cm from the origin. A specialized collection of tubular cells is found at the return point, and these cells provide a channel for information transfer to the arteriole. The collection of cells is called the macula densa, to indicate the morphological differentiation from other, unspecialized cells.

Regulation of the excretion of sodium chloride, a component of the blood plasma and tubular fluid, is the principal means of controlling extracellular fluid volume and blood pressure. Because of the actions of the tubular epithelial cells, the concentration of NaCl falls, and reaches a value only 40% as high in the tubular fluid at the macula densa as in the glomerular filtrate. The concentration of NaCl at this point varies with the local flow rate of tubular fluid, and provides the signal to the macula densa. Increasing the rate of glomerular filtration, for example, increases flow throughout the tubule, and increases the concentration of NaCl in the tubular fluid at the macula densa. Increasing the concentration has the effect of constricting the arteriole, which reduces glomerular pressure and glomerular filtration rate, restoring the concentration of NaCl at the macula densa. These actions, of flow rate on NaCl concentration, and of NaCl concentration on arteriolar diameter, are shown in Figure 4.

The complete feedback system oscillates spontaneously at about 35 mHz (7-12). Each tubule has oscillations of tubular flow, tubular hydrostatic pressure, and of NaCl concentration (7). All variables oscillate at the same frequency, and in constant phase with each other. Representative examples are shown in Figure 5. The oscillations in flow and pressure indicate that glomerular filtration rate is periodic, and the oscillation in NaCl concentration at the

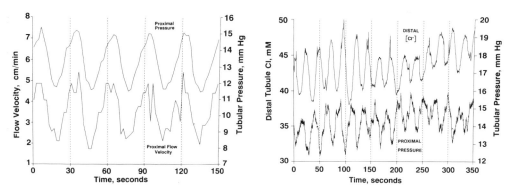

Figure 5. Spontaneous oscillations in proximal tubular pressure and flow (left), and proximal tubular pressure and chloride concentration at the macula densa (right). Different sampling rates were dictated by the temporal resolution of the methods used to measure flow and chloride. From (7).

macula densa indicates that the signal to the feedback system is periodic. The question at issue next becomes to determine whether the oscillation occurs autonomously, or whether it is driven by some pacemaker, examples of which are well known in the body. To answer this question we derived a system of differential equations and boundary conditions to simulate the operation of the tubuloglomerular feedback system(14). If the oscillation is autonomous, the model should have a stable periodic solution; if it fails this test, we are then obliged to identify the pacemaker. The model has a glomerular component and a tubular component, and a feedback component to link the other two. The model is as follows:

GLOMERULAR MODEL

As blood flows through glomerular capillaries, fluid is filtered through membranes that are not permeable to proteins in the plasma. Protein is conserved, and the filtration rate, $Q_T(0)$ can be deduced from the product of the fractional change in protein concentration, C, and the blood flow rate entering the glomerular capillaries , Q_A.

$$Q_T(0) = (1 - \frac{C_A}{C_E})Q_A \qquad (1)$$

The filtration process that causes the change in concentration is proportional to the local pressure difference, which is given by the sum of the capillary hydrostatic pressure, P_G, the tubular pressure, P_T, and the osmotic pressure due to proteins, $\pi(C)$, a nonlinear function that can be approximated as a quadratic from experimental data.

$$\frac{dC}{dx} = \frac{K_f}{LQ_AC_A}(C^2)[P_G - P_T - \pi(C)], \qquad (2)$$

where x is fractional position, K_f is the filtration coefficient, and L is the length of the capillary. The initial condition is determined from the arterial pressure and the computed preglomerular resistance. Time dependent processes in the glomerular capillaries are very rapid compared to tubular processes, and we have therefore assumed that the glomerular system is in a steady state with respect to the rest of the system.

67

TUBULAR MODEL

Equations governing the temporal and spatial behavior of the model are based on local conservation of mass and momentum. Because of the very low Reynolds' number, one dimensional approximations of the Navier - Stokes equations in cylindrical coordinates are suitable.

$$\frac{\partial P_T}{\partial z} = -\frac{\rho}{\pi R^2}\frac{\partial Q_T}{\partial t} - \frac{8\eta}{\pi R^4}Q_T, \quad 0 < z < Z, t > 0 \tag{3}$$

$$\frac{\partial Q_T}{\partial z} = -2\pi R\frac{\partial R}{\partial P_T}\frac{\partial P_T}{\partial t} - J_v(z), \quad 0 < z < Z, t > 0 \tag{4}$$

where P_T is tubular pressure, z is position, t is time, ρ is fluid density, R is tubular radius, Q_T is tubular flow rate, and J_v is the local rate of fluid reabsorption by the renal epithelial cells. Fluid reabsorption is a complex process powered by metabolism, and for our purposes a simple approximation to measured rates of reabsorption is sufficient. One boundary condition, glomerular filtration rate, $Q_T(0)$, is taken from equation (1). The second boundary condition, tubular pressure at the end of the loop of Henle, $P_T(Z)$, is calculated as:

$$P_T(Z) = \frac{Q_T(Z)}{(\alpha P_T(Z) + \beta)^4} \tag{5}$$

where α and β are constants.

The tubular system beyond the macula densa, which is not included in the model, provides a nonlinear hydrodynamic resistance. This nonlinear boundary condition was designed as an approximation to low Reynolds' number flow at low pressure in which the local pressure available to cause axial flow also acts to determine the diameter of the outflow path and therefore its hydrodynamic resistance.

The renal tubules are compliant, and the radius can be calculated from;

$$R(z) = (P_T(z) - P_I)\frac{\partial R}{\partial P_T} + R_0, \tag{6}$$

where R is the radius and P_I is the hydrostatic pressure in the interstitial space surrounding the tubule and assumed constant.

The tubules reabsorb NaCl and water. Experimental measurements show that this process causes no change in NaCl concentration in the proximal tubule. In the loop of Henle, where the NaCl concentration does change:

$$\frac{\partial A C_S}{\partial t} = -\frac{\partial Q_T C_S}{\partial z} - J_S, 0 < z < Z, t > 0, \tag{7}$$

where A is the cross section area of the lumen, C_S is the concentration, and J_S is the local rate of NaCl reabsorption. The boundary condition needed to solve this equation is $C_S = 150mM$ at the end of the proximal tubule.

To calculate NaCl reabsorption in the loop of Henle:

$$J_S = L_S(C_S - C_I) + \frac{V_{max}C_S}{K_m + C_S} \tag{8}$$

The first term on the right hand side represents diffusion across the epithelial cells driven by the local concentration gradient, the second nonlinear term is the conventional Michaelis-Menten expression used to represent active transport that derives its energy from metabolism. Experimental evidence indicates no active transport in the descending limb of Henle's loop, so that Vm=0 for this segment. The interstitial NaCl concentration, C_I, varied linearly from 150 mM at the cortico-medullary boundary to 300 mM at the bend of Henle's loop, to approximate the well known accumulation of NaCl in the renal medulla.

TUBULOGLOMERULAR FEEDBACK FUNCTION AND AFFERENT ARTERIOLAR ACTION

Information about the NaCl concentration in tubular fluid at the macula densa is transmitted to the afferent arteriole. Experimental data(15) shows that the functional relationship between NaCl concentration and arteriolar action is nonlinear and can be approximated by:

$$\phi = \xi_m - \frac{\psi}{1 + exp[k(C_S(\mu) - C_{1/2})]} \tag{9}$$

where ϕ is the action of the feedback, ξ_m is the maximum structural resistance of the afferent arteriole, ψ is the range of the feedback response, and μ is the macula densa.

There is little quantitative information describing the dynamics of the afferent arteriolar response to tubular forcing, and we therefore modeled it as a second order linear response, the simplest model that could simulate the oscillation.

$$\frac{1}{\omega^2}\frac{d^2\xi}{dt^2} + \frac{2\zeta}{\omega}\frac{d\xi}{dt} + \xi = \phi(C_S(\mu)) \tag{10}$$

where ζ is the natural frequency of the system, and ξ is the resistance of the afferent arteriole.

The model equations were solved numerically, and yielded a periodic solution with parameter values consistent with experimentally determined ones. Table 1 provides a quantitative comparison of experimental measures of the oscillation with those predicted by the model. The agreement is good, and there are no discrepancies that would cause the model to be rejected. The model is a reasonably isomorphic respresentation of the nephron, its blood supply, and the control elements that couple them. That the simulation yields a periodic solution with similar properties suggests that the oscillation arises autonomously without a pacemaker. The model results also make clear that their are several lags responsible for the oscillation. The largest lag is due to the accumulation or depletion of NaCl in the tubular fluid of the thick ascending limb of Henle's loop as flow rate changes; this lag arises in the solution of equation 7. The second largest lag is in the signal transmission from the sensing site at the macula densa to the arteriole, which occurs in the solution of equation 10. There is also a small lag that occurs because the tubule walls are compliant, and flow disturbances produce local changes in radius that slow wave propagation.

The model predicts an oscillation in tubular pressure and flow, which has been observed(7); the oscillation involves an action of the arteriole supplying blood to the glomerulus, and the model therefore predicts a corresponding oscillation in blood flow. Blood flow measurements with adequate temporal resolution have only been possible in the renal artery, which supplies blood to 30,000 or more nephrons in the rat kidney, and no oscillations can be found there. As with any measured variable in a biological system, the frequency of oscillation for a population of nephrons shows a distribution of values, and different nephrons are not necessarily in phase. It seemed likely therefore that the failure to find an oscillation

Table 1. Comparison of model output to experimentally obtained values

	Period	Proximal Pressure vs Proximal Flow phase		Proximal vs Distal Pressure phase		Proximal Pressure vs Distal (Cl) phase	
		delay	angle	delay	angle	delay	angle
	sec	sec	rad	sec	rad	sec	rad
Model :	28.0	-2.0	0.45	1.1	-0.25	7.0	-1.57
Experimental Values:							
Mean	29.0	-1.5	0.31	1.1	-0.22	8.9	-2.09
SE	1.0	0.4	0.05	0.4	0.08	0.8	0.17

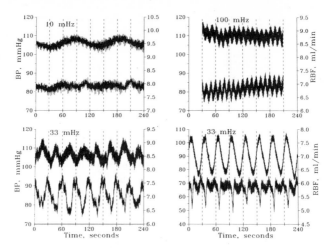

Figure 6. Results of experimental forcing of arterial pressure. Top tracing in each panel is blood pressure, bottom tracing is whole kidney blood flow rate. Upper left: 10 mHz; upper right: 100 mHz; lower left: 33 mHz; lower right: also 33 mHz but higher amplitude than in lower left.

in the whole population could be attributed to the relative incoherence to be expected. It was necessary to test this suggestion, however, before it could be asserted that the oscillation is an appropriate description of nephron dynamics. To provide this test we took advantage of the well known property of autonomous oscillators to become entrained when an external forcing is applied at a frequency near that of the spontaneous oscillation. Figure 6 shows the results of four such forcings, in which arterial blood pressure is made to vary periodically, and whole kidney blood flow is measured. The spontaneous oscillation occurs in the 35 - 50 mHz band. When a forcing is applied at a lower frequency, 10 mHz, the system counters the variation in blood pressure, and blood flow changes little, indicating good autoregulation. At 100 mHz, the mechanical properties of the tubule act as a low pass filter, and there is neither attenuation nor amplification of the forcing. At 33 mHz, when the input amplitude is small, there is a large oscillation in blood flow, with considerable amplification of the input. These results indicate that a large number of nephrons, possibly all, are capable of becoming entrained when a small amplitude forcing is applied near the natural frequency. Since this result can be predicted from the hypothesis that most nephrons have a spontaneous oscillation at nearly the same frequency, we think it is a further indication that the model is a good representation of tubular dynamics. Note that when the amplitude of the 33 mHz forcing was increased, there was a frequency doubling, and a reduction of the amplitude of the blood flow oscillation at 33 mHz. As the input magnitude was increased at this frequency, blood flow spectral power was shifted from 33 to 66 mHz, and there was a phase shift. The most likely explanation for this more complicated pattern is that as the magnitude of the induced flow changes in the ascending limb increased, the lag due to epithelial transport became more pronounced, and produced phase shifts in the signal that caused the frequency doubling. The conclusion from these studies is that the oscillation is a general characteristic of nephron dynamics, that it can be predicted with the model presented above, and that the nonlinearities of the system can lead to qualitative changes in the operation of the system when forcings of sufficient magnitude are applied.

TUBULAR DYNAMICS IN EXPERIMENTAL HYPERTENSION

The results presented to this point are from rats with normal blood pressure. Figure 7 compares the results of tubular pressure measurements from normal rats with those from rats with two different forms of hypertension, or chronic high blood pressure. One group

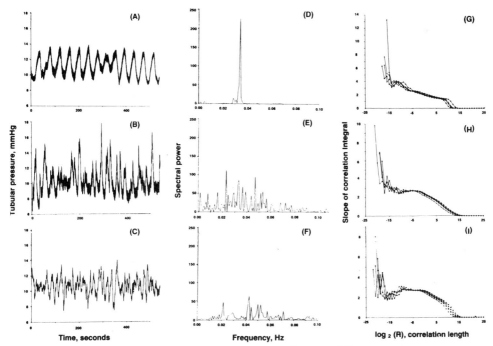

Figure 7. Left column: tubular pressure; top, normal rat; middle, rat with genetic form of hypertension; bottom, rat with renovascular hypertension. Middle column: power spectra. Right column: derivative of log correlation integral with respect to log correlation length, as a function of log correlation length, calculated from time series in left column.

had a genetic form of hypertension, the other had it induced by partial obstruction of one renal artery. Pressure measurements were made from the nonobstructed kidney in the latter group. There is a qualitative change in the tubular pressure record with the development of hypertension; the regular oscillation of normal rats is replaced by an irregular fluctuation. The figure also shows the power spectra. In the spectrum from the normal rat the power is concentrated at a single frequency, but it is more broadly distributed in the spectra from the rats with hypertension. We suspected that there had been a bifurcation to deterministic chaos in hypertension and we therefore estimated the dimension of the attractor, using the Grassberger-Procaccia algorithm(14). Results of these calculations are also shown in the figure. Linear scaling regions were routinely identified. Because of measurement noise and periodicities imposed by experimental circumstances, it was necessary to filter the recorded signals. Estimates of the correlation dimension varied with the cutoff frequency of the filter; the upper limit is approximately 5. The largest Lyapunov exponent was also calculated from each time series, and were were positive in every case. The average largest Lyapunov exponent was $0.10 \pm .01$ in 12 measurements in the spontaneously hypertensive rats, and the same in the renovascular hypertensive group. The presence of a low dimension attractor suggests that the aperiodic time series from hypertensive nephrons arise from an ordered system, while the positive Lyapunov exponents indicate sensitivity to initial conditions, two criteria of deterministic chaos.

The two forms of hypertension yielded similar estimates of attractor dimension and of the largest Lyapunov exponent. These similarities suggest that the bifurcation from the limit cycle oscillation in nephrons from normotensive rats is to similar strange attractors in both forms of hypertension, despite the fact that the underlying causality of the blood pressure rise differs. We suggest that the bifurcation is a consequence of the interaction between blood pressure and renal blood flow control mechanisms. We tested ψ, the magnitude of the feedback action on the afferent arterial, as a bifurcation parameter because experimental results show that it is increased in the genetic model of hypertension we studied. Increasing the value of ψ from an initial low value caused a bifurcation from a steady state solution to a limit cycle oscillation. Some anesthetics act to reduce the magnitude of this parameter, and cause ψ to fall below the bifurcation point; oscillations in tubular pressure are not detected when these anesthetics are used in animals with normal blood pressure(15,16). When anesthetics are used that permit the oscillation to develop, measured values of ψ exceed the level required to induce the oscillation. Still higher values, to levels higher than those found in hypertension, fail to cause further bifurcations. Because we think the model is a good representation of the local system regulating blood flow, we conclude that the change in hypertension that causes the bifurcation to chaos is not yet a part of the simulation. There are structural changes in the afferent arteriole in both types of hypertension and the description of the dynamics of the afferent arteriole in the model is probably oversimplified. Further studies of the dynamic characteristics of the afferent arteriole in both normotensive and hypertensive rats will provide a firmer basis for the modelling of this part of the feedback system.

It has been possible, using an earlier, more simplified model, to simulate chaotic behavior (17). That model had a greatly simplified version of the nephron, and differed by introducing an ad hoc, but not unreasonable, nonlinearity into the differential equation describing afferent arteriolar dynamics. Equation 10 is a second order linear equation; it and the equation from the earlier model were used to provide some form of dynamics for the afferent arteriole, but neither usage can be justified by experimental evidence. This earlier model showed a transition to chaos through a cascade of period doubling bifurcations when either ψ or the total delay in the signal transmission was increased. The prediction that an increase in ψ could be responsible for the bifurcation was initially very attractive, since it could be shown that the gain in the feedback loop was increased in SHR (15). On the other hand, the observation of chaotic fluctuations in the tubular pressure of the unoperated kidney of rats made hypertensive through partial blockade of the other renal artery makes this possibility less likely, because ψ is reduced in the unoperated kidney of these rats, and its change is therefore in the opposite direction from that seen in SHR. To test the prediction that changes in the delay in the transmission of the signal through the feedback loop were of importance in the transition to chaos, a series of experiments were performed where low concentrations of loop diuretics (furosemide and bumetanide) were infused into the tubular fluid (11). The rationale behind these studies was that partial blockade of NaCl transport across the macula densa, a crucial step in the sensing mechanism of the feedback system, would lead to an increase in the response time of the system. When infused in low doses into the tubules of normotensive rats both furosemide and bumetanide caused the oscillations to become irregular. These studies therefore indicate that a change in dynamics of the response of the macula densa/afferent arteriole may be of importance for the transition to chaos. Unfortunately, no studies comparing the dynamics of the macula densa/afferent arteriolar response between normotensive and hypertensive rats have been performed.

The model presented by Jensen et al predicts that the transition from a regular oscillation to a chaotic fluctuation occurs through a cascade of period doubling bifurcations (17). We

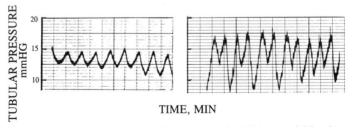

Figure 8. Tubular pressure measurements in an animal with normal blood pressure. Left: normal; right: reduced oxygen tension in inspired air and reduced kidney temperature.

never observed period doublings in rats made hypertensive by partial obstruction of one renal artery. Either the fluctuations were periodic, as in the normotensive animals, or they were irregular with the Lyaponov exponent and correlation dimension as in SHR. On the other hand, it is clear that perturbations of the system in normotensive rats may invoke more complicated dynamics in the tubular pressure. Figure 8 shows an example of a typical period doubling in a normotensive rat provoked by a 2 minute period of ischemia (low oxygen tension) and reduction in temperature to 30° C. The rationale for this experiment was the assumption that low temperature and reduced energy supply might depress the rate of tubular NaCl transport and interrupt signal transmission across the macula densa similar to the action of loop diuretics (furosemide). This observation may not be relevant to an understanding of the bifurcation seen in hypertension, but it does show that the system has the potential for further bifurcations.

We conclude that there is as yet no convincing identification of the parameter change responsible for the bifurcation to chaos in experimental hypertension. More work is needed to identify the changes causing the transition from regular oscillations in normotensive rats to irregular fluctuations in hypertensive rats. The experimental results outlined above lead us to suggest that the changes in the dynamics of the afferent arteriole is of major importance in causing this transition.

In summary, arterial blood pressure has $1/f$ dynamics. These fluctuations in blood pressure, which come from many sources, could cause similar fluctuations in renal blood flow, with deleterious consequences for a number of important renal tasks. The kidney reduces the impact of the blood pressure fluctuations by altering arteriolar blood vessel diameter in response. The mechanism primarily responsible for providing this autoregulation is a local feedback mechanism that uses flow dependent changes of electrolyte concentration at a key tubular site as a signal. The feedback system operates in animals with normal blood pressure as a limit cycle oscillator, but bifurcates in rats with hypertension to deterministic chaos. Preliminary results point to a locus in the afferent arteriole for the bifurcation. One interesting connection that requires further study is between blood pressure dynamics and tubular dynamics. Raising blood pressure acutely does not cause a bifurcation; the oscillation simply increases in amplitude. But raising blood pressure chronically increases the high frequency content of the blood pressure record; the fractal dimension of the blood pressure record is increased. Because of the possibility of long range order implicit in the development of chaos, it will be of considerable interest to determine whether there is a causal link between the change in tubular dynamics and the the change in blood pressure dynamics.

ACKNOWLEDGMENT

Supported by NIH Grants DK15968, DK33729, by a grant from the Whitaker Foundation, and by a grant from the Danish Medical Research Council.

REFERENCES

1. Marsh, D.J., J.L. Osborn, and A.W. Cowley, Jr. $1/f$ fluctuations in arterial pressure and the regulation of renal blood flow in dogs. Am.J.Physiol. 258:F1394 - F1400, 1990.

2. Schlesinger, M F. Fractal time and $1/f$ noise in complex systems. Ann. N.Y. Acad. Sci. 504: 214-228, 1987.

3. Mandelbrot, B.B. Fractals Form, Chance and Dimension. San Francisco, 1977, W.H. Freeman and Co.

4. Blinowska, K and Marsh, D J. Ultra-and circadian fluctuations in arterial pressure and EMG in conscious dogs. Am. J. Physiol. 249:R720-R725, 1985.

5. Livnat, A., Zehr, J E. and Broten, T P. Ultradian oscillations in blood pressure and heart rate in free-running dogs. Am. J. Physiol. 246:R817-R824, 1984.

6. Sakai, T., E. Hallman, and D.J. Marsh. Frequency domain analysis of renal autoregulation in the rat. Am.J.Physiol. 250: F364-F373, 1986.

7. Holstein-Rathlou, N.-H., and D.J. Marsh. Oscillations of tubular pressure, flow, and distal chloride concentration in rats. Am. J. Physiol. 256: F1007-F1014, 1989.

8. Leyssac, P.P. Further studies on oscillating tubulo-glomerular feedback responses in the rat kidney. Acta Physiol. Scand. 58:236-242, 1986.

9. Holstein-Rathlou, N-H. Synchronization of proximal intratubular pressure oscillations: evidence for interaction between nephrons. Pflugers Arch. 408: 438-443, 1987.

10. Holstein-Rathlou, N.-H., and P.P. Leyssac. Oscillations in proximal intratubular pressure: a mathematical model. Am.J. Physiol. 252: F560-F572, 1987.

11. Leyssac, P.P., and N.-H. Holstein-Rathlou. Effects of various transport inhibitors on oscillating TGF pressure responses in the rat. Pflügers Archiv. 407: 285 - 2911, 1986.

12. Holstein-Rathlou, N.-H., and D.J. Marsh. A dynamic model of the tubuloglomerular feedback mechanism. Am.J.Physiol. 258: F1448-F1459, 1990.

13. Briggs, J.P., G. Schubert, and J. Schnermann. Quantitative characterisation of the tubuloglomerular feedback response: effect of growth. Am. J. Physiol. 247: F808-F815, 1984.

14. Grassberger, P., and I. Procaccia. Measuring the strangeness of strange atractors. Physica 9D:189-208, 1983.

15. Leyssac. P.P. and N.-H. Holstein-Rathlou. Tubuloglomerular feedback: enhancement in spontaneously hypertensive rats and effects of anesthetics. Pflügers Archiv. 413: 267-272, 1989.

16. Holstein-Rathlou, N.-H., P. Christensen, and P.P.Leyssac. Effects of halothane-nitrous oxide inhalation anesthesia and Inactin on overall renal and tubular function in Sprague-Dawley and Wistar rats. Acta Physiol. Scand. 114:193-201, 1982.

17. Jensen, K.S., N.-H. Holstein-Rathlou, P.P. Leyssac, E. Mosekilde, and D.R. Rasmussen. Chaos in a system of interacting nephrons. in *Chaos in Biological Systems.*, H. Degn, A.V. Holden, and L.F. Olsen, eds. Plenum Publishing Corp. London, 1987.

ASPECTS OF OSCILLATORY INSULIN SECRETION

Jeppe Sturis,[t][‡] Kenneth S. Polonsky,[t] John D. Blackman,[t]
Carsten Knudsen,[‡] Erik Mosekilde,[‡] and Eve Van Cauter[t]

[t]University of Chicago
Dept. of Medicine, Box 435
Chicago, IL 60637
USA

[‡]Physics Laboratory III
Technical University of Denmark
DK-2800 Lyngby
Denmark

ABSTRACT

Slow oscillations of human insulin secretion have been observed during the last couple of decades. They may be explained either by an unstable insulin/glucose feedback mechanism or in terms of an intrapancreatic pacemaker. In the present paper, a simulation model of the insulin/glucose feedback system is presented which exhibits self-sustained oscillations for realistic parameter combinations when a constant glucose infusion is simulated. Entrainment of the oscillations occurs when an oscillatory rather than a constant glucose infusion is simulated, which suggests that the presence of the oscillations can be explained by the insulin/glucose feedback mechanism. The design and results of clinical experiments performed in non-diabetic humans are described which verify the simulation results, in that entrainment is also experimentally achieved. Finally, additional simulation results are presented which show that the system can have coexisting solutions, and potential clinical implications hereof are discussed.

INTRODUCTION

The complexity of human insulin secretion has become increasingly evident in recent years, in that at least two distinctly different modes of internally generated oscillations have been experimentally observed. Rapid pulses of insulin with a period of 10-15 min have been reported[1,2], and these have been found[3] to be superimposed on slower, larger amplitude oscillations with a period of 80-150 min[3,4,5,6], depending

Complexity, Chaos, and Biological Evolution, Edited by E. Mosekilde and
L. Mosekilde, Plenum Press, New York, 1991

on the individual subject and the experimental conditions. The mechanisms of both types of oscillations have yet to be clearly elucidated, although in vitro experiments using animal pancreases suggest[7] that the rapid pulses originate from an intrapancreatic pacemaker. Here, we shall primarily be concerned with the slow oscillations, trying to provide a plausible explanation for their genesis.

Figure 1 shows examples of the slow oscillations in three different subjects. Certain characteristic dynamic properties have been disclosed for these oscillations: (i) they are self-sustained during continuous delivery of the stimulus (constant glucose infusion[5,6] or continuous enteral nutrition[3]), (ii) they are damped following ingestion of a meal or after a single oral glucose administration[4], (iii) there is a high correlation between glucose and insulin oscillations, (iv) there is a tendency for glucose peaks to precede insulin peaks by a few minutes[6], and (v) an increased delivery rate of the stimulus leads to an increase in amplitude of the oscillations, whereas the period remains unchanged within the limits of detection[6].

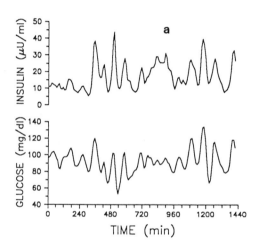

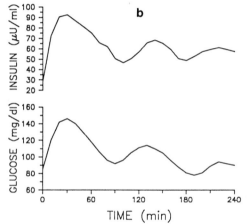

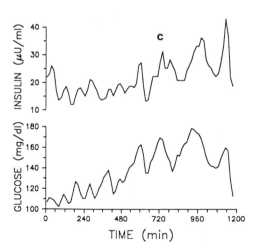

Figure 1. Examples of slow oscillations of insulin and glucose in humans.

a) During continuous enteral nutrition. Redrawn from Simon et al.[3]

b) Following an oral glucose tolerance test. Redrawn from Kraegen et al.[4]

c) During constant glucose infusion. Redrawn from Shapiro et al.[5]

It is of interest to understand how these oscillations are generated. Certain possibilities can be ruled out on the basis of experimental observations. Firstly, the oscillations are not caused by an intermittent uptake of glucose by the gastrointestinal tract, since the oscillations persist during constant intravenous glucose infusion; secondly, preliminary experimental results have shown that the oscillations are present in patients following segmental pancreas transplantation[8], suggesting no dependence on central neurogenic connections; finally, the oscillations do not seem to be generated by counterregulatory hormones, since analyses of simultaneous glucagon and cortisol concentrations have failed to reveal correlations with the insulin and glucose oscillations[5]. Two major hypothetical mechanisms remain: (I) the slow oscillations originate from the activity of an intrapancreatic pacemaker which causes periodic bursts of insulin secretion, forcing the glucose concentration to oscillate as well, or (II) the feedback relationship which exists between glucose and insulin is unstable and produces self-sustained oscillations of glucose and insulin.

Here, we shall pursue the latter hypothesis by presenting first a mathematical model of the insulin/glucose feedback system[9] and then the results of experiments which were designed based upon simulations[10]. Furthermore, various simulation results are presented and discussed.

THE SIMULATION MODEL

Model Description

Insulin/glucose feedback is an important element of the overall metabolic regulation in the human body. Due to the complexity of the involved processes, it would be very difficult to construct a satisfactory model of the complete system. However, it may be possible to isolate the mechanisms underlying the slow oscillations by considering only a relevant part of the metabolic system. Here, we thus hypothesize that the insulin/glucose feedback system suffices to explain the existence of the slow oscillations in pancreatic insulin secretion.

Glucose stimulates insulin secretion, and insulin enhances the uptake of glucose by the cells. Insulin also inhibits endogenous production of glucose. In addition, glucose per se augments its own utilization, representing yet another negative feedback in the system. These are the mechanisms we have chosen to consider in the model.

In figure 2, a flow-diagram of the model is presented. For simplicity and in accordance with other models of insulin/glucose dynamics[12,13], we consider glucose to be uniformly distributed in a single compartment. Insulin, on the other hand, is modeled by a two-compartmental representation, accounting for the slow equilibration between plasma and interstitial fluid for this hormone. Insulin is assumed to act from the compartment representing the interstitium in agreement with recent in vivo

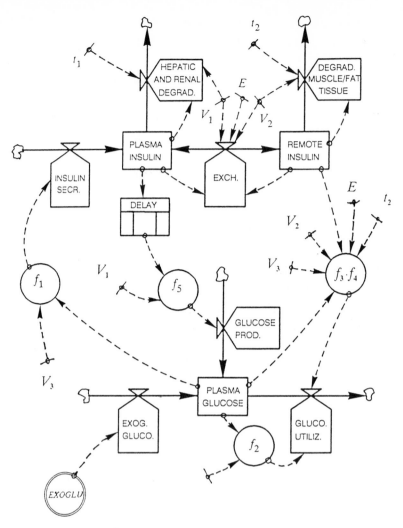

<u>Figure 2.</u> Flow diagram of the model. We have made use of the symbols defined in System Dynamics[11]. Here three of the boxes represent the total amounts of plasma insulin, remote insulin and plasma glucose, respectively, and last box is a third order delay. These are the state variables of the model. The valves represent the rates of production, exchange, infusion or utilization of insulin and glucose. Finally the circles represent the nonlinear controls.

experiments performed in dogs[14] which have shown that glucose utilization is much more highly correlated with lymph insulin than with plasma insulin. In addition to this sluggish insulin action, there is a delay between the appearance of insulin in the plasma and its inhibiting effect upon glucose production[14,15]. Since the exact pathway of this effect is uncertain, we have modeled the delay process as a third order delay with a delay time of 36 min. We tested the plausibility of this approach by using the final

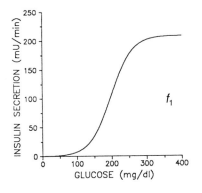

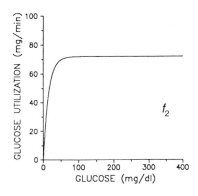

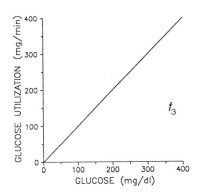

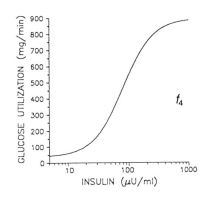

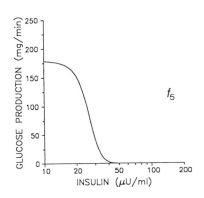

Figure 3. Functional relationships in the model. f_1 is insulin secretion as a function of glucose, f_2 is the insulin-independent term of glucose utilization as a function of glucose concentration, f_3 is the insulin-dependent term of glucose utilization as a function of glucose concentration for an insulin concentration of 18 μU/ml, and f_4 is the insulin-dependent term of glucose utilization as a function of insulin concentration for a glucose concentration of 95 mg/dl, and f_5 is glucose production as a function of insulin.

model to simulate insulin infusion studies which have been carried out experimentally by Prager et al.[15] Among other things, they considered the temporal development of glucose production for various rates of insulin infusion. Both the appearance and disappearance of insulin were investigated. Our simulations revealed dynamics which were in very good agreement with the experimental results. The functional relationships in the model are displayed in figure 3. They were qualitatively and quantitatively estimated via a review of available literature: insulin secretion is a sigmoidal function f_1 of the glucose concentration[16]; glucose utilization depends on both glucose[17] and insulin concentration in the remote compartment[14,18]. This is described by means of the functions f_2, f_3, and f_4. Finally, glucose production is a function f_5 of the delayed plasma insulin concentration[15,18]. To the extent it was possible, the rest of the parameters in the model were also chosen on the basis of available experimental data[19,20]. They are listed in the appendix together with the differential equations and the analytical expressions for the functional relationships.

Simulation Results

When a constant glucose infusion is simulated, self-sustained oscillations of glucose and insulin with a period of 110-120 min occur. A typical result of such a simulation is illustrated in figure 4. Extensive simulations with the model showed that the model can reproduce the properties (listed in the introduction) of the experimentally observed oscillations. In addition, the simulations showed that the existence of oscillations is relatively insensitive to most variations in the different parameter values. Crucial, however, is the delay between plasma insulin and glucose production, and the two-compartmental representation of insulin distribution. Furthermore, the oscillations become damped if the slopes of one or more of the functional relationships are reduced significantly.

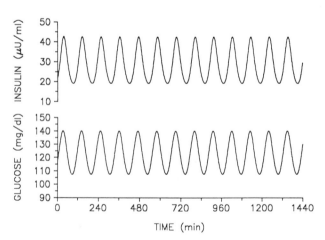

Figure 4. A simulation of a constant glucose infusion (base run) exhibits self-sustained oscillations of glucose and insulin.

Figure 5 shows the result of a simulation of ingestion of an oral glucose load. As in the experimental case (see figure 1), the oscillations are damped.

This relatively simple model thus supports the hypothesis that the slow oscillations are generated purely by the insulin/glucose feedback system, without any necessity for postulating the existence of an intrapancreatic pacemaker.

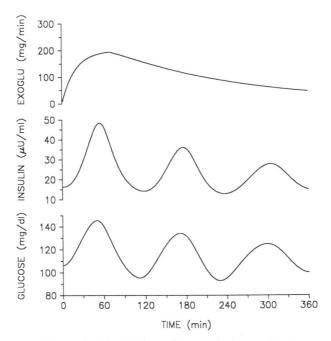

Figure 5. Simulation of an oral glucose load.

With the model in hand, we have a way to further test various properties of the oscillations. Based upon general knowledge of dynamic behavior of nonlinear systems[21], we can predict that it should be possible to alter the periodicity of the oscillations if, instead of a constant glucose infusion, an oscillatory infusion is simulated. If the period of the oscillatory infusion is sufficiently close to the period observed during constant glucose infusion, the system will respond by entraining to the pattern of the glucose infusion.

In figure 6, examples of such simulations are shown. Complete entrainment occurs both when the infusion period is 20% lower (6a) and 20% higher (6b) than the system's endogenous period.

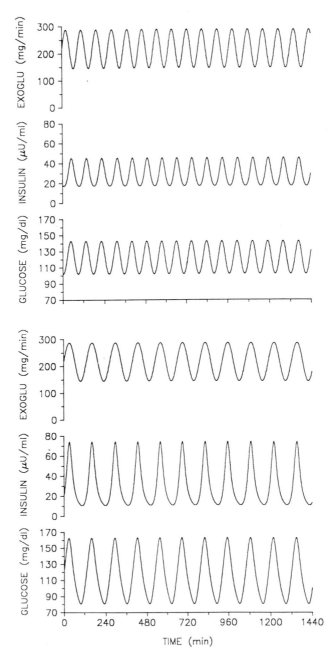

Figure 6. Examples of simulations using oscillatory glucose infusion. Entrainment occurs for exogenous periods shorter and longer than the natural period of the system.

THE CLINICAL EXPERIMENTS

We wanted to test if these results could be experimentally reproduced. If this were the case, we would have an indication that the insulin/glucose feedback system indeed plays an important role in generating the slow oscillations. Therefore we designed a series of experiments in which oscillatory glucose infusion was used.

Experimental Protocol

Seven non-diabetic volunteers participated in the experiments. Each subject underwent a 28-hour glucose infusion at a mean rate of 6 mg/kg/min on three separate occasions. In the first study, a constant glucose infusion was used. This allowed us to estimate the endogenous period in each individual separately. The two subsequent studies were then performed using an oscillatory (sinusoidal, relative amplitude 33% of the mean, see figure 6) glucose infusion pattern. The respective periods of infusion in the two studies were 20% below and 20% above each person's estimated individual period. In all studies, blood sampling was carried out every 10 min during the last 24 hours of the glucose infusion, allowing for measurement of glucose, insulin, and C-peptide.

Due to a relatively large and variable extraction of insulin by the liver, the insulin concentration does not necessarily reflect insulin secretion rates very accurately. C-peptide is secreted in an equimolar fashion with insulin, and since C-peptide is not extracted by the liver to any significant degree, the C-peptide concentration is a better marker of pancreatic insulin secretion[22]. It has previously been shown[23,24] that the actual insulin secretion rates can be accurately calculated from peripheral C-peptide concentration using an open two-compartmental model of C-peptide kinetics and metabolism. This method involves a separate study to estimate the kinetic parameters for C-peptide. We performed such a study on each subject as well, thus allowing us to calculate insulin secretion rates.

Experimental Results

All the studies using a constant infusion of glucose resulted in slow oscillations of glucose, insulin, and C-peptide. In the subsequent studies using oscillatory glucose infusion, the periodicity of the slow oscillations entrained to the infusion pattern. Two examples are illustrated in figure 7. In the two left panels, the outcomes of the constant glucose infusion experiments are plotted. The periodicity was estimated to 160 min for subject A (top) and to 120 min for subject B (bottom). The middle panels depict the results of the experiments using an oscillatory glucose infusion with a periodicity 20% below the estimated endogenous periods (128 and 96 min, respectively).

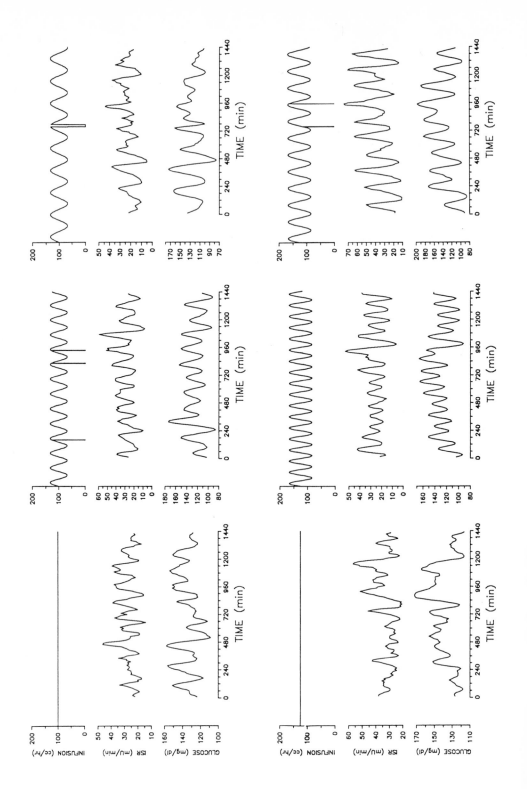

84

Figure 7 (opposite page). Representative experimental results. Subject A is displayed in the top row, and subject B is shown in the bottom row. Left: constant glucose infusion. Middle: oscillatory infusion, period 20% shorter. Right: oscillatory infusion, period 20% longer. (Redrawn from Sturis et al.[10])

Finally, the right panels show the results when the period of the oscillatory infusion was 20% above the period observed during constant glucose infusion (192 and 144 min, respectively). Note that there is no significant oscillatory activity of insulin secretion independent of glucose oscillations, indicating complete entrainment of the β-cell activity.

In subject A, a significant interruption (18 min) of the glucose infusion accidentally occurred during one experiment (right panel). This could be accurately detected because the glucose infusion was computer controlled. As can be seen from the graphs, the interruption resulted in loss of entrainment, which was not restored until several hours later.

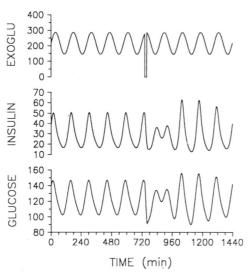

Figure 8. Simulation of the experiment in which the infusion was accidentally interrupted for 18 min.

The impact of an interruption in the glucose infusion was also investigated with the model. A simulation (figure 8) was performed which was designed to mimic the timing and the duration of the interruption. As was the case with the experiment, entrainment was temporarily lost. In light of the fact that the model only represents a partial picture of glucose regulation, it is not surprising that the interruption apparently was of greater disturbance in the experimental situation than in the model simulations.

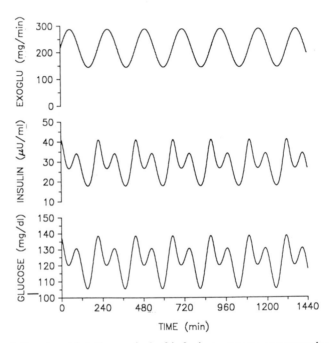

Figure 9. Simulation in which the period of infusion was set at approximately twice the natural period of the system. Frequency-locking occurs in which the system completes exactly two oscillations for each period of the exogenous glucose.

Present understanding of nonlinear dynamics proposes that the simple 1:1 entrainment (one period of the infusion for each resulting period of the system) which we have just observed can occur only in a finite interval around the endogenous period. For instance, there is a limit to how much the period can be stretched. Instead, quasiperiodicity, other modes of entrainment, or deterministic chaos can occur. In addition to the experiments already described, we therefore carried out a study on subject A in which the period of the oscillatory infusion was 320 min. This value was chosen because it is twice the length of the estimated natural period for this subject.

In this way, we hoped to achieve 1:2 entrainment. Before carrying out the experiment, we performed simulations to have an indication of the chance of success for the particular experiment. As shown in figure 9, the model simulation resulted in 1:2 entrainment.

Figure 10 shows the experimental result. It is quite clear that thgre are two large amplitude oscillations of both the glucose concentration and the insulin secretion rate for every one period of the glucose infusion. Simple 1:1 entrainment occurs in an interval around the natural period of the system, and in the same way, 1:2 entrainment can be expected in an interval around <u>twice</u> the natural period.

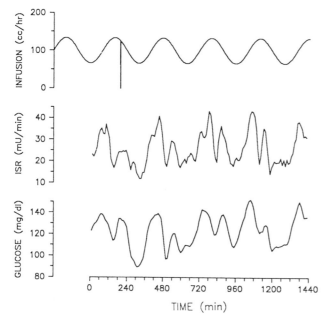

Figure 10. Result of the experiment performed on subject A in which the period of infusion was set at 320 min, approximately twice the subject's natural period as estimated from the constant glucose infusion experiment. (Redrawn from Sturis et al.[10]).

COEXISTING SOLUTIONS

So far, all simulations of oscillatory glucose infusions have been performed with an amplitude of 33%. Of course, other amplitudes could be used, both in simulations and in experimental situations. We performed a large number of simulations using different amplitudes and periods of the exogenous glucose. When the amplitude is set to 70% and the period to 60 min, an interesting situation occurs. In this case, simulations show that there are two <u>different</u> solutions. The initial conditions of the system determine which steady state solution will be obtained. This is potentially very relevant to the similar clinical experimental situation.

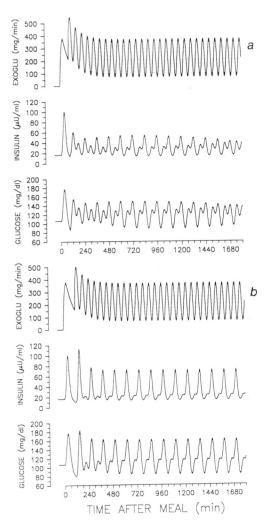

Figure 11. Simulations of how the same experiment can lead to different results if the subject has eaten 1½ hrs (11 a) and 2 hrs (11 b) before initiation of the experiment.

Suppose an experiment is carried out in which the amplitude and period of infusion are 70% and 60 min, respectively. Ideally, the subject has been fasting overnight, but in case he/she has had a meal in the morning, this person's <u>initial conditions</u> (blood sugar, plasma insulin, etc.) are different from the fasting situation. The result of a simulation in which the subject has been fasting as requested leads to entrainment. If, instead, the subject has had a carbohydrate meal 1½ hours before the beginning of the experiment, the picture is quite different, as illustrated in figure 11 a.

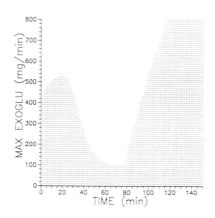

Figure 12. Analysis of which type of solution will be obtained depending on the size of the meal and on the time of meal ingestion relative to the beginning of the experiment. The shaded area shows the parameter combinations which lead to entrainment. Quasiperiodicity is obtained for the other parameter combinations.

Now entrainment is not obtained, and formal pulse analysis would lead to detection of more pulses than in the first case. This is <u>not</u> a matter of transient behavior. Regardless of how long the simulation is extended, entrainment will <u>never</u> occur. If the subject had eaten the same meal <u>2 hours</u> before initiation of the experiment, entrainment would occur, as shown in figure 11 b.

Figure 12 illustrates this dependence on the initial conditions in a different way. Here, we have investigated whether the experiment would lead to entrainment as a function of the size of the meal and the time span between ingestion of the meal and the beginning of the experiment.

DISCUSSION

The multiple ways in which model simulations and experimental results agree with each other strongly suggest that the hypotheses on which the computer model is based are correct. Of course, it is not possible to prove the correctness of any model, but our analyses have indeed suggested that the slow oscillations are generated by the insulin/glucose feedback mechanism. In this study, we found no evidence in support of the alternative hypothesis: that the oscillations could be generated internally in the pancreas via some sort of pacemaker mechanism. This hypothesis has previously been proposed by Marsh et al.[25] From studies performed in dogs, those authors conclude that the slow oscillations in insulin can exist even if the glucose concentration is kept constant. In fact, they claim that the amplitude of the insulin oscillations is reduced by the presence of glucose oscillations. This does not seem to be compatible with our results. We found that by raising the glucose amplitude using oscillatory glucose infusion, the amplitude of insulin secretion was also increased. Although the possibility of a pacemaker still remains, we have experimentally shown that the periodicity of such a pacemaker can be manipulated by applying an oscillatory glucose infusion. Therefore, a potential pacemaker cannot have an <u>independent</u> periodicity.

One property which the model cannot reproduce in its present formulation is the irregular nature of the experimentally observed oscillations. In the construction of the model, we have isolated the insulin/glucose system in order to investigate the possibility that this system is responsible for the slow oscillations. Of course, the insulin/glucose system is coupled to many other hormones and processes in the body, and each of these could cause the oscillations to become irregular. It is quite possible that the present model if modified, for instance by including glucagon, can produce irregular behavior in the form of deterministic chaos. It is not clear, however, that such an extension would capture the instabilities which cause the oscillations to be irregular.

The simulations showing coexisting periodic and quasiperiodic solutions illustrate a interesting problem which may arise in reproducing experimental results. Even though the model is not an accurate representation of the overall metabolic regulation, it is very likely that the same <u>phenomenon</u> could be observed experimentally. Therefore, inability to reproduce a certain experimental result does not necessarily mean that the result is wrong. Acceptance of the fact that two or more experiments can lead to different results, even if the experimental setups appear identical, is an important step towards understanding complex dynamic systems such as the system we have investigated here.

At present, it is not known whether the ultradian oscillations have any physiological significance. In light of the potential implications for solving the mystery of diabetes, this should be an important future subject to be investigated.

APPENDIX

The differential equations of the model are:

$$\frac{dx}{dt}=f_1(z)-E\left[\frac{x}{v_1}-\frac{y}{v_2}\right]-\frac{x}{t_1} \qquad \text{(mU/min)}$$

$$\frac{dy}{dt}=E\left[\frac{x}{v_1}-\frac{y}{v_2}\right]-\frac{y}{t_2} \qquad \text{(mU/min)}$$

$$\frac{dz}{dt}=EXOGLU+f_5(h_3)-f_2(z)-f_3(z)f_4(y) \qquad \text{(mg/min)}$$

$$\frac{dh_1}{dt}=\frac{x-h_1}{t_3/3}$$

$$\frac{dh_2}{dt}=\frac{h_1-h_2}{t_3/3}$$

$$\frac{dh_3}{dt}=\frac{h_2-h_3}{t_3/3}$$

In the equations, x is the amount of plasma insulin, y represents the amount of remote insulin, and z is the amount of glucose in the glucose space. The variables h_1, h_2, and h_3 are used to represent the delay between plasma insulin and glucose production.

The analytical expressions for the functional relationships used in the model (figure 3) are:

$$f_1(z)=\frac{209}{1+\exp\left[-\dfrac{z}{(300\cdot v_3)}+6.6\right]}$$

$$f_2(z)=72\cdot\left[1-\exp\left[-\frac{z}{144\cdot v_3}\right]\right]$$

$$f_3(z)=0.01\frac{z}{v_3}$$

$$f_4(y) = \frac{90}{1+\exp\left\{7.76-1.772\cdot \log\left[y\left(1/v_2+1/\left(Et_2\right)\right)\right]\right\}}+4$$

$$f_5(h_3) = \frac{180}{1+\exp\left[0.29\dfrac{h_3}{v_1}-7.5\right]}$$

The rest of the parameter values were in the base run:

$EXOGLU$ = 216 mg/min, t_1 = 6 min, t_2 = 100 min, t_3 = 36 min, E = 0.2 l/min, v_1 = 3 l, v_2 = 11 l, and v_3 = 10 l.

REFERENCES

1. D. A. Lang, D. R. Matthews, J. Peto, and R. C. Turner, Cyclic oscillations of basal plasma glucose and insulin concentrations in human beings, N. Engl. J. Med. 301:1023 (1979).
2. B. C. Hansen, K. C. Jen, S. B. Pek, and R. A. Wolfe, Rapid oscillations in plasma insulin, glucagon and glucose in obese and normal weight humans, J. Clin. Endocrinol. Metab. 54:785 (1982).
3. C. Simon, G. Brandenberger, and M. Follenius, Ultradian oscillations of plasma glucose, insulin, and c-peptide in man, J. Clin. Endocrinol. Metab. 64:669 (1987).
4. E. W. Kraegen, J. D. Young, E. P. George, and L. Lazarus, Oscillations of blood glucose and insulin after oral glucose, Horm. Metab. Res. 4:409 (1972).
5. E. T. Shapiro, H. Tillil, K. S. Polonsky, V. S. Fang, A. H. Rubenstein, and E. Van Cauter, Oscillations in insulin secretion during constant glucose infusion in normal man: relationship to changes in plasma glucose, J. Clin. Endocrinol. Metab. 67:307 (1988).
6. E. Van Cauter, D. Desir, C. Decoster, F. Fery, and E. O. Balasse, Nocturnal decrease in glucose tolerance during constant glucose infusion, J. Clin. Endocrinol. Metab. 69:604 (1989).
7. J. I. Stagner, E. Samols, and G. C. Weir, Sustained oscillations from the isolated canine pancreas during exposure to a constant glucose concentration, J. Clin. Invest. 65:939 (1980).
8. K. S. Polonsky, J. B. Jaspan, L. Woodle, and R. Thistlethwaite, Alterations in the pattern of insulin secretion and c-peptide kinetics post pancreas transplantation, Diabetes 39, suppl.:15A (1990).
9. J. Sturis, K. S. Polonsky, E. Mosekilde, and E. Van Cauter, The mechanisms underlying ultradian oscillations of insulin and glucose: a computer simulation approach, Am. J. Physiol. (1991 - in press).
10. J. Sturis, E. Van Cauter, J. D. Blackman, and K. S. Polonsky, Entrainment of ultradian pulses of insulin secretion by oscillatory glucose infusion. J. Clin. Invest. (February 1991)

11. J. W. Forrester, "Principles of Systems," Wright-Allen Press, Cambridge, Massachusetts, USA (1968).
12. M. Berger and D. Rodbard, Computer simulation of plasma insulin and glucose dynamics after subcutaneous insulin injection, Diabetes Care 12:725 (1989).
13. R. N. Bergman, Toward physiological understanding of glucose tolerance. Minimal model approach, Diabetes 38:1512 (1989).
14. Y. J. Yang, I. D. Hope, and R. N. Bergman, Insulin transport across capillaries is rate limiting for insulin action in dogs, J. Clin. Invest. 84:1620 (1989).
15. R. Prager, P. Wallace, and J. M. Olefsky, In vivo kinetics of insulin action on peripheral glucose disposal and hepatic glucose output in normal and obese subjects, J. Clin. Invest. 78:472 (1986).
16. G. M. Grodsky, A threshold distribution hypothesis for packet storage of insulin and its mathematical modeling, J. Clin. Invest. 51:2047 (1972).
17. C. A. Verdonk, R. A. Rizza, and J. E. Gerich, Effects of plasma glucose concentration on glucose utilization and glucose clearance in normal man, Diabetes 30:535 (1981).
18. R. A. Rizza, L. J. Mandarino, and J. E. Gerich, Dose-response characteristics for effects of insulin on production and utilization of glucose in man, Am. J. Physiol. 240:E630 (1981).
19. R. Steele, Influences of glucose loading and of injected insulin on hepatic glucose output, Ann. NY Acad. Sci. 82:420 (1959).
20. K. S. Polonsky, B. D. Given, W. Pugh, J. Licinio-Paixao, J. E. Thompson, T. Karrison, and A. H. Rubenstein, Calculation of the systemic delivery rate of insulin in normal man, J. Clin. Endocrinol. Metab. 63:113 (1986).
21. L. Glass and M. C. Mackey, "From Clocks to Chaos: The Rhythms of Life," Princeton University Press, Princeton, New Jersey, USA (1988).
22. K. S. Polonsky and A. H. Rubenstein, C-peptide as a measure of the secretion and hepatic extraction of insulin: pitfalls and limitations, Diabetes 33:486 (1984).
23. R. P. Eaton, R. C. Allen, D. S. Schade, K. M. Erickson, and J. Standefer, Prehepatic insulin production in man: kinetic analysis using peripheral connecting peptide behaviour, J. Clin. Endocrinol. Metab. 51:520 (1980).
24. K. S. Polonsky, J. Licinio-Paixao, B. D. Given, W. Pugh, P. Rue, J. Galloway, T. Karrison, and B. Frank, Use of biosynthetic human c-peptide in the measurement of insulin secretion rates in normal volunteers and type I diabetic patients, J. Clin. Invest. 77:98 (1986).
25. B. D. Marsh, D. J. Marsh, and R. N. Bergman, Oscillations enhance the efficiency and stability of glucose disposal, Am. J. Physiol. 250:E576 (1986).

THE DYNAMIC CODE:

INFORMATION TRANSFER IN HORMONAL SYSTEMS

K. Prank, H. Harms, Chr. Kayser, G. Brabant, L.F. Olsen[*]
and R.D. Hesch

Abteilung für Klinische Endokrinologie
Zentrum Innere Medizin der Medizinischen Hochschule Hannover
Konstanty-Gutschow-Straße 8, D-3000 Hannover 61, FRG

Institute of Biochemistry[*]
University of Odense
Campusvej 55, DK-5230 Odense M, Denmark

In the recent textbook of "Endokrinologie" (Hesch, 1989a) we have formulated a new concept of the evolution of biological information. When we analyzed the current knowledge on the evolution of life it became apparent that its separation into processes is due to the history of western science.

The rapid progress of research on the evolution and selforganization of life as well as molecular biology with its reductionistic approach has mainly led to the concept that life might be encoded by the generation and evolution of structural information. From this approach the **"genetic code"** emerged as the fundamental principle of biology. The impressive success of recombinant gene technology to alter the structure of the genetic code and the expression of the genotypic and phenotypic structural and behavioural appearance of living subjects (Jeffrey, 1990) point to the predominant importance of structure. The recent report of the generation of large combinatorial libraries for designed antibody repertoirs in E. coli offers unlimited combinations of antibody conformers against any ligand-structure (Sastry et al., 1989). The polymerase chain reaction is the ultimate tool to magnify smallest structural information at the level of DNA of even a single cell in an attempt to detect minimal errors in the evolutionary plans to construct life (Eisenstein, 1990). From the predominance of the power of the genetic code for structural information of life it seems to result that functional processes which communicate a biography to the structure may be of a subordinate importance or even result entirely from a preexisting information of structure. Moreover, this view negates that most structures in living systems encode in themself a functional information.

Complexity, Chaos, and Biological Evolution, Edited by E. Mosekilde and
L. Mosekilde, Plenum Press, New York, 1991

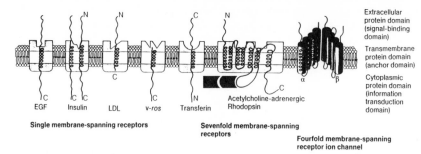

Fig. 1. Topology of receptor structures in membrane: E = effector protein, G = G- protein, C = C-terminus, EGF = epidermal growth factor, LDL = low densisty lipoprotein, v-ros = viral ros-receptor.

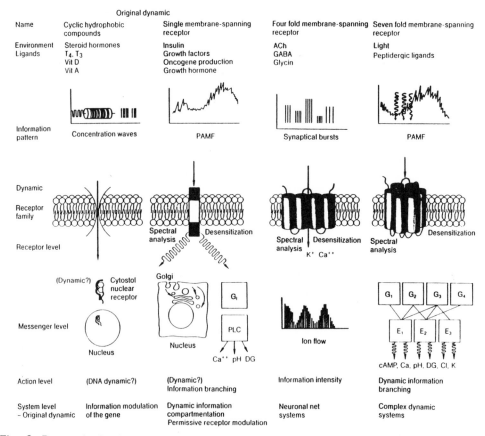

Fig. 2. Dynamical principles of receptor function (receptor families). DG = diacylglycerol, E = effector pathways, G = G-protein, PAM F = pulse amplitude and frequency modulation, PLC = phospholipase C.

To obviate this classical dichotomy in science we have formulated the hypothesis of an inseparable coevolution of structure and function (Hesch, 1989a). Our hypothesis predicts that genetic encoding of a structural theme necessarily implies a simultaneous encoding of negative entropy. This is coding of thermodynamic energy transfer through compartments. We have also postulated that the thermodynamic requirements, i.e. the function, associated to a specific structure are basic components of the smallest biological unity called a "biological system". A biological system in this definition is a quantity of polymeric structures (components) which thermodynamically exhibit a high energy conformation ("Eigenschaft") to form compartments (space). Compartments are connected through complex communications of information by energy transfer (relation, function) against structural gradients to create dysequilibria which appear as biological events in time and space. These events can evolve, construct and maintain the process of life. Using this definition one must be able to formulate a general "Bauplan" in which the constructive laws for both structure and function should be identical. Information communication between different compartments at the structural level of biological systems is performed either through a family of ion channel proteins, substrate transporters or through a family of receptors (Fig.1 and Fig. 2). They are all operated by specific ligands which communicate specific thermodynamic and morphogenetic information to an evolving system or adaptive information to a mature system. This information is permanently transformed into structure and metabolism by deconvolution of the genetic information, i.e. regulation of gene expression (Fig. 3).

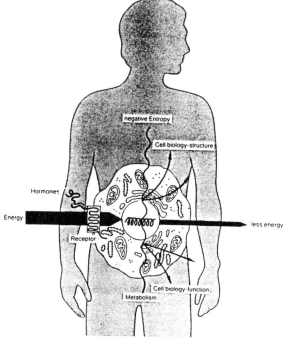

Fig. 3. An organism extracts energy from the environment and increases negative entropy. This is converted into biological information, i.e. cell structure and cell function and driven by metabolism.

The "Bauplan" of biological structures at the organ level follows determinstic construction principles and reveals that organs exhibit a fractal dimension (Sernetz, 1988). We do not know so far if the organization of the genetic code follows the same laws but there is recent evidence that this could be possible (Sheldrake, 1990). Jürgens, Peitgen and Saupe (1989) have recently formulated that systems which create fractal structures are operated by complex dynamics which obey the law of chaos. The common "Bauplan" emerging from a coevolution of structure and function hence means the ensemble of complex information communication in dynamic biological systems where chaos creates fractal structures. We have called this morphogenetic and adaptive encoding of biological information complementary to the genetic code the **"dynamic code"**. The nature of this **"dynamic code"** has remained unclear yet, many speculations have been proposed and even new physical forces have been envisaged recently by (Sheldrake, 1990). There is, however, good evidence from our own research and that of other groups (Goldbeter et al., 1983) that the **"dynamic code"** can now be described on the basis of deterministic encoding of biological information. We have chosen the coding of information within the hormonal system of circulating blood as an example of investigation and we assume the parcrine and autocrine information at organ and cellular level obeys the same laws. This would offer us a more general formulation of the **"dynamic code"** as the essential partner of the genetic code for life (Fig. 3 and Fig. 4).

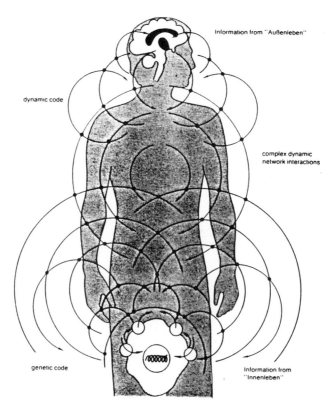

Fig. 4. The genetic code and the dynamic code fuse to complex dynamic networks. The genome ("Innenleben") meets the surrounding environment ("Außenleben") for a limited period of time - life - to exchange and concentrate biological information.

Our next question then is if such a **"dynamic code"** can be found when we analyse the quality of biological information. We can interprete a living organism as a system able to concentrate energy, i.e. information form the environment with negative entropy. The result is the transformation of thermodynamic information into structure, metabolism and cellbiology presenting as action towards the envirnoment. This process separates the outer world from the inner world of organism, the border is called the "Self" (Fig. 5). We interprete the transition from health ("Gesundsein") to disease ("Kranksein") (Hesch, 1988) as a process where Self looses its control. Since these transitions evolve within the dynamic information communication systems of the organism we find the term "dynamic disease" as proposed by Mackey and Milton (1987) appropriate. Information from the environment into Self enters a living organism through a limited amount of entrances: In essence we know of three main entrances with regard to the evolution and maturation of an individual biography and with regard to a subjects health and disease in its enviroment (Fig. 6).

These entrances are formed by ion channels, substrate transporters and receptors. They are operated by various biophysical information properties depending from where they come:

- Information from the outer environment has different physical qualities.

- Information within the body is encoded in hormonal ligands.

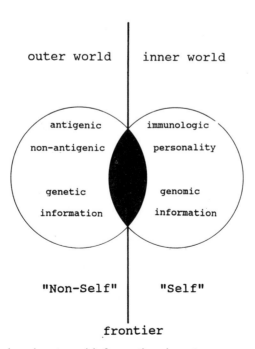

outer world | inner world

antigenic | immunologic

non-antigenic | personality

genetic | genomic

information | information

"Non-Self" | "Self"

frontier

Fig. 5. The internal and external informational system seperate "Self" (Selbst) and "Non-Self" (Nichtselbst) with virtual, individual maturated frontiers.

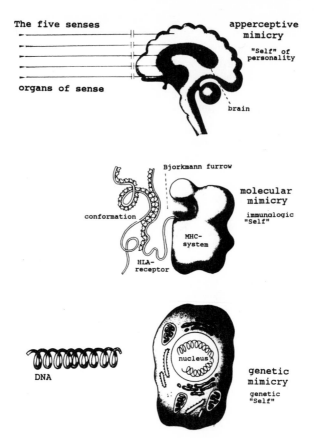

Fig. 6. Diseases perforate frontiers of an individuum (Selbst) towards the surrounding environment by the mechanisms: apperceptive mimikry, molecular mimikry and genetic mimikry.

All informational systems in the body are driven by hormones transcoding information from outside to the cell (Fig. 7 and Fig. 8). Hormonal information in circulating blood presents in two physiocochemical qualities. (1) The first property is a constant concentration within the circulating hormonal pool. (2) On top of this circulating pool all hormones exhibit a dynamic pulse amplitude and frequency modulation of episodic secretory and metabolic events (Fig. 9). Both processes modulate the responsiveness of receptors within the cell membrane by regulation of the receptors synthesis, movement within the membrane layer, coupling to signal transduction proteins and internalization. Constant hormone concentrations seem to preferentially determine the presence of receptors in the membrane whereas the dynamic pulse amplitude and frequency modulation deciphers the specific nature of biological information which enters the intracellular space through the receptor (Fig. 3, Fig. 7, Fig. 9 and Fig. 10).

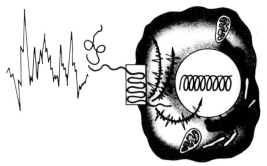

dynamics – hormone – receptor – dynamics

Fig. 7. Coevolution of information transfer of cells. The dynamics of the hormonal information coding, conformation of ligands and receptors have developped togehter.

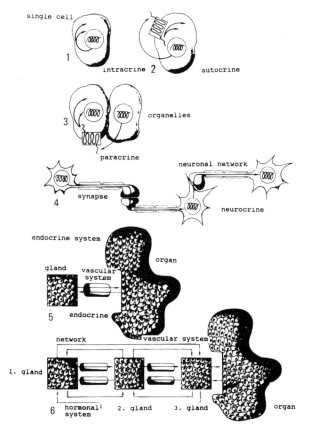

Fig. 8. Principles of information interchange with single cells (1-3), clusters of cells (4), organs and in the system of an adult organism.

101

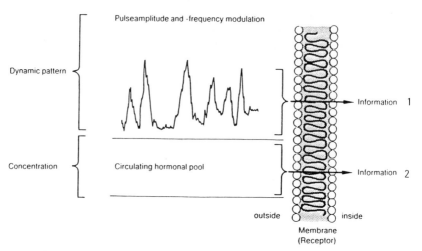

Fig. 9. Biological information encoded into hormones is delivered to a receptor in two forms. The (1) conformational information presents with different concentrations in the circulating pool, the (2) dynamic information is encoded in pulse-amplitude and frequency ("dynamic") pattern of oscillatory hormonal concentrations driving membrane receptor operated pathways.

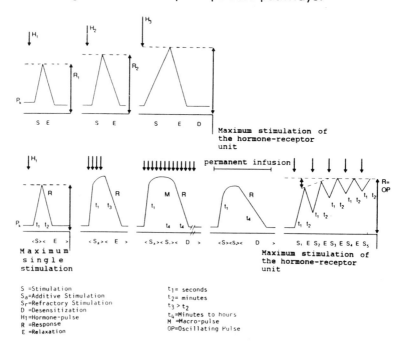

Fig. 10. Modulation of hormone-receptor-interaction. Models of up- and downregulation of the receptor by various deliveries of the ligand.

Continous hormone delivery to form constant hormonal pools corresponds to a steady state system with fixed point attractors. It seems, however, of great interest to delineate the deterministic behaviour and the dimension of attractors responsible for the dynamic pulse amplitude and frequency modulation of the hormone-receptor unit. If dynamic information communication, i.e. the "dynamic code", were, indeed, to have a fundamental importance in biology and human life we should be able to describe their dynamic attractors in limited cycles or other deterministic systems.

There are so far only rare occasions where in the living organisms we can observe how dynamic functions are coupled to the evolution of structural elements. The permanent plasticity of neuronal network-connectivity in the brain is currently the most attractive research project. Increasing amounts of experimental data suggest that changes in the neurotransmitter pattern by information inputs through the apperceptive system ultimately lead to permanent changes in gene expression in brain tissue followed by distinct structural and functional reorganization.

There is, however, another network, that of bone, where we are able to study many components in vivo and in the tissue itself. Alterations within this system allow a good description of a dynamic disease. Bone is a tissue where we can develope our hypothesis with particular strength. Bone undergoes permanent structural and functional changes. During the first time of our life the skeleton maturates to its later composition a process called "modelling". The gene expression necessary for this process is under the plastic control of developping hormonal pattern and growth hormone is the most relevant hormone for this process (Thorner and Holl, 1989), (Fig. 5 and Fig. 6)

In adult life the composition of bone undergoes a permanent plastic adaptation and this concerns hematopoiesis as well as the biomechanic structural network. This process is called "remodelling". Parathyroid hormone (PTH) is the main hormone responsible for bone remodelling and we have therefore, investigated how PTH regulates the architecture of bone and hematopoietic function as well? Since bone exhibits prima facie a typical fractal structure which awaits detailled description one would postulate that the dynamic modulation of this structure should proceed in a deterministic information communication system.

Recent work of Harms et al. (1989) from our group suggests that the normal three dimensional architecture of bone depends upon the dynamics of PTH-pattern in circulating blood. In the following paragraph we present our data on the dynamic PTH pattern.

Normal healthy subjects

In normal healthy subjects PTH circulates in blood in a pulsatile manner. Fig. 11 shows a representative rhythm with intact PTH and its (44-68) peptide. The concentration of intact PTH within the circulating pool is at about 20-40 pg/ml and pulses with a height over 200 pg/ml are not rare. This means that the dynamic power of a peak can be nearly tenfold that of the circulating pool.

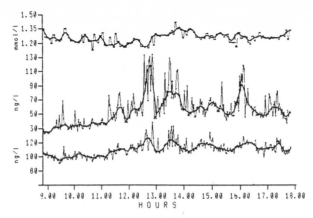

Fig. 11. A representative rhythm of a healthy young man. Depicted from top to bottom are ionized calcium, PTH-(1-84) and PTH-(44-68). A 20-min. moving average in each panel is additionally drawn.

By eye ball-analysis an irregular pattern of oscillations is visible and we have used classical time domain rhythm analyses to describe the pattern.

1. Power spectrum analysis (Fig. 12)

In normal subjects the maximal spectral power for intact and (44-68) PTH is found around 45-min periods. A second peak of interest in the spectrum for both lies around 20-min. periods. Intact PTH additionally expresses spectral power around 12-min. periods. The experimental assay noise spectrum was found to have peaks at 7-, 14,- and 30 min. periods demonstrating harmonic vibrations.

2. Crosscorrelation analysis

The (44-68) PTH is a derivative from intact glandular PTH generated in part by a metabolism of intact PTH through enzymes in various organs of the body. Crosscorrelation (Fig. 13) for non time-delayed comparison shows that either fragmentation occurs within 2 min. which is reasonalble or glandular secretion of peptide fragments.

3. Puls detection analysis

We applied the Pulsar algorithm (Merriam and Wachter, 1982), the method described by Santen and Bardin (1973) and the Cluster analysis (Veldhuis and Johnson, 1986) for pulse rhythm-oscillation analyses (Harms et al., 1989). The results are given in table 1.

Table 1. Summary of pulse rhythm detection of intact PTH and PTH-(44-68) from 10 healthy men

	Cluster	Pulsar	Santen-Bardin
A. Normal PTH-(1-84)			
Peaks/h	3.4 ± 0.6	5.8 ± 0.7	7.1 ± 1.0
Peak width (min)	12.3 ± 1.6	6.4 ± 1.2	
Incremental peak amplitude (ng/L)	18.5 ± 11.0	26.0 ± 16.0	
Interpeak interval (min)	17.1 ± 2.2	10.6 ± 1.2	
B. Normal PTH-(44-68)			
Peaks/h	1.0 ± 0.5	0.8 ± 0.6	2.5 ± 1.0
Peak width (min)	45.5 ± 28.0	6.3 ± 1.9	
Incremental peak amplitude (ng/L)	18.4 ± 8.8	25.4 ± 12.2	
Interpeak interval (min)	56.7 ± 27.0	52.5 ± 24.0	

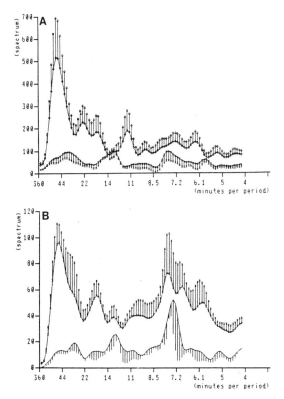

Fig. 12. Spectral estimations in eight healthy men for A) intact PTH (◊) and PTH-(44-68) (X), and for B) PTH-(44-68) (X) and assay noise (◊).

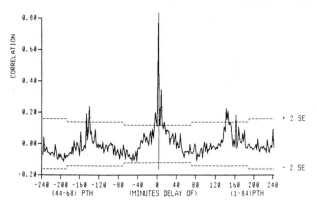

Fig. 13. Cross-correlation of intact and PTH-(44-68) in healthy men.

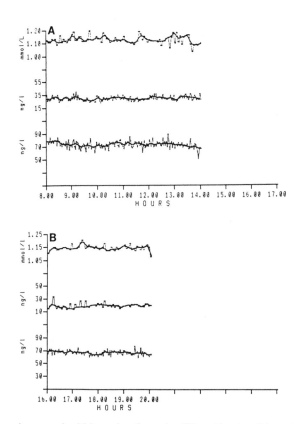

Fig. 14. Rhythms in a male (A) and a female (B) patient with osteoporosis. Depicted from top to bottom are ionized calcium, PTH-(1-84), and PTH-(44-68) in both panels. The decreased pulse amplitude and frequency of PTH secretion are obvious.

Patients with osteoporosis

Vertebral bone is constructed by a three dimensional network of plates. The shape and thickness of the plates, their content of organic and mineral material undergoes a permanent modulation. Osteoporosis is a disease where the three dimensional architecture and the total mass of bone disappears. Normally the action of bone constructing osteoblastic cells and bone resorbing osteoclastic cells is functionally coupled. Parathyroid hormone is the main coupler of both processes and it is reasonable to assume that a disturbed dynamic pattern of PTH results in a decoupling of bone remodelling processes. The consequence of such derangement with more bone resorption than reconstruction is a rapid and progressive perforation of plates. They become porotic, hence the name of the disease osteoporosis. At a certain extension of the perforations they can become irreversible.

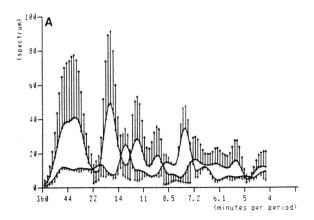

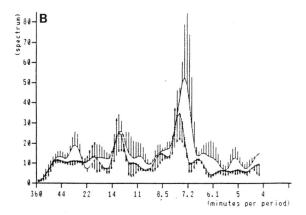

Fig. 15. Depicted are spectral estimations for A) intact (◇) and PTH-(44-68) (X) and for B) PTH-(44-68) (X) and in assay noise spectrum (◇) in two patients with osteoporosis. The maximal spectral power is 10-fold lower than in healthy men.

When we analysed the dynamic PTH pattern in osteoporotic subjects we discovered that the normal pulse amplitude and frequency modulation of intact and (44-68) was completely lost (Fig. 14). Power spectrum analyses revealed a maximum around 45-, 30-, 16- and 12-min. periods (Fig. 15). The pulse detection methods revealed a significant reduction in the number of pulses, their length and their amplitude.

From these results it appears that the bifurcation of the normal PTH-oscillations towards a "low dynamic functional" state is followed by deranged PTH action at the bone cell receptors with a loss of a distinct structure of bone.

Estrogen induced bifurcations of dynamic PTH secretion

We have recently been able to induce artificially such bifurcation in human beings. Estrogen-deficiency is the most frequent cause of osteoporosis and bone mass can be conserved in estrogen-substituted women; the mechanism of estrogen action on bone metabolism, however, remained obscure. We have determined a dynamic oscillatory PTH-pattern in women before and after estrogen substitution (six months 1 mg estradiolvalerate over 25 days combined to 2 mg estriol and levonorgestrel from day 11 to 25). It was found that a low dynamic functional state could be converted to a higher oscillating PTH-pattern (Fig. 16). It is supposed that the restoration of the PTH-oscillatory pattern is responsible for the conservation of bone mass.

Table 2. Pulse detection analysis of two rhythms with hyperparathyroidism

Hyperparathyroidism	Cluster	Pulsar
Normal PTH-(1-84)		
Peaks/h	3.5 ± 0.7	4.8 ± 1.6
Peak width (min)	12.1 ± 3.3	7.4 ± 1.4
mean incremental peak amplitude (ng/L)	84.6 ± 54.3	105.0 ± 72.2
Interpeak interval (min)	15.1 ± 2.2	12.7 ± 3.1

The conversion of a pathological bifurcation in a dynamic system, i.e. dynamic disease ("Kranksein"), back towards a preexisting dynamic state of the respective system is obviously associated with a way back to being healthy ("Gesundsein") (Hesch, 1988).

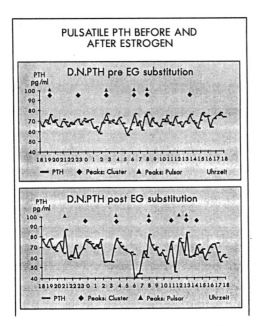

Fig. 16. (taken from Chr. Kayser, unpublished thesis, Medizinische Hochschule Hannover): top: 24h PTH-concentration time series before estrogen substitution; bottom: 24h PTH-concentration time series after estrogen substitution

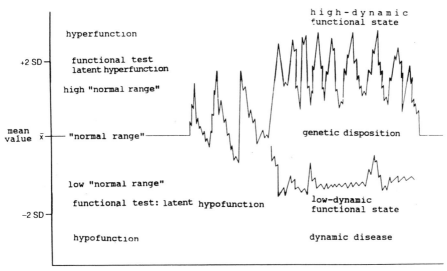

Fig. 17. Left: Current model where diseases are defined by "normal ranges" and states of hyper-or hypofunction; Middle: Dynamic model: Diseases can evolve within the "normal range" as consequence of a disturbed dynamic hormonal pattern; Right: Further bifurcation of dynamic states towards hyperdynamic pattern and low dynamic pattern.

Patients with hyperparathyroidism

Hyperparathyroidism (HPTH) is a disease where the parathyroid glands deliver more PTH to the receptors of PTH-sensitive organs than needed for their physiological demands. Usually the consequence of defective PTH-production is a symptomatic or asymptomatic state where high calcium concentrations are associated with elevated PTH concentrations in blood. In the recent years an increasing number of symptomatic patients presents with hypercalcemia and PTH-concentrations not unequivocally over the upper limit (Hesch, 1989b) (Fig. 6 and Fig. 17)

Harms et al. (unpublished results from our group) have studied the dynamic PTH-secretion in hyperparathyroid patients (Fig. 18). Inspection of the pattern shows a higher oscillating frequency for small pulses. The computer evaluation, however, describes a similar number of peaks with a similar width but with a drastically increased amplitude (Table 2).

It results that the PTH-oscillations in these patients compared to normals results in bifurcation to a "high dynamic functional" state. Whereas low dynamic PTH-receptor function in bone results in a loss of trabecular bone architecture and bone mass. The bone in hyperparathyroid patients can exhibit a higher connectivity of the trabecular network and a higher bone mass (Delling et al., 1987). These new pathophysiological concepts have been used to design the yet most successfull therapy of low turnover osteoporosis by application of synthetic (1-38) PTH in conjunction with calcitonin (Hesch et al., 1989c).

Animal experiments with dynamic PTH

In rat studies Tam et al. (1982) showed that an artificially kept constant circulating hormonal pool of PTH corresponding to fixed point attractor is associated to a disturption of bone apposition and resorption and results in a loss of bone structures. Discontinous injection of PTH corresponding to stable limit cycles creates increased trabecular bone network and enhanced bone mass. Identical results can be found when the experiments were repeated in the greyhound (Podbesek et al., 1983). Our observation in human beings extend these earlier studies by a systematic analysis of the underlying dynamics.

We can now summarize that distinct nonlinaer dynamic oscillatory pattern of PTH which can be mathematically assigned by computer algorithms to distinct PTH-receptor functions and bioarchitectural structures of the trabecular bone network. We can say that the bifurcation from a single fixed point attractor to a stable limit cycle changes bone architecture in an animal model, further that the bifurcation of distinct nonlinear dynamic pattern in human beings forward and backward can be associated to healthy and diseased states of bone.

Evidence for a pulse generator in parathyroid cells ?

Birnbaum et al. (1988) investigated normal and hyperplastic human parathyroid cells in a perfusion incubation system and collected samples every two minutes as we did in human beings. They showed that the perfused cells

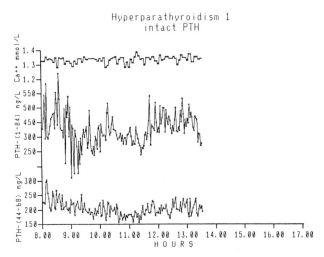

Fig. 18. Rhythms in a patient with hyperparathyroidism. Depicted from top to bottom are ionized calcium, PTH-(1-84) and PTH-(44-68). The increased pulse amplitude and frequency of PTH secretion are obvious.

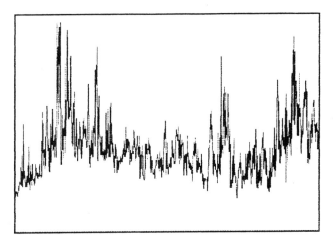

Fig. 19. Time series of PTH concentration in blood. Scales are: x-axis: time 0-813 points (interval 2 min). y-axis: PTH concentration (0-140 ng/l).

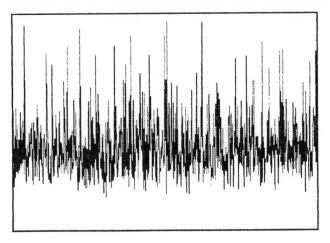

Fig. 20. Time series of PTH concentration in blood in a randomized order. Scales are: x-axis: time 0-813 points (interval 2 min). y-axis: PTH concentration (0-140 ng/l).

were synchronized. A similar pulse amplitude and frequency pattern in normal tissue in vitro as observed in vivo was described. Hyperparathyroid tissue showed more oscillations and higher frequencies. These experiments seem to suggest that the pacemaking pulse generator for dynamic PTH-release can be found within the parathyroid glands. There is, however, not yet a concept about the molecular basis of the selforganization of this hormonal system.

Is the normal PTH secretory pattern due to determinstic chaos ?

Finally, we must now ask the critical questions to the studies presented so far:

- Does bone have a fractal structure ?

- Does the normal secretion of PTH, the main hormone for remodelling bone structure exhibit a deterministic dynamic pattern ?

- Do bifurcations of dynamic pattern where diseases begin lead to chaos ?

There is no immediate answer to these questions. But in cooperation with L.F. Olsen from the University of Odense, Denmark we have performed next maximum maps by plotting $X(n)$ against $X(n+1)$, where X is the PTH concentration in blood. This was done with the original data which consisted of three PTH-rhythms of three healthy subjects (Fig. 19). But this was done also with these data ordered in a randomized manner (Fig. 20). The next maximum maps of the original data in all three cases gave a more deterministic structure than the maps of the randomized data (Fig. 21). Further to this L.F. Olsen has made a preliminary series of embeddings in 3D-phase space of smoothed data (Fig. 22 and Fig. 23), computed the Poincaré-sections and applied the

Fig. 21. Next maximum maps of original (top, data as in Fig. 19) and randomized (bottom, data as in Fig. 20) PTH data.

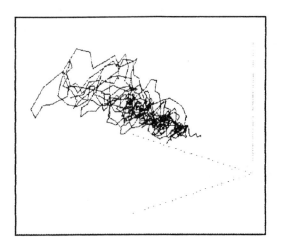

Fig. 22. 3D reconstructed phase portrait of smoothed (5-point) PTH data. Delay = 5 time intervals (data as in Fig. 19).

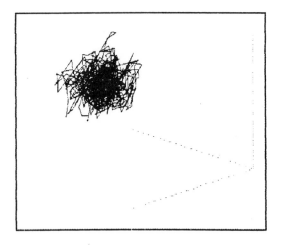

Fig. 23. 3D phase portrait of smoothed (5-point) and randomized PTH data. Delay = 5 time intervals (data as in Fig. 20).

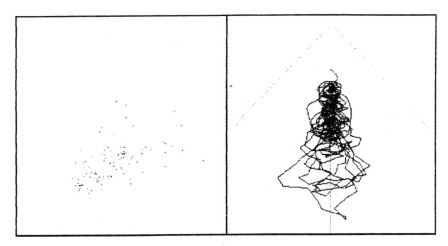

Fig. 24. Right: Reconstructed phase portrait of PTH data seen from above. Line indicates Poincaré section. Left: Intersection points of the Poincaré section (data as in Fig. 19).

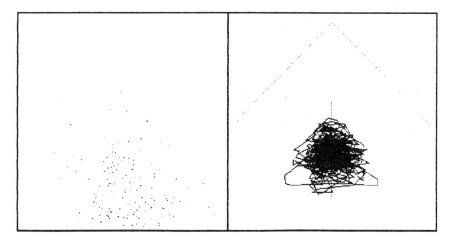

Fig. 25. Right: Reconstructed phase portrait of randomized PTH data viewed from above. Line indicates Poincaré section. Left: Intersection points of the Poincaré section (data as in Fig. 20).

Grassberger-Procaccia method to the points of the Poincaré-section (Fig. 24 and Fig. 25). He then found that the data may be deterministic, the dynamics complex and exhibit an attractor in three degrees of freedom (Fig. 26 and Fig. 27).

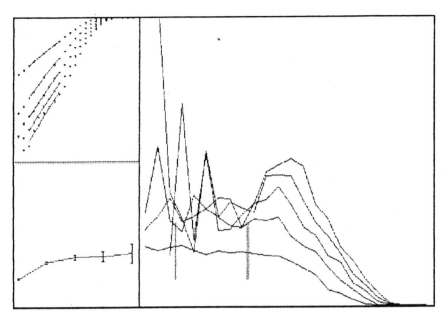

Fig. 26. Computing correlation dimension for PTH data using points on Poincaré section. Scales ln(g) 0-6; ln(C(g)) -8 - 0; dln(C(g))/dln(g) 0-5 (data as in Fig. 19).

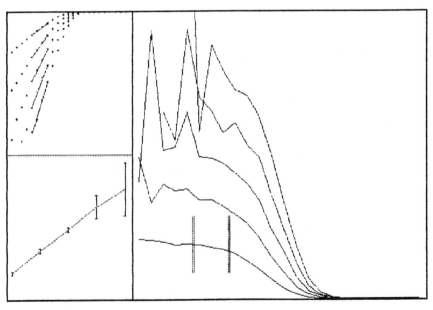

Fig. 27. Estimating correlation dimension for the randomized PTH data. Scales as in Fig. 26 (data as in Fig. 20).

Although this is not a final proof these data encourage us to further hypothesize that oscillatory dynamic PTH-secretion proceeds not following a stochastic principle but obeys deterministic laws, i.e. the **"dynamic code"**, whose understanding ultimately will help to understand bifurcations associated with being healthy and being ill.

References

Birnbaum, J., Klandorf, H., Armando, G. and van Herle, A., 1988, Lithium Stimulates the Release of Human Parathyroid Hormone in Vitro, J. Clin. Endocrinol. Metab., 66:1187-1191.

Delling, G., Dreyer, T., Hesch, R.D., Schulz, W., Ziegler, R. and Bressel, M., 1987, Morphologische Veränderungen der Beckenkammspongiosa beim primären Hyperparathyreoidismus und ihre Bedeutung für die Diagnostik, Klin. Wochenschr., 65:643-653.

Eisenstein, B.I., 1990, The Polymerase Chain Reaction: A New Method of Using Molecular Genetics for Medical Diagnosis. New Engl. J. Med., 322:178-183.

Goldbeter, A., Decroly, O., 1983, Am. J. Physiol., 245:R478-483.

Harms, H.M., Kaptaina, U., Külpmann, W.R., Brabant, G., Hesch, R.D., 1989, Pulse Amplitude and Frequency Modulation of Parathyroid Hormone in Plasma, J. Clin. Endocrinol. Metab., 69:843-851.

Hesch, R.D., 1988, Gesundsein und Kranksein, Futura 1/88, S. 23-27.

Hesch, R.D., 1989a, "Endokrinologie", Teil A, Urban und Schwarzenberg, München.

Hesch, R.D., 1989b, Hyperparathyreodismus in: "Endokrinologie", Teil B, Urban und Schwarzenberg, München.

Hesch, R.D., Rittinghaus, E.F., Harms, H.M., Delling, G., 1989c, Die Frühtherapie der Osteoporose mit (1-38) Parathormon und Calcitonin-Nasalspray, Med. Klin., 84:488-4398 (Nr. 10).

Jeffrey, H.J., 1990, Chaos game representation of gene structure, Nucl. Aci. Res., 18, No. 8:2163.

Jürgens, W., Peitgen, H.O., Saupe, D., 1989, Chaos und Fraktale, Spektrum der Wissenschaft Verlags GmbH, Heidelberg, S. 7.

Mackey, M.C., Milton, J.G., 1987, Dynamical Diseases, Ann. N.Y. Acad. Sci., 504:16-32

Merriam, G.R., Wachter, K.W., 1982, Algorithms for the study of episodic hormone secretion, Am. J. Physiol., 243:E310.

Podbesek, R., Edouard, C., Meunier, P.J., Parsons, J.A., Reeve, J., Stevenson, R.W., Zanelli, J.M., 1983, Effects of Two Treatment Regimes with Synthetic Human Parthyroid Hormone Fragment on Bone Formation and the Tissue Balance of Trabecular Bone in Greyhounds. Endocrinology , 112:1000-1006.

Santen, R.J., Bardin, C.W., 1973, Episodic luteinizing hormone secretion in man. Pulse analysis, clinical interpretation, physiologic mechanisms, J. Clin. Invest., 52:2617.

Sastry, L., Alting-Mees, M., Huse, W.D., Short, J.M., Sorge, J.A.,Hay, B.N., Janda, K.D., Benkovic, S.J., Lerner, R.A., 1989, Cloning of the immunological repertoire in Escherichia coli for generation of monoclonal catalytic antibodies: construction of a heavy chain variable region-specific cDNA library. Proc. Natl. Acad. Sci. USA, 86(15):5728-32.

Sernetz, M., 1988, Spiegel der Forschung, 5:8-11.

Sheldrake, R., 1990, "Das Gedächtnis der Natur", Scherz Verlag, Bern, München, Wien.

Tam, C.S., Heersche, J.N.M., Murray, T.M. and Parsons, J.A., 1982, Parathyroid Hormone Stimulates the Bone Apposition Rate Independently of Its Resorptive Action: Differential Effects of Intermittent and Continous Administration. Endocrinology , 110:506-512.

Thorner, M.O. and Holl, R.W., 1989, Physiologisches Wachstum in: Endokrinologie, Teil A, Urban und Schwarzenberg, München.

Veldhuis, J.D., Johnson, M.L.,1986, Cluster analysis: a simple, versatile and robust algorithm for endocrine pulse detection. Am. J. Physiol., 250:E486.

STRUCTURAL AMPLIFICATION IN CHEMICAL NETWORKS

Erich Bohl

Fakultät für Mathematik der Universität Konstanz
Postfach 5560, D-7750 Konstanz
Federal Republic of Germany

ABSTRACT

The reaction circuit for the cascade of vision is analyzed. We show that the architecture of this circuit is such that it can give rise to a rapid amplification for a moderate stimulus provided that a certain condition on the reaction constants and the concentrations of some of the involved chemicals is satisfied.

INTRODUCTION

In 1986, L. Stryer published the reaction circuit of the cascade of vision in a diagram which we reproduce as Fig. 1. Extracting the main features of this circuit, we arrive at the left-hand picture of Fig. 2 which we call our model circuit. In the transition between these two pictures we have introduced the notation

$$W_1 \stackrel{\wedge}{=} T_\alpha \cdot GTP, \quad W_2 \stackrel{\wedge}{=} [R^* \cdot T \cdot GDP, R^* \cdot T \cdot GTP],$$

$$W_3 \stackrel{\wedge}{=} T \cdot GDP, \quad S \stackrel{\wedge}{=} cGMP, \quad P \stackrel{\wedge}{=} 5' \cdot GMP, \quad U \stackrel{\wedge}{=} R^*, \tag{1}$$

$$Y_1 \stackrel{\wedge}{=} PDE_i, \quad Y_2 \stackrel{\wedge}{=} PDE^* \cdot T_\alpha \cdot GTP, \quad Y_4 \stackrel{\wedge}{=} PDE_i \cdot T_\alpha \cdot GDP.$$

The reaction networks corresponding to the circuits in Fig. 2 are shown in Fig. 3. A central point made by Stryer (1986, 1987) and also by Kühn (1988) is the rapid amplification of W_1 at only moderate levels of U. This suggests that the rate constant

Complexity, Chaos, and Biological Evolution, Edited by E. Mosekilde and
L. Mosekilde, Plenum Press, New York, 1991

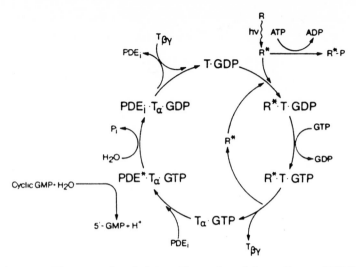

Figure 1. The cascade of vision. Reproduced from Stryer (1986).

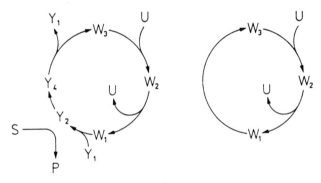

Figure 2. The model circuit (left) and the linearized model circuit (right).

k_{-7} (see Fig. 3) is large enough to secure a substantial formation of W_2 even for moderate U. Note that the rate constant k_{-7} governs the transition $W_3 \to W_2$ catalyzed by U.

In this paper we show that the architecture of the model circuit can produce a rapid structural amplification effect on W_1 for moderate U if $k_{-7} k_{-6} / \bar{Y} >> 1$. Here $\bar{Y}$ denotes the sum of all concentrations of the Y's in (1) (which is conserved in the reaction process). Note that our condition is in accordance with a large rate constant

k_{-7}. Our analysis shows that the structural amplification of W_1 comes from the interaction between the Y's and the W's and is hence a consequence of the overall architecture of the circuit.

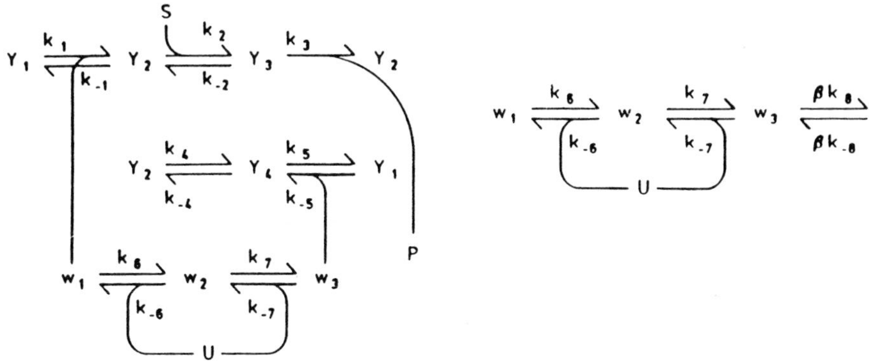

Figure 3. The reaction network for the model circuit (left) and for the linearized model circuit (right).

What really happens is best seen in the simplest reaction scheme

$$U + W_3 \xrightarrow{k} W_1 \xrightarrow{\beta} W_3 , \tag{2}$$

which is able to produce amplification. This scheme leads to the two differential equations $\dot{w}_1 = kuw_3 - \beta w_1$ and $\dot{w}_3 = -kuw_3 + \beta w_1$ which we can solve to find the steady state vector

$$w_1 (u,\beta) = \alpha \{1 - \beta / (ku + \beta)\} , \ w_3 (u,\beta) = \alpha\beta / (ku + \beta) ,$$

$$\alpha : = w_1 (0) + w_3 (0) (=w_1 (t) + w_3 (t) \ \text{for} \ t \geq 0) . \tag{3}$$

The function $w_1 (u,\beta)$ increases monotonously from $w_1 (0,\beta) = 0$ to $w_1 (\infty,\beta) = \alpha$, and it assumes values close to α very rapidly if $k / \beta >> 1$. This condition basically means that the last transition governed by the rate constant β is much slower than the first reaction defined by the rate constant k. All that follows in this paper is just a variation on this theme for more complicated systems.

The model circuit involves a nonlinear action at the two points where the W's and the Y's meet. A further reduction which uses the W's alone retains the structural amplification effect: It is the linearized model circuit to the right in Fig. 2 which is composed of only first order (linear) transitions. The corresponding reaction network is shown in Fig. 3. As in the case of (2), it turns out that the steep rise of w_1 in the linearized model circuit for small $U > 0$ is increasingly pronounced as a rational expression like k / β for (2) becomes very large. In particular, this occurs if the parameter β tends to zero. Indeed, β in the linearized model circuit plays an analogous role to the sum $\bar{Y}$ of the concentration of the Y's in the model circuit: Both parameters control the amplification effect on the chemical W_1. Our discussion will be based on this similarity.

In section 1 we discuss the linearized model circuit. We transform its reaction network into dynamical equations and perform a steady state analysis. This yields the steady state vector

$$w(u,\beta) \; = \; (w_1 \, (u,\beta), \, w_2 \, (u,\beta), \, w_3 \, (u,\beta))$$

as a function of the outside stimulus u and the parameter $\beta > 0$. We then discuss a possible dramatic rise of the function $w_1 \, (u,\beta)$, which describes an activating action of the network. In section 2 we turn to the model circuit on the basis of the results of the linearized model circuit. Special features of the cascade of vision require restrictions of the various rate constants in Fig. 3. This, along with some comments on more general networks exhibiting similar features of structural amplification, is discussed at the end of section 2. We illustrate all theoretical findings by numerical experiments. For a more mathematical approach and a discussion of numerical difficulties with the resulting nonlinear systems, we refer the reader to Bohl (1990).

THE LINEARIZED MODEL CIRCUIT

The transformation of the right-hand network in Fig. 3 into a system of differential equations uses in a standard way the law of mass action with the rate constants k_j, k_{-j} as indicated. The result is the system

$$\dot{w}_1 \; = \; -k_6 \, u w_1 + k_{-6} \, w_2 - \beta \, (k_{-8} \, w_1 - k_8 \, w_3) \,, \tag{4a}$$

$$\dot{w}_2 \; = \; k_6 \, u w_1 - (k_{-6} + k_7) \, w_2 + k_{-7} \, u w_3 \,, \tag{4b}$$

$$\dot{w}_3 \; = \; k_7 \, w_2 - k_{-7} \, u w_3 + \beta \, (k_{-8} \, w_1 - k_8 \, w_3) \,. \tag{4c}$$

We are interested in the steady state solution to these equations. To simplify the notation we define the polynomials

$$L(u) = k_6 k_{-7} u + k_6 k_7 + k_{-6} k_{-7},\tag{5a}$$

$$Q(u) = (k_{-8} k_{-7} + k_8 k_6) u + (k_8 + k_{-8}) (k_{-6} + k_7)\tag{5b}$$

and the constant

$$\gamma = k_{-8} k_{-6} k_{-7} - k_8 k_6 k_7.\tag{6}$$

Furthermore, we apply the conservation law

$$w_1(t) + w_2(t) + w_3(t) \equiv w_1(0) + w_2(0) + w_3(0) =: \alpha\tag{7}$$

which follows from (4) if we add up both sides of the three equations. With this simplification, the steady state solution to (4) reads

$$w_1(u,\beta) = \frac{\alpha}{L(u)} \left\{ k_{-6}k_{-7} - \frac{\gamma\beta(k_{-7}u + k_{-6} + k_7)}{uL(u) + \beta Q(u)} \right\},\tag{8a}$$

$$w_2(u,\beta) = \frac{\alpha u}{L(u)} \left\{ k_6 k_{-7} + \frac{\gamma\beta(k_{-7} - k_6)}{uL(u) + \beta Q(u)} \right\},\tag{8b}$$

$$w_3(u,\beta) = \frac{\alpha}{L(u)} \left\{ k_6 k_7 + \frac{\gamma\beta(k_6 u + k_{-6} + k_7)}{uL(u) + \beta Q(u)} \right\}.\tag{8c}$$

Note that the constant α is defined in (7). The formulae (8a-c) describe the response of the linearized model circuit to a stimulus u. Since this response depends sensitively on the constant $\beta > 0$, we represent the solution as a function of the two variables u, β.

A discussion of the formulae (8a-c) is in order: Let us concentrate on the function given in (8a), which is the steady state concentration of the chemical W_1 experiencing the dramatic increase in response to a moderate stimulus U. The rational function

$$R(u) = \frac{\beta P(u)}{uL(u) + \beta Q(u)} = \frac{P(u)}{u\beta^{-1}L(u) + Q(u)},\tag{9}$$

where

$$P(u) = k_{-7} u + k_{-6} + k_7$$

123

is responsible for this behavior. It drops from $R(0) = P(0) / Q(0)$ at $u = 0$ rapidly to almost zero in a small positive neighborhood of $u = 0$ if $L(0) / \beta >> 1$. We can see this if we calculate the derivative

$$R'(0) = - \frac{L(0)}{\beta} \frac{P(0)}{Q(0)} + \frac{Q(0)P'(0) - P(0)Q'(0)}{Q(0)^2} .$$

The absolute value of the first term on the right-hand side of this expression tends to infinity if $L(0) / \beta$ does. Consequently, the function (8a) increases very rapidly from $w_1 (0, \beta) = \alpha L(0)^{-1} (k_{-6} k_{-7} - \gamma R(0))$ to $w_1 (u, \beta) \sim \alpha L(u)^{-1} k_{-6} k_{-7}$ $(0 < u,$ small$)$. It describes mathematically the dramatic change in the concentration of the chemical W_1 in response to a (positive but possibly very small) stimulus U: Hence, the structural amplification inherent in the architecture of the linearized model circuit is obtained if

$$\gamma = k_{-8} k_{-6} k_{-7} - k_8 k_6 k_7 > 0 . \tag{10}$$

This inequality implies that the reactions in the clockwise direction of the linearized model circuit are faster than the counter-clockwise. Therefore, (10) requires a continuous supply of energy pumped into the linearized reaction circuit.

In passing, we note that for $\beta = 0$, there is no amplification of the chemical W_1: on the contrary, (8a) reduces to the monotonously decreasing expression $w_1 (u, \beta) = \alpha k_{-6} k_7 (L(u))^{-1}$ so that the concentration of W_1 is reduced in response to an injection of U (if $k_{-7} k_6 > 0$). Furthermore, note that the condition $L(0) / \beta >> 1$ for amplification directly corresponds to the condition $k / \beta >> 1$ which we arrived at in the discussion of the simple reaction scheme (2) in the introduction. Since $L(0) = k_6 k_7 + k_{-6} k_{-7}$, it is enough to require $k_{-7} k_{-6} / \beta >> 1$, which is in accordance with a large value of the reaction constant k_{-7} as mentioned in Stryer's and Kühn's work. However, we see that k_{-7} alone is not responsible for the steep rise in the concentration of W_1. It is rather its ratio to β. We can produce instant amplification by choosing β very small relative to k_{-7}. In the next section we will learn that β of the linearized model circuit directly corresponds to the sum of the concentrations of the Y's in the model circuit.

We illustrate these results with the numerical example given in Fig. 4. The rate constants are $k_6 = 10$, $k_7 = 0.1$, $k_8 = 3$, $k_8 = k_{-6} = k_{-7} = 1$ and $\alpha = 1$. In Fig. 4 the function $w_1 (u, \beta)$ is drawn for three different values of the parameter β. The picture shows clearly how the steep ascent develops in a positive vicinity of the origin $u = 0$ if $\beta > 0$ is small enough. The picture also shows the possibility that the function $w_1 (u, \beta)$ may decrease after passing through a maximum. But this behavior is only due to the choice of the various constants. For different (and probably more appropriate) choices, the function may be monotonously increasing with the steep ascent near the origin leveling off outside a neighborhood of $u = 0$. We return to this point in the discussion of the model circuit in the next section. If β is kept constant and k_{-7} is

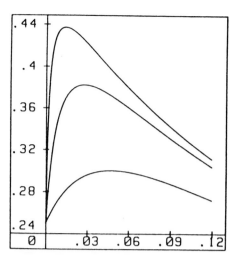

Figure 4. The function $w_1 (u, \beta)$ for β = 0.03, 0.005, 0.001 (from bottom to top curve); all other parameters are given in the text (u-axis horizontal, w_1-axis vertical).

increased, the resulting pictures look very similar to Fig. 4: The amplification grows as k_{-7} / β increases.

THE MODEL CIRCUIT

In this section we consider the model circuit of Fig. 2 which is transformed into the chemical network given as the left-hand picture of Fig. 3. Here, the chemical Y_3 is added to the list (1) to model the enzyme action occurring in the lower left corner of the model circuit. Taking a closer look at both networks in Fig. 3, we observe that their main parts are identical. The direct connection between W_1 and W_3 in the simpler network appears in the more complicated case as the path from W_1 via Y_2, Y_4 to W_3 : hence, the reversible reactions with the rate constants k_1, k_4 and k_5. This way of reasoning leads to the hypothesis that we can understand the model circuit on the basis of the linearized model if we identify the constants according to

$$k_8 \stackrel{\wedge}{=} k_{-5} \, k_{-4} \, k_{-1}, \quad k_{-8} \stackrel{\wedge}{=} k_1 \, k_4 \, k_5. \tag{11}$$

Note that k_8 and k_{-8} occur in the linearized model circuit, whereas all other rate constants in (11) enter in the more complicated model circuit. Next, we must find the analog to the parameter β of the linearized model circuit. We note that, in the case of the left-hand network in Fig. 3, the conservation law

$$y_1(t) + y_2(t) + y_3(t) + y_4(t) \equiv$$

$$y_1(0) + y_2(0) + y_3(0) + y_4(0) =: \bar{Y} \qquad (t \geq 0)$$

(12)

holds for the concentrations $y_i(t)$ of the chemical Y_i. (12) defines a constant $\bar{Y} > 0$, and it turns out that this constant plays the role of the parameter $\beta > 0$ in the linearized model circuit. Therefore, the analogy is complete if we use the identifications (11) and set

$$\beta \hat{=} \bar{Y}. \tag{13}$$

Of course, we cannot expect that the formulae (8a-c) for the steady state concentrations of the W's apply to the model circuit as well. However, we can expect, and this is indeed true, that the qualitative implications drawn from (8) for the linearized model circuit apply word for word to the model circuit itself: Using the identifications (11), the crucial constant γ defined in (6) becomes

$$\gamma = k_1 k_4 k_5 k_{-6} k_{-7} - k_{-1} k_{-4} k_{-5} k_6 k_7. \tag{14}$$

According to (10), this constant should be positive:

$$k_1 k_4 k_5 k_{-6} k_{-7} > k_{-1} k_{-4} k_{-5} k_6 k_7. \tag{15}$$

As for the linearized model circuit, this means that the reactions in the clockwise direction for the model circuit must be faster than the counter-clockwise; a thermodynamically feasible assumption if we recall that energy is continuously pumped into the reaction circuit of the cascade of vision.

Next, we apply the analogy (13) and recall that, for the spontaneous amplification effect in the linearized model circuit, the condition $k_{-7} k_{-6} / \beta >> 1$ was sufficient. For the model circuit we therefore expect an amplification effect if $k_{-7} k_{-6} / \bar{Y} >> 1$. Note that, by (12), $\bar{Y}$ measures the total concentration of all the chemicals Y_i ($i = 1,2,3,4$).

In fact, a rigorous analysis of the model circuit taking into account all nonlinearities shows that the qualitative description for the linearized model circuit applies word for word to the model circuit itself if we strictly use (11) and (13).

Let us again support our findings with numerical examples as presented in Fig. 5. In all four pictures we have used

$$k_j = k_{-j} = 1 \quad (j = 1,2,3,4), \quad \text{and } G = 1. \tag{16}$$

Note that $w_1(t) + w_2(t) + w_3(t) - y_1(t) =: G$ remains constant during the process,

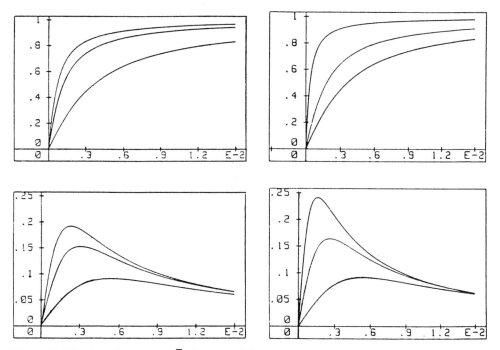

Figure 5. The function $w_1 (u, \bar{Y})$ for two different sets of parameters (see (17)) and the following values for $\bar{Y}$ and k_{-7}:

upper left: $\bar{Y} = 0.01$, $k_{-7} = 1,3,5$ (bottom to top curve)
lower left: $\bar{Y} = 0.1$, $k_{-7} = 1,3,5$ (bottom to top curve)
upper right: $k_{-7} = 1$, $\bar{Y} = 0.01, 0.005, 0.001$ (bottom to top curve)
lower right: $k_{-7} = 1$, $\bar{Y} = 0.1, 0.025, 0.01$ (bottom to top curve)
(u-axis horizontal, w_1-axis vertical)

and it is this constant G which appears in (16). The other constants are as follows:

upper pictures: $k_5 = 1000$, $k_{-5} = 0$ (17a)
                               $k_6 = 0$, $k_{-6} = 1000$,
                               $k_7 = 1$, $s = 1$,

lower pictures: $k_5 = 1$, $k_{-5} = 0.001$,   (17b)
                               $k_6 = 1000$, $k_{-6} = 1$,
                               $k_7 = 10^{-13}$, $s = 1$.

In all pictures we see how the steep ascent builds up in the vicinity of the origin if the quotient $k_{-7} / \bar{Y}$ increases. The lower pictures of Fig. 5 are very similar to Fig. 4 for the linearized model circuit. In producing the upper pictures, we have used quantitative information provided by N. Bennett (1990) for the cascade of vision (compare Fig. 3, left-hand picture): Firstly, the reaction with the rate constant k_5 is irreversible so that $k_{-5} = 0$, and secondly, the concentration of W_1 for $u = 0$ must vanish: $w_1 (0,\bar{Y}) = 0$. A more detailed discussion of the model circuit shows the implication $k_{-1} k_{-4} k_{-5} = 0 \Rightarrow w_1 (0,\bar{Y}) = 0$. Therefore, the requirement that w_1 must vanish for $u = 0$ and $k_{-5} = 0$ is consistent. Furthermore, $k_{-5} = 0$ is in accordance with (15).

A closer look at the left-hand network in Fig. 3 reveals that it consists basically of two parts: a network of the Ys and one of the Ws. Both operate in the style of the MWC-theory (Monod et al. 1965) and are linked together at two points. This structure is probably the next level of complexity after the networks resulting from the MWC-theory. It seems to arise quite frequently. As a simple example, we turn to our model circuit splitting W_2 into two chemicals W_2, W_2' as suggested by the original circuit of Fig. 1. It turns out that this little change in our model circuit does not alter its behavior: The new network shows the described amplification effect for W_1 if the sum of the concentrations of the Ys (which is conserved during the process) is positive but small enough.

ACKNOWLEDGMENTS

I thank Professor B. Wurster for stimulating discussions and helpful comments as well as Dr. N. Bennett for her comments on the subject discussed in this paper.

REFERENCES

Bennett, N., 1990, private communication.

Bohl, E., A boundary layer phenomenon for linear systems with a rank deficient matrix. To appear in the ZAMM.

Kühn, H., 1988, Die lichtaktivierte Enzymkaskade, <u>Aus Forschung und Medizin</u>, Jahrgang 3, 2:63-74.

Monod, J., Wyman, J., Changeux, J.-P., 1965, On the nature of allosteric transitions: A plausible model, <u>J. Mol. Biol.</u>, 12:88-118.

Stryer, L., 1986, Cyclic AMP cascade of vision, <u>Ann. Rev. Neurosci.</u>, 9:87-119.

Stryer, L., 1987, Die Sehkaskade, <u>Spektrum der Wissenschaft</u>, 9:86-95.

Section III
Membrane Activity and Cell to Cell Signalling

It appears that pulsatile signalling is more effective than constant stimuli in eliciting a sustained physiological response.

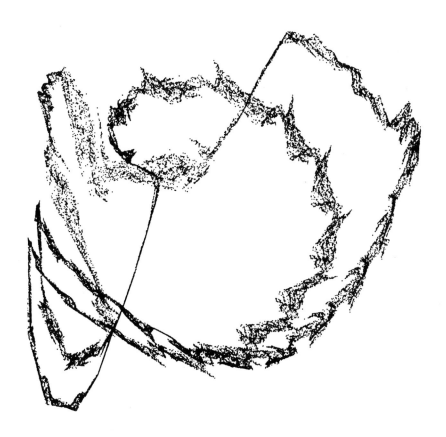

PERIODICITY AND CHAOS IN cAMP, HORMONAL, AND Ca^{2+} SIGNALLING

A. Goldbeter, Y.X. Li, and G. Dupont

Faculté des Sciences
Université Libre de Bruxelles
Campus Plaine, C.P. 231
B-1050 Brussels, Belgium

INTRODUCTION

Rhythmic, pulsatile signals are repeatedly encountered in intercellular communication (Goldbeter, 1988, 1990). Besides neurons and muscle cells which communicate by trains of electrical impulses, examples range from the generation of cyclic AMP (cAMP) pulses in the slime mold *Dictyostelium discoideum* (Gerisch, 1987) to the pulsatile release of a large number of hormones (Crowley and Hofler, 1987; Wagner and Filicori, 1987). While in all these instances the oscillatory dynamics characterizes the extracellular signal, recent observations indicate that signal transduction itself may be based on oscillations of intracellular messengers. Thus, many hormones or neurotransmitters act on target cells by triggering a train of intracellular Ca^{2+} spikes (Berridge et al., 1988).

The purpose of this paper is to briefly discuss the role of periodic or chaotic processes in various modes of intercellular communication. We shall deal with pulsatile signalling in *Dictyostelium* cells and in endocrine systems, before turning to signal transduction based on intracellular Ca^{2+} oscillations. The efficiency of pulsatile signalling will be assessed in a general model based on the ubiquitous process of receptor desensitization. In this model, it appears that periodic signals are more effective than constant, stochastic or chaotic stimuli in eliciting a sustained physiological response in target cells.

PULSATILE SIGNALS OF cAMP IN *DICTYOSTELIUM* CELLS

In view of the relative simplicity of its pattern of differentiation, the cellular slime mold *Dictyostelium discoideum* represents a particularly useful model in developmental biology (Loomis, 1975). This system also provides a prototype of pulsatile signalling in intercellular communication. After starvation, these amoebae indeed aggregate by a chemotactic response to pulses of cAMP emitted by cells which behave as aggregation centers. The periodicity of the phenomenon is of the order of 5 to 10 min (Gerisch and Wick, 1975).

The analysis of a model (Martiel and Goldbeter, 1987) based on cAMP-induced cAMP synthesis and on receptor desensitization through reversible phosphorylation (Vaughan and Devreotes, 1988) accounts for the phenomena of autonomous oscillations of cAMP and of relay of suprathreshold signals of cAMP observed both in cell suspensions and during

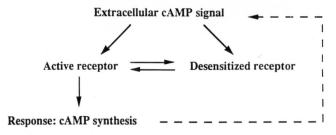

Fig. 1.   Scheme for the autocatalytic synthesis of cyclic AMP (cAMP) in the slime
          mold *Dictyostelium discoideum* in the presence of receptor desensitization.

aggregation on agar (Gerisch and Wick, 1975; Roos et al., 1975). The model considered is schematized in Fig.1; it is described by a set of three nonlinear, ordinary differential equations which govern the time evolution of intracellular and extracellular cAMP, and the fraction of active cAMP receptor (see Martiel and Goldbeter, 1987; Goldbeter, 1990). The core mechanism of the model for cAMP relay and oscillations reduces to a two-variable system (Martiel and Goldbeter, 1987) which has been used to simulate wavelike propagation of cAMP signals in the course of slime mold aggregation (Tyson et al., 1989). This wavelike process has also been simulated with an alternative model for cAMP signalling based on a putative inhibition of adenylate cyclase by Ca^{2+} (Monk and Othmer, 1990).

Besides their role in controlling aggregation, the chemotactic signals govern differentiation during the hours that follow starvation. Pulses of cAMP applied every 5 min are indeed found to accelerate development in contrast to constant cAMP stimuli which are without effect (Darmon et al., 1975; Gerisch et al., 1975). More surprisingly, pulses of cAMP delivered every 2 min also fail to have any physiological effect (Wurster, 1982), as is the case for signals delivered at random (Nanjundiah, 1988).

The finding that the frequency of cAMP pulses directly controls their effect on cellular differentiation raises the intriguing question as to what is the molecular basis of frequency encoding of pulsatile signals of cAMP in *Dictyostelium*. This question has been investigated theoretically in the receptor-desensitization model for cAMP signalling (Li and Goldbeter, 1990).The differential equations of the model were integrated numerically for various square-wave stimuli of extracellular cAMP. The stimuli differed by the duration (τ_1) of each pulse and/or by the interval (τ_0) between successive pulses (see Fig. 2 below).

In this study, the response, i.e., synthesis of intracellular cAMP, is determined for square-wave stimuli characterized by the same pulse duration (e.g., $\tau_1 = 3$ min) and increasing time intervals between successive pulses (e.g., $\tau_0 = 1, 2, 5, 10$ min). The results show that the mean quantity of cAMP synthetized increases with the value of the interval τ_0. A comparison of the different situations indicates that this is due to the fact that the fraction of active receptor decreases during stimulation – owing to cAMP-induced receptor desensitization – and increases as soon as the pulse ceases. When the interval is short, the receptor has not enough time to recover significantly from desensitization and cAMP synthesis is damped. As the interval becomes larger, more receptor returns to the active state so that a stronger response is triggered by the next stimulus. If the interval τ_0 is very large, each response is nearly maximum but the total number of responses will be reduced. Therefore there exists a particular pattern of periodic stimulation that maximizes the number of significant responses over a given amount of time (Li and Goldbeter, 1990).

These results provide a molecular explanation for the effectiveness of pulses delivered at 5 min intervals and for the failure of constant or stochastic cAMP stimuli and of pulses delivered at 2 min intervals in promoting cell differentiation in *Dictyostelium*. Only when the interval is sufficiently (but not too) long would the mean level of intracellular cAMP be high enough to bring about differentiation. This explanation relies on the assumption that developmental changes require a significant synthesis of intracellular cAMP upon external stimulation. Such an assumption is corroborated by experimental observations which show that intracellular cAMP mediates some developmental responses (Simon et al., 1989). Similar conclusions on the efficiency of cAMP pulses of appropriate frequency hold for other intracellular messengers, such as diacylglycerol or inositol 1,4,5-trisphosphate, whose synthesis is also elicited by the cAMP signal and which appear to play a role in some of the subsequent cellular changes (Ginsburg and Kimmel, 1989).

The model discussed above does not take explicitly into account the role of various GTP-binding proteins (G-proteins) in the activation of adenylate cyclase and in the adaptation process (Snaar-Jagalska and Van Haastert, 1990). That receptor modification is not the only process involved in adaptation of the cAMP relay response to constant stimuli (Snaar-Jagalska and Van Haastert, 1990) is reflected by the partial failure of the receptor desensitization model to account for some experiments based on a series of rapid step increases in stimulation (Monk and Othmer, 1990). As more information becomes available from biochemical experiments, the reaction steps involving G-proteins are being incorporated into the model with the aim of extending the repertoire of experimental observations accounted for by the theory. It is expected that the conclusions reached above on the existence of an optimal pattern of periodic stimulation by cAMP pulses will continue to hold, in a large measure, regardless of the detailed molecular mechanism of cAMP signalling and adaptation in *Dictyostelium*.

Besides providing a prototype for pulsatile signalling in intercellular communication, cAMP oscillations in *Dictyostelium* also yield an example of autonomous chaos resulting from the interplay between two endogenous oscillatory mechanisms present within the same cellular system (Goldbeter, 1990). For some parameter values, cAMP oscillations predicted by the receptor desensitization model are indeed chaotic (Martiel and Goldbeter, 1985). Such a result could account for the aperiodic nature of aggregation on agar in the *Fr17* mutant of the slime mold (Durston, 1974). Studies of this mutant in cell suspensions showed regular instead of chaotic oscillations (Goldbeter and Wurster, 1989). One reason might be that the mutant cells had evolved toward the domain of periodic behavior which is much larger than the domain of chaos in parameter space. Alternatively, the presence of a certain proportion of cells oscillating periodically could destroy the chaotic properties of other cells present within the mixed suspension. The latter conjecture has recently been verified in a theoretical model for cAMP signalling in *Dictyostelium*. Besides indicating the existence of a period-doubling route to chaos upon mixing a periodic population with an increasing proportion of chaotic cells, this analysis shows that a tiny proportion of periodic cells suffices to suppress chaos within the mixed suspension owing to the strong coupling of cells through extracellular cAMP (Halloy et al., 1990; Li et al., 1990).

PULSATILE HORMONE SECRETION

Intercellular signalling by cAMP pulses in *Dictyostelium* may be viewed as a primitive example of hormonal communication. In multicellular organisms, processes of pulsatile communication are also observed, particularly in endocrinology. Thus, many hormones are secreted in a rhythmic manner into the circulation (Wagner and Filicori, 1987; Crowley and Hofler, 1987). What is remarkable is that the physiological effect of hormones, in most cases, is found to depend on the frequency of their pulsatile secretion.

The prototype of pulsatile hormone release is that of the hormone GnRH which induces the periodic secretion of the gonadotropic hormones LH and FSH by the pituitary (Knobil, 1980). In man and in the rhesus monkey, GnRH is released by the hypothalamus with a frequency close to one pulse every hour.

Experiments performed by Knobil and coworkers have shown that constant GnRH stimuli fail to establish appropriate levels of LH and FSH; a similar failure is found for GnRH pulses delivered 2 or 3 times per hour, or once every two hours (Belchetz et al., 1978; Knobil, 1980). Therefore, as in *Dictyostelium* amoebae, a major part of the information carried by the hormo-nal stimulus resides in its frequency. To paraphrase Knobil (1981), the temporal pattern of the hormone may be of greater importance than its level in the blood.

What are the molecular bases of frequency encoding of the hormonal signal? This question has been investigated (Li and Goldbeter, 1989) in a theoretical model (Segel et al., 1986) whose general form ensures wide applicability. Once again, as in the specific model for cAMP signalling discussed in the preceding section, a receptor existing in an active and a desensitized state is subjected to a square-wave stimulus (Fig. 2). For the sake of generality, the response is taken as proportional to an *activity* defined as a linear combination of the concentrations of the liganded and unliganded receptor states, each state being characterized by an *activity coefficient* that weights its contribution to the overall response (Segel et al., 1986).

Target cell responsiveness is measured by a function that takes into account both the magnitude of each response and the number of significant responses that can be triggered by

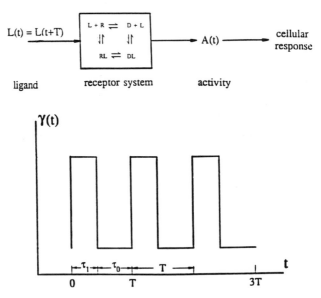

Fig. 2.   Pulsatile, square-wave stimulus (lower panel) applied to a receptor undergoing ligand-induced desensitization (upper panel). The normalized concentration of the ligand L is denoted by γ; the period T of the pulsatile stimulus is equal to the sum of pulse duration (τ_1) and pulse interval (τ_0). The receptor exists in active (R, RL) or desensitized (D, DL) states whose interconversion occurs through conformational transitions or through covalent modification (Li and Goldbeter, 1989).

stimulation in a given amount of time (Li and Goldbeter, 1989). Cellular responsiveness is computed as a function of the two parameters that characterize the pulsatile signal, namely, the interval τ_0 and its duration τ_1 (the amplitude of the signal is held at a constant value). The results of this analysis indicate (Li and Goldbeter, 1989; Goldbeter and Li, 1989) that there exists an *optimal pattern* of periodic stimulation, corresponding to a pair of values (τ_0*, τ_1*), which allows for maximum cellular responsiveness (Fig. 3). It appears, therefore, that desensitization in target cells permits the efficient encoding of pulsatile hormonal signals.

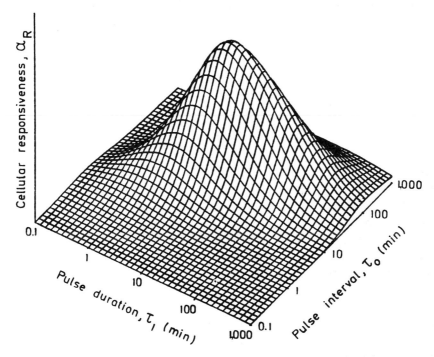

Fig. 3.   Cellular responsiveness as a function of pulse duration (τ_1) and pulse interval (τ_0) of the square-wave stimulus considered in Fig. 2 (lower panel). Cellular responsiveness provides a measure of the number and magnitude of integrated responses in a given amount of time. The data, obtained for the model and stimulus of Fig. 2, indicate the existence of a pattern of pulsatile stimulation yielding optimal cellular responsiveness (Li and Goldbeter, 1989). Receptor parameters have been chosen so as to yield an optimal pattern of stimulation similar to the optimal pattern observed for the hypothalamic hormone GnRH, i.e., one 6 min-pulse per h.

The questions arise as to what is the effect of stochastic variations in the pulsatile signal, and as to whether chaotic signals are more (or less) efficient than periodic ones in eliciting cellular responses. These issues have been addressed in the receptor-desensitization model of Fig. 2, by subjecting parameters τ_0 and τ_1 to stochastic (Gaussian or white-noise) variations around the optimal values τ_0*, τ_1*, or to chaotic variations generated by the logistic map (Li and Goldbeter, manuscript in preparation). The results are exemplified by the data in Fig. 4 which indicate that the optimal, periodic signal is always more efficient in this model than the signals subjected to stochastic or chaotic variations. The conclusion on the superior efficacy of periodic versus chaotic stimulation depends in a certain measure on the type of mechanism used to generate aperiodic variations. However, we may expect that in this receptor system the qualitative result holds for chaos-generating processes other than the logistic map. With chaos, indeed, some time intervals between pulses will be too short to allow for significant resensitization, and will therefore lead to smaller responses, while longer intervals will decrease the number of significant responses in a given amount of time.

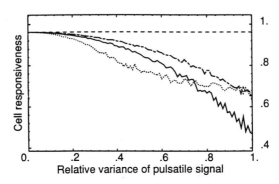

Fig. 4. Comparison of cell responsiveness to the optimal periodic stimulus (dashed line) with responsiveness to square-wave stimuli subjected to Gaussian (dotted line), white-noise (dashed-dotted line), and chaotic (solid line) variation in pulse duration τ_1 and pulse interval τ_0. The chaotic variation is generated by the logistic map $r_{n+1} = \mu\, r_n (1 - r_n)$ with $\mu = 4$ (Li and Goldbeter, in preparation). For model equations and definition of cell responsiveness, see Li and Goldbeter, 1989).

SIGNAL TRANSDUCTION BASED ON CALCIUM OSCILLATIONS

Frequency encoding is not restricted to the external signal used in intercellular communication. In an increasing number of instances, it is found that hormones or neurotransmitters act on target cells by triggering a train of cytosolic Ca^{2+} spikes. The oscillations occur in a variety of cells, with periods ranging from less than 1 second to minutes (see Berridge and Galione, 1988; Berridge et al., 1988; and Cuthbertson, 1989, for recent reviews). Here, therefore, the oscillations occur at the level of the second rather than first messenger.

The mechanism of Ca^{2+} oscillations appears to involve the synthesis of inositol 1,4,5-trisphosphate (InsP3) that follows binding of the stimulatory ligand to the membrane receptor. According to Berridge, the rise in InsP3 in turn triggers the release of a certain amount of Ca^{2+} from an InsP3-sensitive intracellular store (Berridge and Galione, 1988; Berridge et al., 1988). The analysis of a two-variable model taking into account the Ca^{2+} input from the external medium, the extrusion of Ca^{2+} from the cell, and the exchange of Ca^{2+} between the cytosol and an InsP3-insensitive store (see scheme in Fig. 5), shows that sustained oscillations may arise through a mechanism of Ca^{2+}-induced Ca^{2+} release (Endo et al., 1970; Fabiato, 1983) from the InsP3-insensitive store once the signal-induced rise in cytosolic Ca^{2+} triggered by InsP3 has reached a sufficient level (Dupont and Goldbeter, 1989; Goldbeter et al., 1990).

The model accounts for the observation that the frequency of the Ca^{2+} spikes increases with the extent of stimulation; the latter is measured by the saturation function of the InsP3 receptor, given that the level of InsP3 rises with the external signal. Also accounted for by this model is the observed linear correlation between the period of Ca^{2+} oscillations and their latency, i.e., the time required for the appearance of the first Ca^{2+} peak after stimulation (Dupont et al., 1990). In contrast with an alternative model based on a feedback of Ca^{2+} on

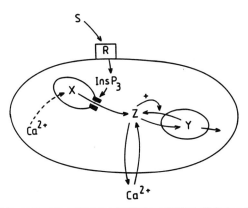

Fig. 5. Scheme of the model considered for signal-induced, intracellular Ca^{2+} oscillations. The stimulus (S) triggers the synthesis of InsP$_3$; the latter intracellular messenger elicits the release of Ca^{2+} from an InsP$_3$-sensitive store (X) at a rate proportional to the saturation function (β) of the InsP$_3$ receptor. Cytosolic Ca^{2+} (Z) is pumped into an InsP$_3$-insensitive intracellular store; Ca^{2+} in the latter store (Y) is released into the cytosol in a process activated by cytosolic Ca^{2+}. This feedback, known as Ca^{2+}-induced Ca^{2+} release, plays a primary role in the origin of Ca^{2+} oscillations. In its simplest version, the model contains only two variables, i.e., Y and Z (Dupont and Goldbeter, 1989; Dupont et al., 1990; Goldbeter et al., 1990).

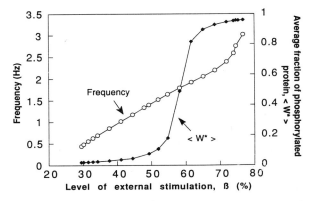

Fig. 6. Dependence of frequency of Ca^{2+} oscillations on magnitude of stimulus in the two-variable model for signal-induced Ca^{2+} oscillations based on Ca^{2+}-induced Ca^{2+} release (see Fig.5). Also indicated is the mean fraction of a putative protein phosphorylated by a Ca^{2+}-dependent kinase (see also Goldbeter et al., 1990).

InsP$_3$ production (Meyer and Stryer, 1988), a specific prediction of the analysis is that sustained oscillations in cytosolic Ca^{2+} may occur in the absence of a concomitant, periodic variation in InsP$_3$. On the other hand, the model is closely related to that proposed by Kuba and Takeshita (1981) for membrane potential oscillations in sympathetic neurons treated with caffeine; there also, oscillations were due to Ca^{2+} spikes resulting from the mechanism of Ca^{2+}-induced Ca^{2+} release, but the role of InsP$_3$, discovered since, was not considered.

How the oscillations in cytosolic Ca^{2+} might control cellular responses through their frequency is shown by the analysis of the situation where a Ca^{2+}-dependent kinase phosphorylates a protein substrate (Goldbeter et al., 1990). The mean level of phosphorylated protein is shown to increase with the frequency of Ca^{2+} transients which rises in turn with the level of the stimulus (Fig. 6). An efficient encoding depends, however, on the kinetic parameters of the kinase and phosphatase. In particular, the dependence of the mean fraction of phosphorylated protein on the frequency of Ca^{2+} spikes — and, hence, on the level of external stimulation — is steeper when the converter enzymes involved in covalent modification become saturated by the target protein.

The dynamics of cytosolic Ca^{2+} does not always possess a simple periodic nature. Thus, upon stimulation, some cells exhibit complex transients which resemble bursting oscillations (Berridge et al., 1988; Cuthbertson, 1989). It is not possible to generate such complex behavior when the model schematized in Fig. 5 contains only two variables, namely, Y and Z. To allow for complex oscillations, one has to consider the interplay between two (or more)

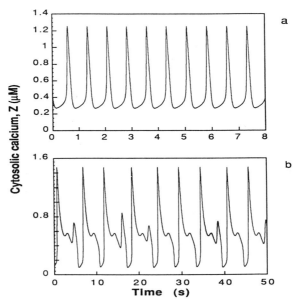

Fig. 7. Periodic (a) and chaotic (b) dynamics of cytosolic Ca^{2+} following sustained stimulation by an external signal in versions of the model based on Ca^{2+}-induced Ca^{2+} release containing two and five variables, respectively. The aperiodic oscillations in (b) are obtained when the model of Fig. 5 is supplemented with additional regulatory processes (Dupont and Goldbeter, in preparation; see text).

endogenous oscillatory mechanisms (Decroly and Goldbeter, 1982; Goldbeter et al, 1988) involving these and additional variables. Thus, upon incorporating into the model based on Ca^{2+}-induced Ca^{2+} release the activation of $InsP_3$ synthesis by cytosolic Ca^{2+} — which process is at the core of the oscillatory mechanism proposed by Meyer and Stryer (1988) —, and desensitization of the $InsP_3$ receptor — which process leads to a progressive decrease in the release of Ca^{2+} from the $InsP_3$-sensitive store —, we have recently obtained evidence, in a five-variable system, for complex periodic oscillations as well as chaos (Dupont and Goldbeter, manuscript in preparation). An example of the latter type of behavior is shown in Fig. 7.

FREQUENCY CODING IN INTERCELLULAR COMMUNICATION

We have discussed successively three different examples of oscillatory dynamics in intercellular communication. In *Dictyostelium* as in hormonal regulation, it appears that frequency encoding of the pulsatile extracellular signal is based on ligand-induced desensitization of target cells. This process may involve the covalent modification of receptors, as in the case of the slime mold (Vaughan and Devreotes, 1988), or some post-receptor mechanism as in the case of GnRH where desensitization of pituitary cells appears to involve inactivation of a Ca^{2+} channel (Stojilkovic et al., 1989). Whenever frequency coding relies on desensitization in target cells, the response of the latter increases as the time interval between successive stimuli augments and cells recover from refractoriness. There exists, however, an optimal frequency of periodic stimulation, since cellular responsiveness should take into account not only the magnitude but also the number of significant responses that can be elicited in a given period of time. The optimal frequency of pulsatile stimulation depends on the kinetics of desensitization and resensitization in target cells, as shown by the specific analysis of cAMP and GnRH signalling (Li and Goldbeter, 1989, 1990; Goldbeter and Li, 1989).

On the other hand, oscillatory dynamics in intercellular communication also plays a role in signal transduction within the cell. The third situation discussed above indeed pertains to the rhythmic evolution of cytosolic Ca^{2+} in cells stimulated by a nonperiodic, extracellular signal such as a step increase in hormone or neurotransmitter. The frequency of Ca^{2+} oscillations rises with the level of external stimulation (Berridge and Galione, 1988; Berridge et al., 1988; Cuthbertson, 1989). The analysis of a minimal model shows how the oscillations may originate from the well-known process of Ca^{2+}-induced Ca^{2+} release, and how they may be encoded in terms of their frequency through phosphorylation of a cellular substrate by a Ca^{2+}-dependent protein kinase (Dupont and Goldbeter, 1989; Dupont et al., 1990; Goldbeter et al., 1990).

Frequency encoding of signals in intercellular communication is thus not restricted to the nervous system (see Cazalis et al., 1985, and Whim and Lloyd, 1989, for examples of neural responses modulated by the frequency of incoming action potentials in neurosecretory cells). The advantages of frequency- versus amplitude-encoding are numerous. Periodic stimuli clearly allow for increased sensitivity and modulation as both their amplitude and frequency can be tuned. Control of the frequency may also prove more accurate than that of the amplitude (Rapp et al., 1981; Rapp, 1987). Other advantages may be more specific; thus, in the case of signal-induced Ca^{2+} oscillations, the characteristic time for the rise in Ca^{2+} during a spike may add a further parameter for control of the subsequent Ca^{2+}-dependent responses, according to the time scale on which the latter reactions occur (Woods et al., 1987).

The examples discussed here illustrate how pulsatile signals are used in intercellular communication between unicellular organisms and, even more, between distinct cells within higher organisms. An increasing number of hormones appear to be released periodically into the circulation and are optimally active when secreted in a pulsatile rather than constant manner (Wagner and Filicori, 1987; Crowley and Hofler, 1987). Besides the episodic release of LH and FSH induced by GnRH pulses (Belchetz et al., 1978; Knobil, 1980), other examples include insulin and glucagon (Weigle, 1987), GH (Bassett and Gluckman, 1986) and its releasing factor (Borges et al., 1984), TSH and PTH (Hesch, 1990). A similar efficiency of pulsatile delivery has recently been observed for a growth factor in the lens, where constant stimulation has no effect (Brewitt and Clark, 1988). These observations support the view that pulsatile signalling in intercellular communication represents a major function of biological rhythms (Goldbeter and Li, 1989; Goldbeter, 1990).

As pointed out by Hesch (1990), a *dynamic code* thus complements the genetic code. The latter carries information in the form of sequences of amino acids within the various proteins, and thereby determines their catalytic and regulatory properties. These properties, in turn, govern the various modes of dynamic behavior at the cellular and supracellular levels, including chaos and periodic oscillations. In comparing chaotic with periodic rhythms in cell to cell signalling, one could argue that the latter possess a richer information content; indeed, the kinetics of the receptor-mediated responses is often such that periodic stimuli of appropriate frequency will lead to optimal responsiveness in target cells subjected to desensitization. Thus, for hormones, as argued above on theoretical ground, it would seem more appropriate to base a dynamic code on periodic rather than chaotic release. That the episodic patterns of hormonal secretion often look irregular when revealed by high-frequency sampling should not eclipse the fact that what may be of physiological significance is the *roughly periodic* nature of pulsatile hormone secretion rather than the details of the episodic pattern.

The conclusions on the increased efficiency of periodic, pulsatile signals bear on the rapidly developing field of chronopharmacology (Lemmer, 1989), since the concept of optimal frequency of stimulation should hold for a number of drugs whose continuous delivery would result in target cell desensitization. Most studies in chronopharmacology presently focus on the important role of circadian rhythms. However, besides taking into account circadian rhythms which indeed influence most if not all physiological functions, the timing of drug delivery should also take into account the kinetics of desensitization and resensitization in target cells. These parameters will likely be specific to the drug and to the target cell, so that a fine tuning of the pulsatile delivery program might be required to optimize the effectiveness of a particular drug. Such strategy is already used successfully for hormones, as exemplified by the treatment of reproductive disorders by means of GnRH pulses of appropriate frequency (Leyendecker et al., 1981). Similarly, optimal pulsatile therapy could apply to drugs in the treatment of a number of physiological disorders.

Acknowledgments: *This work was supported by the Belgian National Incentive Program for Fundamental Research in the Life Sciences (Convention BIO/08), launched by the Science Policy Programming Services of the Prime Minister's Office (SPPS).*

REFERENCES

Bassett, N.S., and Gluckman, P.D., 1986, Pulsatile growth hormone secretion in the ovine fetus and neonatal lamb, J. Endocr., 109:307-312.

Belchetz, P.E., Plant, T.M., Nakai, Y., Keogh, E.J., and Knobil, E., 1978, Hypophysial responses to continuous and intermittent delivery of hypothalamic gonadotropin-releasing hormone, Science, 202:631-633.

Berridge, M.J., and Galione, A., 1988, Cytosolic calcium oscillators, FASEB J., 2:3074-3082.

Berridge, M. J., Cobbold, P. H., and Cuthbertson, K.S.R., 1988, Spatial and temporal aspects of cell signalling Phil.Trans. R. Soc. Lond. B., 320:325-343.

Borges, J.L.C., Blizzard, R.M., Evans, W.S., Furlanetto, R., Rogol, A.D., Kaiser, D.L., Rivier, J., Vale, W., and Thorner, M.O, 1984, Stimulation of growth hormone (GH) and somatomedin C in idiopathic GH-deficient subjects by intermittent pulsatile administration of synthetic human pancreatic tumor GH-releasing factor, J. Clin. Endocr. Metabol., 59:1-6.

Brewitt, B., and Clark, J.I., 1988, Growth and transparency in the lens, an epithelial tissue, stimulated by pulses of PDGF, Science, 242:777-779.

Cazalis, M., Dayanithi, G., and Nordmann, J.J., 1985, The role of patterned burst and interburst interval on the excitation-coupling mechanism in the isolated rat neural lobe, J. Physiol. Lond., 369:45-60.

Crowley, W.F., and Hofler, J.G., eds, 1987, "The Episodic Secretion of Hormones," Wiley, New York.

Cuthbertson, K.S.R., 1989, Intracellular calcium oscillators, in: "Cell to Cell Signalling: From Experiments to Theoretical Models", A. Goldbeter, ed., Academic Press, London, pp. 435-447.

Darmon, M., Brachet, P., and Pereira da Silva, L.H., 1975, Chemotactic signals induce cell differentiation in Dictyostelium discoideum, Proc. Natl. Acad. Sci. USA, 72:3163-3166.

Decroly, O., and Goldbeter, A., 1982, Birhythmicity, chaos, and other patterns of temporal self-organization in a multiply regulated biochemical system, Proc. Natl. Acad. Sci. USA, 79:6917-6921.

Dupont, G., and Goldbeter, A., 1989, Theoretical insights into the origin of signal-induced calcium oscillations, in: "Cell to Cell Signalling: From Experiments to Theoretical Models", A. Goldbeter, ed., Academic Press, London, pp. 461-474.

Dupont, G., Berridge, M.J., and Goldbeter, A., 1990, Latency correlates with period in a model for signal-induced Ca^{2+} oscillations based on Ca^{2+}-induced Ca^{2+} release, Cell Regul., 1: in press.

Durston, A.J., 1974, Pacemaker mutants of Dictyostelium discoideum, Dev. Biol., 38:308-319.

Endo, M., Tanaka, M., and Ogawa, Y., 1970, Calcium induced release of calcium from the sarcoplasmic reticulum of skinned skeletal muscle fibers, Nature, 228:34-36.

Fabiato, A., 1983, Calcium-induced release of calcium from the cardiac sarcoplasmic reticulum, Amer. J. Physiol., 245:C1-C14.

Gerisch, G., 1987, Cyclic AMP and other signals controlling cell development and differentiation in Dictyostelium, Annu. Rev. Biochem., 56:853-879.

Gerisch, G., and Wick, U., 1975, Intracellular oscillations and release of cyclic AMP from Dictyostelium cells, Biochem. Biophys. Res. Commun., 65:364-370.

Gerisch, G., Fromm, H., Huesgen, A., and Wick, U., 1975, Control of cell contact sites by cAMP pulses in differentiating Dictyostelium cells, Nature, 255:547-549.

Ginsburg, G., and Kimmel, A.R., 1989, Inositol trisphosphate and diacylglycerol can differentially modulate gene expressiion in Dictyostelium, Proc. Natl. Acad. Sci. USA, 86:9332-9336.

Goldbeter, A., 1990, "Rythmes et Chaos dans les Systèmes Biochimiques et Cellulaires," Masson, Paris.

Goldbeter, A., Decroly, O., Li, Y.X., Martiel, J.L., and Moran, F., 1988, Finding complex oscillatory phenomena in biochemical systems. An empirical approach, Biophys. Chem., 29:211-217.

Goldbeter, A., and Li, Y.X., 1989, Frequency coding in intercellular communication, in: "Cell to Cell Signalling: From Experiments to Theoretical Models", A. Goldbeter, ed., Academic Press, London, pp.461-474.

Goldbeter, A., Dupont, G., and Berridge, M.J., 1990, Minimal model for signal-induced Ca^{2+} oscillations and for their frequency encoding through protein phosphorylation, Proc. Natl. Acad. Sci. USA, 87:1461-1465.

Goldbeter, A., and Wurster, B., 1989, Regular oscillations in suspensions of a putatively chaotic mutant of *Dictyostelium discoideum*, Experientia, 45:363-365.

Halloy, J., Li, Y.X., Martiel, J.L., Wurster, B., and Goldbeter, A., 1990, Coupling chaotic and periodic cells results in a period-doubling route to chaos in a model for cAMP oscillations in *Dictyostelium* suspensions, Phys. Lett. A, in press.

Hesch, R.D., 1990, Thyroid growth — An example of modulation of cell function, Hormone and Metab. Res., Suppl. Ser. Vol. 23: 47-50.

Knobil, E., 1980, The neuroendocrine control of the menstrual cycle, Recent Progr. Horm. Res., 36:53-88.

Knobil, E., 1981, Patterns of hormone signals and hormone action, New Engl. J. Med., 305:1582-1583.

Kuba, K., and Takeshita, S., 1981, Simulation of intracellular Ca^{2+} oscillations in a sympathetic neurone, J. Theor. Biol., 93:1009-1031.

Lemmer, B., ed., 1989, "Chronopharmacology: Cellular and Biochemical Interactions," M. Dekker, New York and Basel.

Leyendecker, G. L., Wildt, L., and Hansmann, M., 1980, Pregnancies following intermittent (pulsatile) administration of GnRH by means of a portable pump ("Zyklomat"): a new approach to the treatment of infertility in hypothalamic amenorrhea, J. Clin. Endocr. Metab., 51:1214-1216.

Li, Y.X., and Goldbeter, A., 1989, Frequency specificity in intercellular communication: The influence of patterns of periodic signaling on target cell responsiveness, Biophys. J., 55:125-145.

Li, Y.X., and Goldbeter, A., 1990, Frequency encoding of pulsatile signals of cAMP based on receptor desensitization in *Dictyostelium* cells, J. Theor. Biol., 146:355-367.

Li, Y.X., Halloy, J., Martiel, J.L., Wurster, B., and Goldbeter, A., 1990, Suppression of chaos by a tiny proportion of periodic cells in a model for cAMP signalling in *Dictyostelium*, submitted for publication.

Loomis, W.F., 1975, "*Dictyostelium discoideum:* A Developmental System," Academic Press, New York.

Martiel, J. L., and Goldbeter, A., 1985, Autonomous chaotic behaviour of the slime mould *Dictyostelium discoideum* predicted by a model for cyclic AMP signalling, Nature, 313:590-592.

Martiel, J. L., and Goldbeter, A., 1987, A model based on receptor desensitization for cyclic AMP signaling in *Dictyostelium* cells, Biophys. J., 52:807- 828.

Meyer, T., and Stryer, L., 1988, Molecular model for receptor-stimulated calcium spiking, Proc. Natl. Acad. Sci. USA, 85:5051-5055.

Monk, P.B., and Othmer, H.G., 1990, Wave propagation in aggregation fields of the cellular slime mould *Dictyostelium discoideum*, Proc. R. Soc. Lond. B, 240:555-589.

Nanjundiah, V., 1988, Periodic stimuli are more successful than randomly spaced ones for inducing development in *Dictyostelium discoideum*, Biosci. Rep., 8:571-577.

Rapp, P.E., 1987, Why are so many biological systems periodic? Progr. Neurobiol., 29:261-273.

Rapp, P.E., Mees, A.I., and Sparrow, C.T., 1981, Frequency encoded biochemical regulation is more accurate than amplitude dependent control, J. Theor. Biol., 90:531-544.

Roos, W., Nanjundiah, V., Malchow, D., and Gerisch, G., 1975, Amplification of cyclic-AMP signals in aggregating cells of *Dictyostelium discoideum*. FEBS Lett., 53:139-142.

Segel, L.A., Goldbeter, A., Devreotes, P.N., and Knox, B.E., 1986, A mechanism for exact sensory adaptation based on receptor modification, J. Theor. Biol., 120:151-179.

Simon, M.N., Driscoll, D., Mutzel, R., Part, D., Williams, J., and Veron, M., 1989, Overproduction of the regulatory subunit of the cAMP-dependent protein kinase blocks the differentiation of *Dictyostelium discoideum*, EMBO J., 8:2039-2034.

Monk, P.B., and Othmer, H.G., 1990, Wave propagation in aggregation fields of the cellular slime mould *Dictyostelium discoideum*, Proc. R. Soc. Lond. B, 240:555-589.

Snaar-Jagalska, B.E., and Van Haastert, P.J.M., 1990, Pertussis toxin inhibits cAMP-induced desensitization of adenylate cyclase in *Dictyostelium discoideum*, Mol. Cell. Biochem., 92:177-189.

Stojilkovic, S.S., Rojas, E., Stutzin, A., Izumi, S.-I., and Catt, K.J., 1989, Desensitization of pituitary gonadotropin secretion by agonist-induced inactivation of voltage-sensitive calcium channels, J. Biol. Chem., 264:10939-10942.

Tyson, J.J., Alexander, K.A., Manoranjan, V.S., and Murray, J.D., 1989, Cyclic-AMP waves during aggregation of *Dictyostelium* amoebae, Physica, 34D:193-207.

Vaughan, R., and Devreotes, P.N., 1988, Ligand-induced phosphorylation of the cAMP receptor from *Dictyostelium discoideum*, J. Biol. Chem., 263:14538-14543.

Wagner, T.O.F., and Filicori, M., eds, 1987, "Episodic Hormone Secretion: From Basic Science to Clinical Application," TM-Verlag, Hameln (FRG).

Weigle, D.S., 1987, Pulsatile secretion of fuel-regulatory hormones, Diabetes, 36:764- 775.

Whim, M.D., and Lloyd, P.E., 1989, Frequency-dependent release of peptide cotransmitters from identified cholinergic motor neurons in *Aplysia*, Proc. Natl. Acad. Sci. USA, 86: 9034-9038.

Woods, N.M., Cuthbertson, K.S.R., and Cobbold, P.H., 1987, Agonist-induced oscillations in cytoplasmic free calcium concentration in single rat hepatocytes, Cell Calcium, 8:79-100.

Wurster, B., 1982, On induction of cell differentiation by cyclic AMP pulses in *Dictyostelium discoideum*, Biophys. Struct. Mecha., 9:137-143.

FRACTAL, CHAOTIC, AND SELF-ORGANIZING CRITICAL SYSTEM:

DESCRIPTIONS OF THE KINETICS OF CELL MEMBRANE ION CHANNELS

Larry S. Liebovitch and Ferenc P. Czegledy

Department of Ophthalmology
Columbia University
College of Physicians & Surgeons
630 West 168th Street
New York, NY 10032 USA

INTRODUCTION

Channels are proteins in the cell membrane that spontaneously fluctuate between conformational shapes that are closed or open to the passage of ions. The kinetics of these changes in conformational state can be described in different ways, that suggest different physical properties for the ion channel protein. We describe kinetic models based on: 1) random switching between a few independent states, 2) random switching between many states that are cooperatively linked together, 3) deterministic, chaotic, nonlinear oscillations, amplifying themselves until the channel switches states, and 4) deterministic local interactions that self-organize the fluctuations in channel structure near a phase transition, switching it between different states.

ION CHANNELS

Ion channels are proteins of several subunits that span the cell membrane. They have an interior hole connecting the inside of the cell with the exterior solution. Ions such as potassium, sodium and chloride, can enter or exit the cell through this hole. These subunits, or pieces of these subunits, can rearrange themselves to close off this hole. We want to understand the nature of this transition from conformations that are open to the flow of ions to conformations that are closed to the flow of ions.

We can learn about these transitions between closed and open conformational states from the patch clamp experiment developed by Neher, Sakmann, Sigworth, Hamill, and others (Sakmann and Neher 1983). A small piece of cell membrane is sealed in the approximately 1 μm opening of a glass micropipette. The patch can even be ripped completely off the cell. Such a patch will contain only a small number of channels, sometimes only one channel. An electrochemical gradient is imposed across the patch and

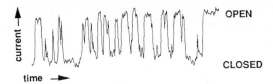

OPEN

CLOSED

current →

time →

Fig. 1. The current recorded through a single ion channel protein from a cell in the cornea.

the current through the channel is measured. Thermal fluctuations cause the channels to spontaneously change between their closed and open states. As shown in Fig. 1, the picoamp currents that flow through an individual ion channel when it is open can be resolved. Thus, the kinetics, the changes in conformation between closed and open states, can be studied in a single protein molecule.

HODGKIN-HUXLEY DESCRIPTION

Hodgkin and Huxley (1952) first proposed that channel kinetics can be described as a Markov process with a few states. That is, the switching between the states is random and the probability of leaving a state is constant and does not depend on the time already spent in a state or the history of previous states. Physically, this implies that the channel protein has a few, distinct conformational states, which are separated by significant activation energy barriers, that are constant in time. Although such models have been widely used in analyzing electrophysiological data, it is now know that proteins do not have these physical properties.

FRACTAL DESCRIPTION

Extensive experiments and simulations over the last decade have led to a new understanding of protein dynamics and kinetics (Austin et al. 1975, Karplus and McCammon 1981, Welch 1986). Proteins have many energy minima and thus many conformational states. There are many ways of going from one conformational state to another over many parallel activation energy barriers. Because the protein structure is flexible, these energies vary in time. For example, the unbinding and rebinding kinetics of ligands (Austin et al. 1975), or the flipping of a tryptophan ring as reported by its fluorescence (Alcala et al. 1987), show a broad, continuous distribution of activation energy barriers, rather than a few discrete (delta functions) predicted by the few state Markov model. The energy barrier determined from the static crystallographic structure of heme proteins, is so large that one would predict that oxygen would never bind to hemoglobin (Karplus and McCammon 1981). But oxygen does bind to hemoglobin because fluctuations in the structure open up a passageway for oxygen to get in and out. This passageway is slightly different every time, and thus rather than discrete states, there is a broad continuum of many slightly different states. Reactions in proteins crucially depend on such time dependent motions.

These physical properties of proteins are also also supported by a new analysis of the patch clamp data. We found that the time course of the current through ion channels is fractal (Liebovitch et al. 1987, Liebovitch and Tóth 1990a). That is, there are bursts within bursts within bursts of channel closings and openings. The data is self-similar. As we examine the current at ever finer time resolution, we see ever shorter closed and open durations. This suggests a new quantitative measure of channel kinetics. The most used measure has been the kinetic rate constant, which is the probability per unit time that the channel changes states. However, the channel must remain in a state long enough for us to detect it in state. Thus, we defined the effective kinetic rate constant k(eff) as the conditional probability per unit time that the channel exits a state, given that it has been in that state for a time t(eff), which is the effective time scale at which the measurement is done. We measured how the effective kinetic rate constant varied with the time scale at which it is measured for channels recorded from cells in the cornea (Liebovitch et al. 1987) and neurons in culture (Liebovitch and Sullivan 1987). The effective kinetic rate constant was larger when measured at finer time resolution. That is, the faster we can look, the faster we see the channel close and open. Our data was well fit by the fractal scaling

$$k(eff) = A \ t(eff)^{1-d}$$

where A is a constant and d is the fractal dimension which was between 1 and 2.

We showed that this fractal scaling of the effective kinetic rate constant implies that the distribution of durations of closed and open times will be a stretched exponential or a power law (Liebovitch et al. 1987). The patch clamp data often has these distributions (Blatz and Magleby 1986, Liebovitch and Sullivan 1987, McGee et al. 1988, French and Stockbridge 1988) rather than the sum of a few exponentials predicted by a Markov model having a few conformational states. These fractal distributions imply that channels have a broad continuum of many different conformational states separated by small energy differences. This is consistent with the new understanding of protein structure determined from other experiments and simulations. Since this continuum can be described by a fractal scaling with a small number of parameters, these many states must be linked together by a simple physical mechanism that causes this fractal scaling. Mechanisms that could produce such a linkage were proposed by Millhauser et al. (1988), Läuger (1988), Croxton (1988), and Levitt (1989). We developed models having many closed states connected over activation energy barriers to an open state (Liebovitch 1989, Liebovitch and Tóth 1990b). We found that a uniform distribution of such barriers was required to fit our experimental data. Thus, the fast processes over the small barriers are linked to the slow processes over the large barriers. Since each activation energy barrier represents a different physical process, this simple linkage between the activation energy barriers means that the physical processes are cooperatively linked together. Thus, the closing and opening of the channel involves a cooperative interaction between many different parts and processes in the ion channel, rather than the few independent processes predicted by the few state Markov model.

CHAOTIC DESCRIPTION

Two similar looking time series that have approximately the same mean, variance, and power spectra are shown in Fig. 2. However, they are vastly different. The one on the left is random, but the one on the right is not random at all, the value at the (n+1)-th time point is a simple, quadratic, deterministic function of the value at the n-th point time. Such deterministic systems, having a small number of independent variables, that mimic random behavior, are now called chaotic (Gleick 1988, Guckenheimer and Holmes 1983, Moon 1987).

We asked if the seemingly random switching of the channel between closed and open states might also be due to deterministic chaotic, rather than random processes. First, we developed a deterministic chaotic model that can fit the patch clamp data (Liebovitch and Tóth 1990c). In this model, x(n) is the current through the channel at the n-th time point. The current at the next time point x(n+1) is a function of the current at the previous time point x(n). This function is shown in Fig. 3.

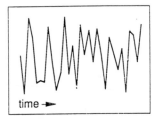 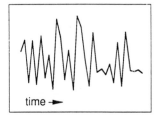

Fig. 2. Two time series with approximately the same mean, variance, and power spectra, that appear to be random. The series on the left is random where the n-th point was generated by x(n)=RND. The series on the right is not random, but deterministic chaos, where x(n+1)=3.95x(n)[1-x(n)].

What is the physical meaning of such a model? Since, the slope of the plot of x(n+1) vs. x(n) is greater than 1, whatever the channel is doing now, it will be doing more of it later. An example of such a system, at a slightly larger scale than a protein, was the failure of the Tacoma Narrows Bridge (Scanlan and Vellozi 1980). Contrary to what is often taught in elementary physics classes, this failure was not due to resonance. In resonance, a periodic driving force matches a natural resonance. That is not what happened because the wind did not blow periodically. The bridge was destroyed by a nonlinearity called "wind-induced flutter" (Moon 1987). The wind blew over the bridge causing it to twist in a way that changed the airflow, that caused the bridge to twist even more. This positive feedback destroyed the bridge. In the same way, perhaps, the channel amplifies its own motions until it kicks itself into a new state. Another viewpoint is to note that the bridge acted as a nonlinear oscillator, organizing the nonperiodic motion of the wind into the periodic motion of the bridge. Perhaps, the ion channel protein, acting as a nonlinear mechanical oscillator just like the bridge, is able to organize the nonperiodic thermal fluctuations in its structure into coherent motions. This happens because the positions and motions of the atoms in the channel at the next instant are a deterministic function of the positions and motions at the previous instant. This is very different than the Markov model description, where the channel sits in a state, and sits in a state, and one time, by magic, without any prior physical antecedent, the channel jumps across activation energy barriers from one conformational state to another.

Although it can be given the physical interpretation described above, the iterated map model was based primarily on mathematical principles. We also developed a deterministic chaotic model based on a more physical description of the dynamics in an ion channel protein (Liebovitch and Czegledy 1990). We assume a potential function $V(x) = \alpha x^2 + \beta x^4$ that has two energy minima corresponding to the closed and open states. The acceleration is proportional to the sum of three forces: a velocity dependent friction due to a viscosity caused by the relative motions within the protein, a force proportional to the gradient of the potential function, and a driving force from the environment which we chose to be periodic. Thus, we solve the modified Duffing's equation

$$m\, d^2x/dt^2 = -\gamma m\, dx/dt - dV(x)/dx + f_0 \cos(\omega t).$$

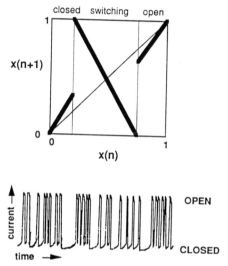

Fig. 3. Chaotic model of ion channel kinetics based on an iterated map (Liebovitch and Tóth 1990c). The next value of the current through the channel $x(n+1)$ is a function of the previous value $x(n)$ given by the heavy lines in the plot at the top. The sequence of current values predicted by this model is shown at the bottom.

The current $x(t)$ through the channel predicted by this model is shown in Fig. 4.

For the parameters that we used, the two well potential always had distributions of closed or open times that were approximately single exponentials. We had hoped that if the system was chaotic enough that it might show the stretched exponential or power law distributions seen in the experimental data. Only potentials with many local minima or time variation could produce such distributions. This is good, however, because other studies of proteins have shown that the energy structure of proteins has exactly such multiple minima and time variations. We also found subconductance states, where the current is less than the fully open value, corresponding to subharmonic resonances. That is, there are long lived states that are due to dynamic resonances rather than local minima in the potential energy function. Thus, some long lived protein states may be a result of dynamical properties rather than energy minima.

Fig. 4. Chaotic model of ion channel kinetics based on a modified Duffing's equation (Liebovitch and Czegledy 1990). The sequence of current values predicted by this model is shown.

We were not able to tell if our experimental data had a low dimensional attractor and was therefore chaotic, because we are not able to construct the phase space set from the time series of the current. This is because it has bursts within bursts, within bursts...that is, it is fractal, and it is therefore not differentiable. The embedding theorems, such as Takens' theorem, require that the time series be differentiable. When the time series is differentiable, the space of the time lagged signal, for example, {x(t), x(t+Δt)} is linearly related to the phase space of the derivatives {x(t), dx(t)/dt} because dx(t)/dt $\approx$ [x(t+Δt)-x(t)]/Δt. When the derivatives do not exist, we can still construct the space of the time lagged signal, but its relationship to the real phase space is not clear. This point is highlighted by the work Osborne and Provenzale (1989). They studied a fractal time series, that is not differentiable, that had a power law power spectrum. They randomized the phases of the components so that the system was infinite dimensional. Thus, the true phase space was infinite dimensional. However, the space generated by the lagged components was low dimensional, sometimes as low as one. We do not know how to determine the dimension of the attractor of a fractal time series.

SELF-ORGANIZING CRITICAL SYSTEM DESCRIPTION

In conservative systems where no energy enters and no energy leaves, the energy level remains constant. In dissipative systems where no energy enters, energy leaves until the system falls to an energy minimum. However, in systems where energy enters and energy leaves, a very interesting thing happens. An example of such a system is a pile of sand, that has been studied by Bak et al. (1988). One grain of sand at a time is added onto a random location of the sandpile. If the additional grain causes the slope to be greater than a critical value, then sand will move downhill. This avalanche will cascade down the sandpile until all the local slopes are less than the critical value. Any system will have the characteristic properties of the sandpile if: 1) energy is continually supplied by adding stress to the system and 2) when the stress at a point exceeds a critical value, it is relieved by being spread to its neighboring points.

150

The state of such a sandpile is not at an energy minimum. Instead, it is at the edge of instability, because each avalanche continues until the slope at that point is once again only marginally stable. Such a system has properties similar to those of a critical system, that is, one undergoing a phase transition, such as a substance changing from a liquid to a gas. Because the sandpile lives at the edge of instability, it self-organizes itself to be very close to a phase transition. Critical systems near a phase transition, have many properties that scale as power laws. That is, the self-organizing critical system of the sandpile generates power laws and fractals.

How is an ion channel protein like a pile of sand? The same mechanisms are at work. Thermal fluctuations are always adding sand, that is energy, to the protein. This increases the strain in the molecule, by stressing the dihedral angles between the amino acid residues. The strain at any residue is lessened by being spread to its neighboring amino acid residues. Thus (with lots of easy handwaving to be followed later by lots of difficult numerical calculations), we conclude that the distribution of the probability of the different conformational shapes will have the form of a power law, rather than an exponential Boltzmann distribution. Simple systems with discrete activation energies will have an exponential Boltzmann distribution. Complex, cooperative systems, such as the sandpile or ion channel, will have power law distributions. The importance of this fact, is that a power law distribution has a much higher probability of something unusual happening than an exponential distribution. Thus, often enough, the sidechains in hemoglobin will swing out of the way, opening up a passageway for oxygen to get into, or out of the binding site. Similarly, there will be a high enough probability for local fluctuations to induce a conformational change of state in an ion channel protein, to produce the power law distributions of the durations of the closed and open times.

SUMMARY

We have described several different types of models of ion channel kinetics.

Hodgkin-Huxley

Hodgkin and Huxley first proposed that ion channel kinetics can be described as a Markov process with a few states. The switching between states is assumed to be inherently random. Physically, this corresponds to a channel protein having a few conformational states separated by large, constant activation energy barriers. Although this model is widely used to interpret electrophysiological experiments, it is now contradicted by extensive biophysical evidence which demonstrates that proteins have many states, separated by small activation energy barriers, that are changing in time.

Fractal

These models have many Markov states, but do not need to be described by many adjustable parameters, because there is a fractal scaling that can be described by a small number of parameters. This scaling is due to cooperative behavior that links different physical processes together. Thus, the closing or opening of the channel is a result of a cooperative or global interaction between many parts and processes in the channel. The switching between different states is still inherently random.

Chaotic

In chaotic models, the switching between different states is not inherently random, but rather the result of a deterministic process that can be described by a small number of independent variables. The positions and motions of the atoms in the channel at the next instant in time are uniquely determined by their positions and motions in the previous instant in time. The channel acts like a nonlinear oscillator that organizes its random thermal fluctuations into coherent motion that kicks it closed or open.

Self-Organizing Critical

Lastly, we described self-organizing critical systems such as the sandpile. Like the chaotic models, these models are also deterministic. However, unlike the chaotic model, these models can only be described by a very large number of independent variables. In these models, interactions between local motions of the atoms in the channel cause the entire protein to self-organize near a phase transition. This yields a power law distribution of the probability of conformational states, that has a high probability for an unusual conformation that can close or open the channel.

ACKNOWLEDGMENTS

This work was done during the tenure of an established investigatorship from the American Heart Association (L.S.L.) and was also supported in part by grants from the Whitaker Foundation and from the National Institutes of Health, EY6234.

REFERENCES

Alcala, J. R., Gratton, E., and Prendergast, F. G., 1987, Interpretation of fluorescence decays in proteins using continuous lifetime distributions, *Biophys. J.*, 51:925-936.

Austin, R. H., Beeson, L., Eisenstein, H., Frauenfelder, H., and Gunsalus, I. C., 1975, Dynamics of ligand binding to myoglobin, *Biochem.*, 14:5355-5373.

Bak, P., Chao, T., and Wiesenfeld, K., 1988, Scale invariant spatial and temporal fluctuations in complex systems, *in* "Random Fluctuations and Pattern Growth: Experiments and Models," H. E. Stanley and N. Ostrowsky, eds., Kluwer, Boston.

Blatz, A. L. and Magleby, K. L., 1986, Quantitative description of three modes of activity of fast chloride channels from rat skeletal muscle, *J. Physiol. (Lond.).*, 378:141-174.

Croxton, T. L., 1988, A model of the gating of ion channels, *Biochem. Biophys. Acta*, 946:19-24.

French, A. S. and Stockbridge, L. L., 1988, Fractal and Markov behavior in ion channel kinetics, *Can. J. Physiol. Pharm.*, 66:967-970.

Gleick, J., 1987, "Chaos, Making a New Science," Viking, New York.

Guckenheimer, J. and Holmes, P., 1983, "Non-linear Oscillations, Dynamical Systems, and Bifurcations of Vector Fields," Springer-Verlag, New York.

Hodgkin, A. L. and Huxley, A. F., 1952, A quantitative description of membrane current and its application to conductance and excitation in nerve. *J. Physiol. (Lond.).*, 117:500-544.

Karplus, M. and McCammon, J. A., 1981, The internal dynamics of globular proteins, *CRC Crit. Rev. Biochem.*, 9:293-349.

Läuger, P., 1988, Internal motions in proteins and gating kinetics of ionic channels, *Biophys. J.*, 53:877-884.

Levitt, D. G., 1989, Continuum model of voltage dependent gating, *Biophys. J.*, 55:489-498.

Liebovitch, L. S., 1989, Analysis of fractal ion channel gating kinetics: kinetic rates, energy levels, and activation energies, *Math. Biosci.*, 93:97-115.

Liebovitch, L. S. and Czegledy, F. P., 1990, A model of ion channel kinetics based on deterministic chaotic motion in a potential with two local minima, submitted to *Ann. Biomed. Engr.*

Liebovitch, L. S., Fischbarg, J., and Koniarek, J. P., 1987, Ion channel kinetics: a model based on fractal scaling rather than multistate Markov processes, *Math. Biosci.*, 84:37-68.

Liebovitch, L. S. and Sullivan, J. M., 1987, Fractal analysis of a voltage-dependent potassium channel from cultured mouse hippocampal neurons, *Biophys. J.*, 52:979-988.

Liebovitch, L. S. and Tóth, T. I., 1990a, Fractal activity in cell membrane ion channels, *in* "Mathematical Approaches to Cardiac Arryhthmias," J. Jalife, ed., Ann. New York, Acad. Sci., 591:375-391.

Liebovitch, L. S. and Tóth, T. I., 1990b, Distributions of activation energy barriers that produce stretched exponential probability distributions for the time spent in each state of the two state reaction $A \rightleftharpoons B$, *Bull. Math. Biol.*, in press.

Liebovitch, L. S. and Tóth, T. I., 1990c, A model of ion channel kinetics using deterministic rather than stochastic processes, *J. Theor. Biol.*, in press.

McGee Jr., R., Sansom, M. S. P., and Usherwood, P. N. R., 1988, Characterization of a delayed rectified K^+ channel in NG108-15 neuroblastoma x glioma cells, *J. Memb. Biol.*, 102:21-34.

Millhauser, G. L., Salpeter, E. E., and Oswald, R. E., 1988, Diffusion models of ion-channel gating and the origin of the power-law distributions from single-channel recording, *Proc. Natl. Acad. Sci., U.S.A.*, 85:1503-1507.

Moon, F. C., 1987, "Chaotic Vibrations," Wiley, New York.

Osborne, A. R. and Provenzale, A., 1989, Finite correlation dimension for stochastic systems with power-law spectra, *Physica D*, 35:357-381.

Sakmann, B. and Neher, E., eds., 1983, "Single-Channel Recording," Plenum, New York.

Scanlan, R. H. and Vellozzi, J. W., 1980, Catastrophic and annoying responses of long-span bridges to wind action, *in* "Long-Span Bridges," E. Cohen and B. Birdsall, eds., Ann. New York, Acad. Sci., 352:247-263.

Welch, G. R., ed., 1986, "The Fluctuating Enzyme," Wiley, New York.

A DISCONTINUOUS MODEL FOR MEMBRANE ACTIVITY

Henrik Østergaard Madsen[1], Morten Colding-Jørgensen[2]
and Brian Bodholdt[1]

[1] Physics Laboratory III
The Technical University of
Denmark
DK-2800 Lyngby
Denmark

[2] Department of General
Physiology and Biophysics,
The Panum Institute
DK-2200 Copenhagen N
Denmark

ABSTRACT

A model for the excitable membrane is described. The model exhibits many types of firing patterns seen in excitable cells. The model includes potassium, sodium and calcium ions. These are affected by an active calcium pump, by potential dependent calcium, sodium and potassium conductances, and a calcium dependent potassium conductance. The cytosolic calcium concentration is buffered. Inactivation or time-delays of the conductances are not included, and the membrane capacity is zero. This leads to a discontinuous model, which is nevertheless capable of showing stable behavior, slow or fast spontaneous oscillations or bursting. To illustrate the generality of the model, it is adopted to mimic two specific cell types: a pancreatic β-cell and a cerebellar Purkinje cell.

INTRODUCTION

Variations in membrane potential are of great importance for living cells and their interactions. Experimental research has revealed that the underlying mechanisms are a multiplicity of ionic currents passing the cell membrane through channels, via carrier mechanisms or by active ion pumps; all with a strongly non-linear dependency on potential, time and chemical substances. Previous models have therefore been equally complex. Our approach is to extract the basic mechanisms needed to model the characteristic behavior of the excitable membrane.

This paper describes a model of the potential activity including three ions, sodium, potassium and calcium. The model is capable of issuing slow and fast oscillatory behavior and periodic bursts as well as stable and excitatory behavior. As the model does not include a membrane capacity, the potential becomes a discontinuous function of time (Colding-Jørgensen et al., 1990a, b). This simplistic approach can be regarded as a combination of non-linear dynamics and catastrophe theory (Zeemann, 1977, and Poston & Steward, 1978).

Complexity, Chaos, and Biological Evolution, Edited by E. Mosekilde and
L. Mosekilde, Plenum Press, New York, 1991

155

The model is first adopted to simulate the behavior of pancreatic β-cells, which exhibit stable bursts (Atwater et al., 1977).

Another application of the scheme is to simulate cerebellar Purkinje cells, where the spiking mechanisms are situated in different parts of the cell (Llinás & Sugimori, 1979). The cell body (soma) contains potential dependent sodium and potassium channels, whereas the dendrite part contains calcium dependent potassium channels and potential dependent calcium channels. The slow, calcium dominated spikes in the dendrites are overlaid by the faster, sodium-potassium spikes from the soma. The fast soma spikes, however, stop during the slow dendrite spike. The two parts are connected via the dendritic tree and can be modeled by two cavities connected by a resistance. This allows the two parts to have different membrane potentials.

THE MODEL

The model includes three ions: sodium, potassium and calcium, with high external calcium and sodium and high internal potassium concentrations. Sodium and calcium enter the cell passively through sigmoidal potential dependent conductances of the form :

$$g_X = g_{X,min} + \frac{g_{X,max} - g_{X,min}}{1 + \exp(-\gamma_X(V_m - V_{0,X}))} \qquad (1)$$

where X stands for Na, Ca or K. Calcium is pumped out by an active and electrically neutral pump; the pump rate is proportional to the intracellular concentration of free calcium ions, $[Ca^{2+}]_i$. The potassium conductance consists of a calcium dependent and a potential dependent part, $g_{K,Ca}$ and $g_{K,V}$. The calcium dependent $g_{K,Ca}$ is proportional to $[Ca^{2+}]_i$, and $g_{K,V}$ follows first-order kinetics with the time-constant τ_V with a potential dependent steady state value as in equation (1). The model includes a simple buffer system, so $[Ca^{2+}]_i$ is a fixed fraction of the total intracellular calcium concentration, $[Ca^{2+}]_{total}$.

The time delays associated with membrane capacity and activation of the conductances are ignored (Colding-Jørgensen, 1990). This allows the membrane potential to jump instantaneously and thereby to become discontinuous in time. The model regards only Na^+, Ca^{2+} and K^+ ions; other currents are assumed constant and are contained in I_m. The Na^+- and K^+-concentrations are relatively large and regulated and are therefore, like the external Ca^{2+}-concentration, considered fixed. Finally, no inactivation of the conductances is included.

This gives the following expression for the currents :

$$I_m = I_{Na} + I_{Ca} + I_K = g_{Na}(V_m - E_{Na}) + g_{Ca}(V_m - E_{Ca}) + (g_{K,Ca} + g_{K,V})(V_m - E_K) \qquad (2)$$

where E_K, E_{Na} and E_{Ca} are the equilibrium potentials. This transcendental equation controls the membrane potential, and by fulfilling this equation, V_m becomes discontinuous. The previous assumption leads to the two equations of motion for the system; one from the time delays of $g_{K,V}$ and one from the calcium ion balance (a detailed derivation can be found in Colding-Jørgensen et al., 1990 a, b) :

$$\tau_V \frac{dg_{K,V}}{dt} = G_K(V_m) - g_{K,V} \quad \text{and} \quad \tau_{Ca} \frac{dg_{K,Ca}}{dt} = -\phi I_{Ca} - g_{K,Ca} \quad (3)$$

Equation (2) can be explained graphically as the intersection between the *curve* $I_{Na} + I_{Ca} - I_m = 0$ and the *line* $g_K(V_m - E_K) = 0$ through $(I=0, V_m = E_K)$ in the I-V phase diagram. As $g_{K,Ca}$ and $g_{K,V}$ change, the line sweeps up or down, and because of the humps on the curve, some of the intersections may disappear, forcing the potential to jump to another intersection (cf. Fig. 1). This leads to a hysteresis-like behavior also described as cusp-catastrophes (Zeemann, 1977, and Poston & Steward, 1978). By applying this technique, one gains simplicity and clearness compared to the otherwise very complicated models consisting solely of differential equations.

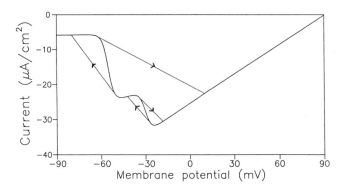

FIG. 1. The sum $I_{Na} + I_{Ca} - I_m$ as a function of the membrane potential. The thin lines represent the line $g_K(V_m - E_K)$ through $(I=0, V_m = E_K)$ for four different g_K corresponding to the jumping points. The arrows indicate the direction of the jumps. The parameters are : $I_m = 5 \ \mu A/cm^2$; $V_{0,Ca} = -55$ mV; $E_{Ca} = 100$ mV; $\tau_{Ca} = 0.5$ mV^{-1}; $g_{Ca,max} = 0.125$ mS/cm^2; $g_{Ca,min} = g_{Ca,max}/15$; $V_{0,Na} = -30$ mV; $E_{Na} = 50$ mV; $\tau_{Na} = 0.5$ mV^{-1}; $g_{Na,max} = 0.15$ mS/cm^2; $g_{Na,min} = g_{Na,max}/15$; $\phi = 0.05$ mV^{-1}.

THE PANCREATIC β-CELL

Depending on the parameters, this model exhibits many different types of behavior (Colding-Jørgensen *et al.*, 1990a). Use of parameters typical for the pancreatic β-cell yields the results shown in Fig. 2. The cell exhibits fast spikes governed by $g_{K,V}$ for a short while, until the slow spikes governed by $g_{K,Ca}$ temporarily force the potential down and cut off the fast spikes. The result is a bursting behavior, very similar to that found by experiments (Atwater *et al.*, 1977). When blocking the potential dependent potassium conductance $g_{K,V}$ by lowering $g_{K,max}$, the model only shows slow, calcium dominated spikes. Blocking the calcium dependent potassium conductance $g_{K,Ca}$ by lowering ϕ results in only fast, potassium-sodium spikes at high levels, as also seen in experiments (Chay & Keizer, 1983).

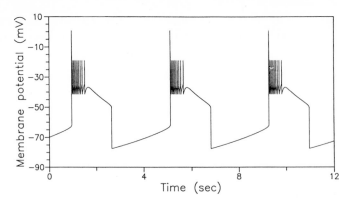

FIG. 2. The membrane potential as a function of time when using parameters typical for the pancreatic β-cell. The parameters are $\tau_{Ca}=0.25$ sec^{-1}; $\tau_V=\tau_{Ca}/30$; $V_{0,K}=-25$mV; $E_K=-90$ mV; $\tau_K=0.5$ mV^{-1}; $g_{K,max}=1$ mS/cm^2; $g_{K,min}=g_{K,max}/50$, and the rest as in Fig. 1.

THE CEREBELLAR PURKINJE CELL

The cerebellar Purkinje cell contains essentially the same mechanisms as the β-cell above, but the conductances are situated in different parts of the cell. The potential dependent potassium and sodium channels are in the main cell body, the soma, and the potential dependent calcium and calcium dependent potassium channels are in the dendritic extensions of the cell (Llinás & Sugimori, 1979). In this way, the cell can spike with slow, calcium dominated spikes in the dendrites, and with fast, sodium-potassium spikes in the soma. The fast, somatic spikes are superimposed on the slow spikes in the dendrites, and the slow dendrite spikes, when on a high potential, can stop the somatic spikes (Hounsgaard & Midtgaard, 1987).

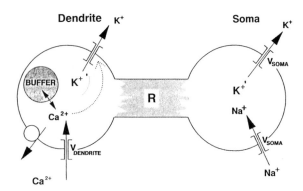

FIG. 3. Diagram of the mechanisms included in the model of a cerebellar Purkinje cell.

This leads to a division of the model into two separate parts, one with the calcium system and the calcium dependent potassium conductance, and one with the potential dependent potassium and sodium conductances. Because of the physical distance, the two parts are connected through a resistance coupling (Fig. 3). The model therefore includes *two* different membrane potentials : V_{SOMA} and $V_{DENDRITE}$.

This will not change the equations for $g_{K,V}$ and $g_{K,Ca}$, but equation 2 will be divided into two parts :

$$I_{m,DENDRITE} + I_{Ca} + g_{K,Ca}(V_{DENDRITE} - E_K) + \frac{V_{DENDRITE} - V_{SOMA}}{R_{DENDRITE}} = 0$$

$$I_{m,SOMA} + I_{Na} + g_{K,V}(V_{SOMA} - E_K) + \frac{V_{SOMA} - V_{DENDRITE}}{R_{SOMA}} = 0$$

(4)

The first equation represents the dendrite and the second the somatic part. In both equations, the last term represents the coupling resistance. Because $R_{DENDRITE}$ and R_{SOMA} are dependent on the surface areas, they do not have to be equal.

Setting $R_{DENDRITE}$ and R_{SOMA} to infinity results in two completely separated systems which will behave as seen in Colding-Jørgensen (1990) and Colding-Jørgensen et al. (1990b), respectively. Setting $R_{DENDRITE}$ and R_{SOMA} to nearly zero forces $V_{DENDRITE}$ and V_{SOMA} to be equal, and the result will be like the model described for the β-cell.

As in the β-cell model, each potential is an intersection between a curve (which now includes a coupling term) and a line. The coupling term makes the jumping points dependent on the other potential. The jumping points are characterized by dV/dt becoming infinite (for both V's), which means that the determinant of the Jacobian matrix for the system (4) becomes zero.

The paths by which the system reaches the jumping points can be seen in the $V_{DENDRITE}$-V_{SOMA} phase diagram (Fig. 4). The parameters for a Purkinje cell are applied to the model, and the areas of negative Jacobian determinant are marked gray. When the trajectory reaches the bounds of zero determinant, it jumps to another solution of Equation (4). In the diagram, the curves for steady-state $g_{K,V}$ and $g_{K,Ca}$ are also shown. The system will tend to follow the steady-state curve for $g_{K,V}$, as it has the smallest time-constant, τ_V. The corresponding time series for $V_{DENDRITE}$ and V_{SOMA} are shown in Fig. 5. This behavior complies with the results found experimentally (Hounsgaard & Midtgaard, 1987). Blocking the potential dependent potassium channels in the soma will give only large, slow, calcium dominated spikes in the dendrites. An additional blocking of the calcium dependent potassium channels in the dendrites will stop all spiking behavior, in accordance with the experimental findings (Llinás & Sugimori, 1979).

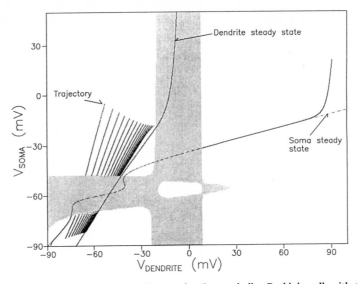

FIG. 4. The $V_{DENDRITE}$-V_{SOMA} phase diagram for the cerebellar Purkinje cell, with the steady-state curve for the dendrite and the soma part. The gray area represents negative Jacobian determinant. The potentials have to jump when reaching the bounds of zero determinant. The parameters are : $R_{DENDRITE}=R_{SOMA}=10\ \Omega/cm^2$; $\Phi=0.05\ mV^{-1}$; $\tau_{Ca}=0.5\ sec^{-1}$; $\tau_V=\tau_{Ca}/10$;

$V_{0,Ca}=0\ mV$; $E_{Ca}=100\ mV$; $\tau_{Ca}=0.3\ mV^{-1}$; $g_{Ca,max}=1.5\ mS/cm^2$; $g_{Ca,min}=g_{Ca,max}/100$;

$V_{0,Na}=-55\ mV$; $E_{Na}=50\ mV$; $\tau_{Na}=0.2\ mV^{-1}$; $g_{Na,max}=0.1\ mS/cm^2$; $g_{Na,min}=g_{Na,max}/1000$;

$V_{0,K}=-60\ mV$; $E_K=-90\ mV$; $\tau_K=0.5\ mV^{-1}$; $g_{K,max}=0.2\ mS/cm^2$; $g_{K,min}=g_{K,max}/25$.

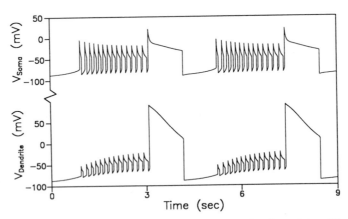

FIG. 5. The two membrane potentials $V_{DENDRITE}$ and V_{SOMA} as functions of time. The results are from the same simulation as in Fig. 4.

CONCLUSION

With this simple model scheme, it has been possible to treat models of two specific cell types; the pancreatic β-cell and the cerebellar Purkinje cell. The models mimic experimental observations, also with modification like drug addition (channel blockers), with appropriate parameter changes. By similar modifications of the model, it will be possible to simulate many other types of cells.

The models' great advantage lies in their simplicity, which makes detailed analyzation of behavior possible. This enables identification of the mechanisms necessary to give rise to a certain behavior. Furthermore, for appropriate parameters, this model can show aperiodic spikes. This is an important intrinsic capability of an excitable membrane, resulting in the possibility of irregular and chaotic behavior for a single nerve cell, or for the transmission between nerve cells (Colding-Jørgensen, 1991).

REFERENCES

ATWATER, I., DAWSON, C. M., RIBALET, B. & ROJAS, E. (1977). Potassium Permeability Activated by Intracellular Calcium Ion Concentration in the Pancreatic β-cell. *J. Physiol.* **288**, 575-588.

CHAY, T. R. & KEIZER, J. (1983). Minimal Model for Membrane Oscillations in the Pancreatic β-Cell. *Biophys. J.* **42**, 181-190.

COLDING-JØRGENSEN, M. (1990). Fundamental Properties of the Action Potential and Repetitive Activity in Excitable Membranes Illustrated by a Simple Model. *J. theor. Biol.* **144**, 37-67.

COLDING-JØRGENSEN, M., MADSEN, H. Ø., BODHOLDT, B. & MOSEKILDE, E. (1990a). A simple Model for Ca^{2+}-dependent Oscillations in Excitable Cells. *Modelling and Simulation*, Proceedings of the 1990 European Simulation Multiconference, 630-635.

COLDING-JØRGENSEN, M., MADSEN, H. Ø., BODHOLDT, B. & MOSEKILDE, E. (1990b). Minimal Model for Ca^{2+}-dependent Oscillations in Excitable Cells. *J. theor Biol.* Submitted.

COLDING-JØRGENSEN, M. (1991). Chaos in Coupled Nerve Cells. *This volume.*

HOUNSGAARD, J. & MIDTGAARD, J. (1987). Intrinsic Determinants of the Firing Pattern in Purkinje Cells of the Turtle Cerebellum *in vitro*. *J. Physiol.* **402**, 731-749.

LLINÁS, R. & SUGIMORI, M. (1979). Electrophysiological Properties of *in vitro* Purkinje Cell Somata in Mammalian Cerebellar Slices. *J. Physiol.* **305**, 171-195.

POSTON, T. & STEWARD, I. (1978). Catastrophe Theory and its Applications. Pitman, London.

ZEEMANN, E. C. (1977). Catastrophe Theory; Selected Papers, 1972-1977. Addison-Wesley, London.

CHAOS IN COUPLED NERVE CELLS

Morten Colding-Jørgensen

Department of General Physiology and Biophysics
The Panum Institute
DK-2200 Copenhagen N
Denmark

A model for the impulse transmission between nerve cells is presented. The model is compared with experiments on nerve cell bodies of the large land snail *Helix pomatia*. Synaptic input is simulated by equal-sized square current pulses applied at a constant rate. The output pattern consists of a mixture of spikes and dropouts, where the response fraction depends on stimulus strength and frequency in a non-trivial manner. In the periodic model, the output locks to the input at simple ratios (1:1, 2:1, 3:2, etc.) resulting in the fractal relation called the "Devil's staircase" between response fraction and stimulation strength or rate. In the chaotic model the near-threshold behavior of the nerve cell is included, resulting in a breakdown of the staircase into a mixture of regions with regular behavior and regions with chaotic behavior. In the chaotic regions, the mean output frequency differs from the mean frequency of nearby regions. The behavior of this model is close to the behavior of the nerve cell. Coupling is mimicked by applying the output from one cell to another cell with the same model parameters. Even in very simple systems, the resulting output depends not only on the strength of the coupling, but also on the precise timing of the incoming impulses. The signal processing of a nerve cell is therefore not just a function of the mean firing frequency. It is a result of subtly timed mixtures of regular and chaotic firing patterns.

Introduction

Nerve impulses are brief events with a relatively constant size and shape. Consequently, the information contained in a given nerve signal is embedded in the timing or pattern of several impulses. As the signal is transmitted from cell to cell, the pattern is changed, and so is the information content. The rules for these changes are mostly unknown. A way to study them is to apply simple input patterns to a cell and observe its output. This has been done for many types of cells, but mostly on cells where the mechanisms underlying the firing pattern were unclear or unknown (Colding-Jørgensen, 1977, 1990; Glass *et al.*, 1983; Tuckwell *et al.*, 1984; Matsumoto *et al.*, 1987).

Complexity, Chaos, and Biological Evolution, Edited by E. Mosekilde and
L. Mosekilde, Plenum Press, New York, 1991

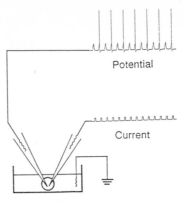

Fɪɢ. 1. The experimental setup. Two electrodes are impaled in a nerve cell. Current pulses are passed through one electrode and the potential changes are recorded with the other.

In many excitable cells, the firing is governed by a Ca^{2+}-dependent K^{+}-conductance (Gorman & Thomas, 1980; Brown & Griffith, 1983; Benham *et al.*, 1986), so the information processing of a cell depends on its submembraneal Ca^{2+}-concentration. An increase in Ca^{2+}-concentration increases the K^{+}-conductance and thereby the firing threshold, so the decision whether to fire or not on a synaptic input depends on the previous movements of Ca^{2+}-ions. In this way the dynamics of the Ca^{2+}-concentration (Akaike *et al.*, 1978; Rasmussen & Barrett, 1984; Ahmed & Connor, 1988) becomes one of the main determinants of the signal processing of a single nerve cell. With a relatively simple model for the Ca^{2+}-movements (Colding-Jørgensen *et al.*, 1990), it is possible to elucidate some of the rules for signal processing and to compare them with the experimental findings.

Experiments were performed on nerve cell bodies of the large land snail *Helix pomatia*, which have a large Ca^{2+}-dependent K^{+}-conductance (Lux & Hofmeier, 1982). Synaptic input was mimicked by applying equal sized current pulses with a constant frequency (Fig. 1). Only the strength of the current pulses and/or the application frequency were varied. Even this simple input resulted in output patterns with a complex mixture of spikes and dropouts. These experimental findings are compared with two slightly different models: The periodic model and the chaotic model.

The periodic model is a modification of a simple model presented previously (Colding-Jørgensen, 1983), where the effect of the stimulus itself on the firing is included. As a result, the cell only responds to the stimuli up to a certain stimulation frequency. Above this frequency no spikes are elicited, no matter how large the stimuli are. In the chaotic model, the near-threshold behavior of the nerve cell is included. With the stimulus close to the threshold, the membrane stays depolarized for some time before it decides whether to fire an impulse. During this period, an extra amount of Ca^{2+} flows into the cell, resulting in an additional threshold increase. The near-threshold phenomena gives a strong possibility of chaotic behavior, and because it is present in all nerve cells, the chaotic firing appears to be a natural part of the signal processing rather than the exception.

Coupling between nerve cells is mimicked by applying the output from one cell to another cell with the same model parameters. Even with very simple couplings, it is shown that the resulting output depends not only on the strength of the coupling, but also on the precise timing of the incoming impulses. Thus the signal processing of the single nerve cell is not a function of the mean firing frequency but depends on a subtly timed mixture of regular and chaotic firing.

The periodic model

The cell is stimulated by repetitively applied current pulses of strength I_S and with a period T or a frequency $f = 1/T$. The influx of Ca^{2+}-ions during an impulse increases the firing threshold I_{th} by the amount i_I and during the stimulus by the amount i_S. Between stimuli, I_{th} decreases exponentially towards the resting value I_0 with the time constant τ. Both the impulse and the stimulus are considered to be short compared with τ. When the stimulus I_S is larger than (or equal to) the actual threshold I_{th}, an impulse is elicited; when it is smaller, there is no impulse.

To give dimensionless notation, the currents are given in units of i_I. The dimensionless stimulus S is then

$$S = (I_S - I_0)/i_I \tag{1}$$

and the dimensionless threshold x is

$$x = (I_{th} - I_0)/i_I. \tag{2}$$

To give rise to an impulse, it is necessary that $S \geq x$. With the simplification that i_S/i_I is proportional to S, this gives the following iteration scheme for the n'th stimulus:

$$x_{n+1} = r \cdot [x_n + \epsilon S + \delta_n], \tag{3}$$

where

$$\delta_n = \begin{cases} 0 & \text{for } S < x_n \\ 1 & \text{for } S \geq x_n, \end{cases} \tag{4}$$

$$r = \exp(-T/\tau), \tag{5}$$

and the factor ϵ is a constant.

The return map for eqns (3) and (4) is a linear map with a gap and with a slope $r < 1$. This results in a periodic impulse pattern. Of interest is the response fraction Q, which is the number m of impulses elicited divided by the number N of stimuli for a long series of stimuli, so

$$Q = m/N. \qquad (6)$$

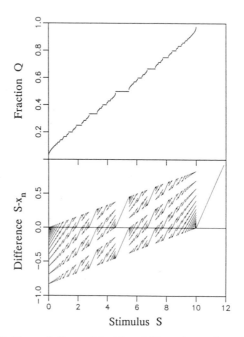

FIG. 2. The periodic model with $r = 0.83$, $\epsilon = 0.1$, and $N = 5000$. *Upper curve:* A complete "Devils staircase" as the relation between the response fraction Q and the stimulus strength S. *Lower curve:* The difference $S-x_n$ plotted for the same parameters. A positive difference corresponds to a spike and a negative to a dropout.

An analysis of the impulse pattern has been given previously (Colding-Jørgensen, 1983). In the periodic model, S and T (or f) are the independent variables, and in Fig. 2 (upper curve), the relation between Q and S is shown. The curve is a variant of the fractal relation called the "Devil's staircase." The name is due to the fact that there is an infinite number of steps interposed between any two steps, so with a finite step rate, it will take an infinite time to cover the distance. To see the variations in x_n underlying the firing pattern, the difference $S-x_n$ is shown in the lower part of Fig. 2 for a long series of stimulations. It is in the principle an analog to the bifurcation tree for chaotic models, but bifurcations are absent, because the behavior is periodic for all S. Correspondingly, the Liapunov exponent (Cf. Schuster, 1988) is always negative and independent of S. However, there is a fractal structure with a mixture of spaced and dense regions. As seen in the upper curve, the most spaced regions correspond to the broadest steps on the staircase, and the densest regions correspond to the end-points of these steps.

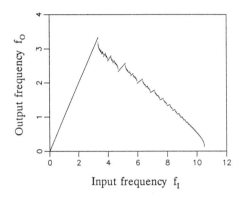

FIG. 3. Input-output relation for the periodic model. Mean output frequency f_O as function of the stimulus frequency f_I in units of $1/\tau$ with $S=4$ and $\epsilon=0.1$.

The difference $S-x_n$ is also an indicator of spike generation. A positive difference corresponds to a spike and a negative to a dropout. It is typical for the staircase that the rational fractions $Q=m/N$ with a small denominator occupy the largest part of the abscissa, and that the size of the steps with an irrational Q is zero. When the fraction $Q=m/N$ is rational and reduced so that N and m have no common factor, the length L of a step is proportional to r^N for large N. The fractal dimension D of the staircase can be found as $D=1-D_{sp}$, where D_{sp} is the dimension of the space between the steps (Bak, Bohr & Jensen, 1985). The number K of steps larger than L approaches $0.3 \cdot N^2$ in the limit, so D_{sp} becomes:

$$D_{sp} = \lim_{L \to 0} \frac{\log(K)}{\log(L)} \approx \lim_{N \to \infty} \frac{2 \cdot \log(N)}{N \cdot \log(r)} = 0.$$

Consequently, D equals 1, and the staircase is complete. The relation between S and Q is thus a twisted, but unbroken, line of horizontal segments with varying length. The relation is therefore an increasing, continuous function with $dQ/dS=0$ everywhere it is defined.

Varying the stimulation frequency with constant S yields also a "Devil's staircase" of almost the same nature, but asymmetric. The effect of the stimulus on the threshold is better seen when the input-output relation is considered. This is demonstrated in Fig. 3, where the output frequency $f_O=Q \cdot f_I$ is plotted as a function of the input frequency f_I. In the previous model (Colding-Jørgensen, 1983) without the stimulus effect ($\epsilon=0$), the f_O rose linearly with f_I until a certain limit f_{max}, where it changed to a sawtooth-like pattern approaching a constant value. With the present model, the relation is the same in the beginning, but above f_{max} the sawtooth pattern decreases towards zero at a limiting frequency f_{lim}.

The transition from linear to sawtooth pattern at f_{max} depends on S. With δ equal to 1 for all n and $x_n = x_{n-1} = S$, eqn (3) gives

$$r = \exp(-T_{max}/\tau) = 1/(1+\epsilon+1/S), \tag{7}$$

$$T_{max} = \tau \cdot \ln(1+\epsilon+1/S), \tag{8}$$

with $T_{max} = 1/f_{max}$. In the sawtooth region the stimulus-dependent increase in x will diminish the firing rate (at $S = 1/\epsilon$ the stimulus has the same effect as a spike), and above f_{lim} no spikes are elicited. This corresponds to $\delta = 0$ and $x_n = x_{n-1} \geq S$. From eqn (3) with a similar argument as in eqn (7)

$$r = \exp(-T_{lim}/\tau) \geq 1/(1+\epsilon),$$
or
$$T_{lim} = \tau \cdot \ln(1+\epsilon), \tag{9}$$

so in the periodic model the limiting frequency $f_I = f_{lim}$ is independent of the stimulus strength and depends only on ϵ and τ. The cell is able to fire in the interval $0 < f_I < f_{lim}$ for all S. The linear part is small for small S and larger for S large. With $\epsilon = 0.1$, f_{lim} equals $10.5/\tau$. For $S = 1$, the linear part is 13% of this interval, but for $S = 10$, it is 52%.

The form of the curve is peculiar. It consists of line segments with a slope $df_O/df_I = Q$, which is positive, where it is defined. Nevertheless, the curve is both continuous and decreasing towards zero for $f_I > f_{max}$. Close to f_{lim} the curve becomes very steep, so the apparent slope approaches $-\infty$.

Experimental evidence for the periodic model

Experimental methods

The experimental methods have been described in detail previously (Colding-Jørgensen, 1977). The ventral ganglion of *Helix pomatia* was isolated and the covering connective tissue removed after softening with pronase-E. The ganglion was placed in a solution of the following composition (mM): NaCl 80; KCl 4; $CaCl_2$ 7; $MgCl_2$ 5; Tris-Cl 5 (pH 7.5). The cells were impaled with two glass micro-electrodes (Fig. 1) filled with 3M KCl and with resistance of 5-15 MΩ.

Only cells without spontaneous firing were used. Square current pulses were passed through the current electrode, and the cell's response was recorded with the potential electrode. For a given current strength and frequency, the response fraction was registered over 30-100 stimulations.

Results

The upper recording of Fig. 4 is an experimental "Devil's staircase." Current pulses of duration 0.2 s were applied at a period of 1.08 s and a varying strength. The resolution (0.1 nA) suffices to show a clearly non-linear structure with "steps"

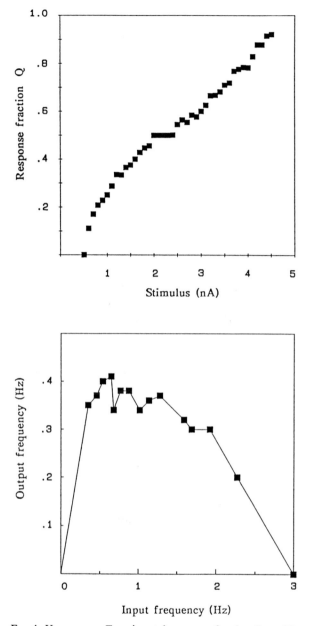

Fig. 4. *Upper curve:* Experimental response fraction $Q = m/N$ for a single nerve cell as a function of the stimulus strength. *Lower curve:* The relation between the output frequency $f_O = Q \cdot f_I$ and the input frequency f_I. Stimulus is 2 nA.

at $Q = 1/2$, $1/3$, $2/3$, and $3/4$. A better resolution is difficult to obtain, because the cell has to remain stable for a long time (several hours). The time constant τ for the cell was 6 s, so r was close to the 5/6 used in the periodic model. From eqn (7), ϵ is estimated as

$$\epsilon = (1\text{-}r)/r - 1/S_{max} = 0.2 - 1/S_{max},$$

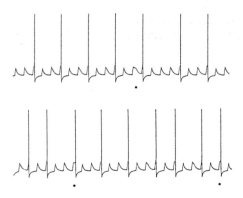

FIG. 5. Near-threshold behavior of the nerve cell.
The cell hesitates sometimes before a dropout
(upper recording) or a spike (lower recording).

where S_{max} is the stimulus value, where $Q = 1$. To find S_{max}, it is necessary to take into account that the potential changes are delayed by the membrane capacity, that the effect of the potential on the Ca^{2+}-influx is non-linear (Colding-Jørgensen, unpublished), and that the threshold current I_0 to some extent depends on the stimulus (Colding-Jørgensen, 1990). The value of I_S in eqn (1) is therefore only 30-50% of the stimulus strength shown in Fig. 4 or around 2 nA. With an i_I around 0.2 nA/spike, this gives $S_{max} = 10$ or $\epsilon = 0.1$ as used in the periodic model.

The input-output relation $f_O = Q \cdot f_I$ is shown in Fig. 4 (lower curve). With a stimulus strength $I_S = 2$ nA, the relation is linear up to $f_I = 0,35$ Hz followed by a sawtooth part and ending with a f_{lim} less than 3 Hz. This gives an estimate of ϵ from eqn (9) of 0.06-0.07, which, with the same corrections as before, is reasonably close to the $\epsilon = 0.1$ used in the model. The break away from the linear part probably takes place too soon. From eqn (8) with an effective S around 3 the f_{max} is 0.46 Hz. However, the very acute maximum of Fig. 3 is vulnerable, so any fluctuation will decrease Q and flatten the maximum.

In conclusion, the experimental results support the periodic model qualitatively and, with the above reservations, also quantitatively.

The chaotic model

The previous section demonstrates that the inclusion of the Ca^{2+}-influx during the stimulus results in a model that can mimic the experimental findings reasonably well. It also demonstrates that the value of ϵ is critical. The variation in ϵ with S makes the staircase in Fig. 4 asymmetrical in contrast to the staircase in Fig. 2 and modifies the input-output relation. A detailed analysis of the variations in ϵ is outside the scope of this paper except for one decisive detail. When the stimulus is very close to the threshold, the potential stays depolarized for some time before the decision whether to fire or not is taken. This hesitation can be several times the normal duration of the depolarization, so a noticeable extra amount of Ca^{2+}-ions enter the cell. Consequently, the threshold is increased both in the case where the cell fires an impulse and when it does not. An example of this near-threshold behavior is shown in Fig. 5. It is present in all nerve cells and is also included in

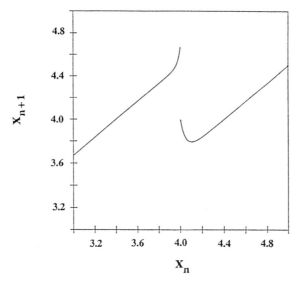

FIG. 6. Part of the return plot, eqn (9), for the chaotic model with $S=4$, $\beta_1=0.4$, $\beta_2=0.2$, $\gamma_1=20$, and $\gamma_2=40$. Other parameters as in Fig. 2.

most nerve models (Hodgkin & Huxley, 1952). The effect is an extra increase in x_n, when $|S-x_n|$ is small. The increase modifies the linear return plot for the model, so an upward bending of the line appears both before and after the gap (Fig. 6). The crucial factor is the bending of the right-hand branch. If the effect is so large that dx_{n+1}/dx_n in the return plot becomes zero (or negative), chaotic behavior becomes a distinct possibility.

The near-threshold behavior is exceedingly fleeting, so experimental results are sparse. The best approximation has been to assume that the effect has a fast exponential decay with $|S-x_n|$. This modifies eqns (3) and (4) as

$$x_{n+1} = r \cdot [x_n + \epsilon S + \delta_n], \tag{9}$$

with

$$\delta_n = \begin{cases} 0 + \beta_1 \cdot \exp(\gamma_1(S-x_n)) & \text{for } S < x_n \\ 1 + \beta_2 \cdot \exp(-\gamma_2(S-x_n)) & \text{for } S \geq x_n, \end{cases} \tag{10}$$

where β_1, β_2, γ_1, and γ_2 are positive constants. A part of this return plot (x_{n+1} as a function of x_n) is shown in Fig. 6 for $S=4$. In the chaotic model, $\beta_1=0.4$, $\beta_2=0.2$, $\gamma_1=20$, and $\gamma_2=40$. Other parameters are as in the periodic model.

The effect is most pronounced for dropouts, so $\beta_1 > \beta_2$ and $\gamma_1 < \gamma_2$. For dropouts ($S < x_n$) the slope is

$$\frac{dx_{n+1}}{dx_n} = r \cdot [1-\beta_1 \cdot \gamma_1 \cdot \exp(\gamma_1(S-x_n))],$$

which is zero for $x_n = S + \ln(\beta_1\gamma_1)/\gamma_1$. Because $S < x_n$, the condition for zero slope, and thereby a possibility for chaos, is $\beta_1\gamma_1 > 1$.

The relation between S and Q is shown in the upper curve of Fig. 7. It demonstrates the breakdown of the "Devil's staircase" into regions with chaotic firing interposed between regular regions. It is typical that the largest steps (1/2, 1/3, 2/3, etc.) survive the breakdown, while the smaller steps disappear. Notice that in the beginning of the curve there is no firing, even though S is positive, so the threshold increase caused by the stimulus is larger than the effect of the stimulus itself. If the series of stimuli is applied to a resting membrane, it will, of course, in the beginning respond with spikes, but the spikes will disappear after some time.

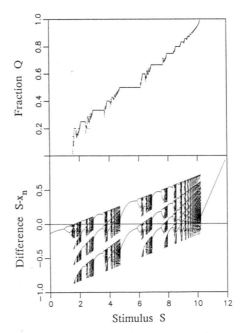

Fig. 7. The chaotic model. Parameters as in Figs. 2 and 6. *Upper curve:* A broken-down "Devil's staircase." The relation between the response fraction Q and the stimulus strength S. *Lower curve:* The bifurcation diagram. The difference $S - x_n$ plotted for the same parameters. A positive difference corresponds to a spike and a negative to a dropout.

The underlying variations in x_n are demonstrated in the lower part of Fig. 7. The overall structure is much like the structure of Fig. 2, but the details reveal a typical bifurcation diagram. The first bifurcations take place below the line $S - x_n = 0$, and the chaotic behavior is present before the first impulse is fired. The Liapunov exponent λ varies throughout the diagram between large negative values and up above $+0.5$. In the most chaotic region, the region where the first spike comes, a

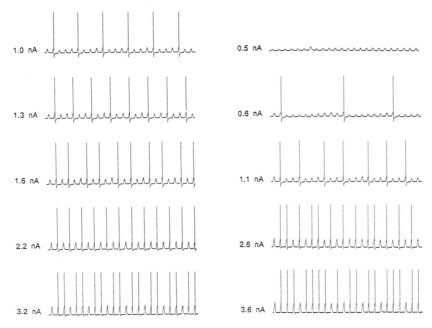

Fɪɢ. 8. Examples of regular (left) and chaotic (right) firing. The corresponding stimulus strength I_S in nA indicated at the recordings. For further description see text.

small difference is therefore amplified $e^{0.5N}$ or 10^6 to 10^{20} times during the 30-100 impulses counted. The dense regions in Fig. 2 are changed to either bifurcations cascading into the chaotic regions or to dense, intermittent regions, where the transition from chaotic to periodic behavior takes place. In the intermittent regions the regularity is only occasionally broken, and λ is still positive but close to zero. Exactly at $\lambda = 0$, the behavior is called quasiperiodic, and as the firing becomes regular, λ becomes negative.

Because the near-threshold effect always results in an increased threshold, the introduction of the effect diminishes the chances for firing. This is seen for small values of S and in the fact that Q=1 first is reached for an S around 10.5. Another effect is that whenever the firing is chaotic, the near-threshold situations are frequent, while they are almost absent in regular firing. As a consequence, the value of Q jumps up and down in a discontinuous manner, so even very small differences in S can result in dramatic differences in Q and thereby in the mean output frequency.

Experimental evidence for the chaotic model

The present technique is insufficient to measure the minute changes of the Ca^{2+}-dependent K^+-conductance that underlies the chaotic model. The behavior of the cell is, however, so close to the behavior of the chaotic model that it not only represents strong evidence for the presence of chaotic firing, but also indicates that chaotic firing represents a normal function of a nerve cell.

173

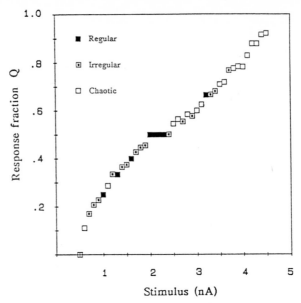

F<small>IG</small>. 9. Same relation as Fig. 4, but with indication of regular, irregular, and chaotic firing.

This is demonstrated in Fig. 8, where the firing patterns arising from different values of the stimulus strength are shown. All patterns are from the same cell, but they are sorted, so the left column contains regular (periodic) patterns and the right column patterns which are called chaotic. It is assumed that experimental error, biological changes, etc. occasionally can break a regular pattern, so only patterns with multiple fluctuations are catagorized as chaotic. The intermediate patterns with one or a few deviations from regularity are named irregular to indicate that they on one hand might be deviations from regular patterns and on the other hand intermittent, chaotic patterns.

The pattern for $S = 0.5$ nA is without spikes. However, the large fluctuations in the potential changes caused by the stimuli indicate a strong possibility of an underlying chaotic variation. This is in accordance with the chaotic model, where the first bifurcations and transition to chaos take place before any spikes are elicited, and where the first spikes appear in the chaotic regime. In Fig. 8, the first appearance of spikes takes place for a stimulus somewhere between 0.5 and 0.6 nA, which is 10-15% of S_{max}. In the chaotic model, the figure was 15%. The threshold stimulus for a single spike was for this cell very small - around 0.1 nA - so even though the chaotic model is very simplified, its quantitative behavior is close to the cell behavior.

The mixture of regular, irregular, and chaotic firing is shown in Fig. 9, which is the same as in Fig. 4 except for the partition into the three categories defined above. As for the chaotic model (Fig. 7, upper curve), the largest steps survive the breakdown. Also the main form of the curve is similar, with an abrupt start in the chaotic region and a pronounced chaotic region after $Q = 0.5$. Small differences are found in the upper part of the curve, but here the stimuli are so large that the chaotic model may be too simple.

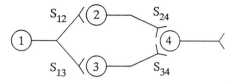

Fɪɢ. 10. Coupling diagram. Four cells are coupled, so cell 1 stimulates cells 2 and 3 with the strengths S_{12} and S_{13}, respectively, and cells 2 and 3 stimulate cell 4 with the strengths S_{24} and S_{34}, respectively. Same model parameters (chaotic model) for the four cells. Only variation the coupling strengths vary.

Coupled nerve cells

In the previous sections, the input to the nerve cell is a regular series of stimuli, and it is shown how such series can result in chaotic output patterns from the cell. Most of the signal processing in the nervous system is, however, a cooperation between many nerve cells. With the demonstrated complex behavior of a single nerve cell, it is obvious that a collection of even a few interconnected nerve cells has a noticeable signal processing capacity. It is also clear that the function of such a network is quite different from the function of conventional neural nets, both with respect to the complexity of the steady state behavior and even more with respect to the adaptability and the power of on-line dynamic performance.

In the present analysis, only the steady state behavior is considered, and to emphasize the complexity of the signal processing, only one net version is considered, as shown in Fig. 10. The first cell is always firing with a constant frequency (100 spikes shown). This gives a "clock frequency" f_1 that is used to syncronize cells 2 and 3. Cell 1 stimulates cells 2 and 3 with the stimulus strengths S_{12} and S_{13}, respectively, mimicking EPSP's (Excitatory postsynaptic potentials) to the cells. With different values of S_{12} and S_{13}, the cells can be brought to fire with different frequency in regular or chaotic patterns. The outputs from cells 2 and 3 are fed into cell 4, and again variations in the EPSP's are mimicked by varying the stimulus strengths S_{24} and S_{34}. This is a very simple network, and by observing how the output from cell 4 depends on the input patterns from cells 2 and 3, some of the basic network rules can be revealed.

Two examples are shown in Fig. 11. In the left recording, both cells 2 and 3 are firing in a chaotic region. If the clock frequency f_1 (cell 1) is set to 100, the mean frequency f_2 of cell 2 is 44, and f_3 for cell 3 is 20. In pattern (a), where only cell 2 stimulates cell 4, the result is a chaotic pattern with a frequency $f_4 = 10$, and in (b), where only cell 3 stimulates, $f_4 = 5$. Finally, when both cells 2 and 3 are active, the pattern is rather complex, and $f_4 = 14$. It is tempting to conclude that cell 4 is a simple adding component, whose mean output frequency is the sum of the two mean input frequencies. This is in agreement with the conventional view on synaptic transmission, where it is supposed that addition of excitatory input to a nerve cell results in an increased output frequency.

175

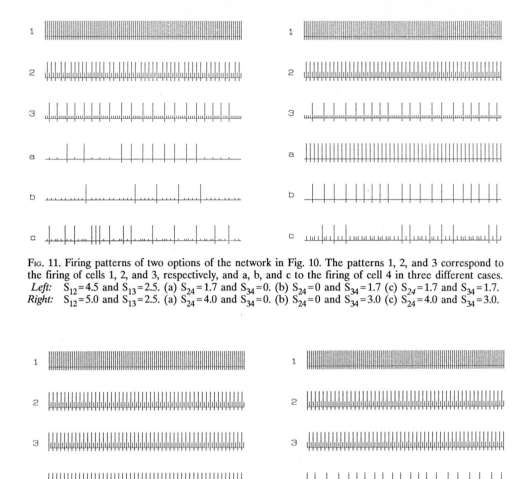

F_{IG}. 11. Firing patterns of two options of the network in Fig. 10. The patterns 1, 2, and 3 correspond to the firing of cells 1, 2, and 3, respectively, and a, b, and c to the firing of cell 4 in three different cases.
Left: $S_{12}=4.5$ and $S_{13}=2.5$. (a) $S_{24}=1.7$ and $S_{34}=0$. (b) $S_{24}=0$ and $S_{34}=1.7$ (c) $S_{24}=1.7$ and $S_{34}=1.7$.
Right: $S_{12}=5.0$ and $S_{13}=2.5$. (a) $S_{24}=4.0$ and $S_{34}=0$. (b) $S_{24}=0$ and $S_{34}=3.0$. (c) $S_{24}=4.0$ and $S_{34}=3.0$.

F_{IG}. 12. The effect of timing. In both cases, $S_{12}=S_{13}=5$ and $S_{24}=S_{34}=2.5$.
Left: Cells 2 and 3 synchronized. *Right:* Cells 2 and 3 out of phase.

This conception is invalidated in the other recording of Fig. 11 (right). Here, cells 2 and 3 are firing with almost the same frequencies as before ($f_2=50$ and $f_3=20$), but S_{24} and S_{34} are different. In case (a), S_{24} is strong enough to synchronize cell 4 and cell 2 when cell 3 is disconnected, and in (b), the same is the case when cell 2 is disconnected and 3 is active. However, when both are connected to cell 4, the firing of cell 4 is strongly inhibited, so $f_4=10$, while the sum should give 70. Considering the change from (a) to (c), the addition of the strong, excitatory input from cell 3 results in a drop in f_4 from 50 to 10. A similar, but much weaker, effect is present, if the periodic model is used in the network.

A simple demonstration of the importance of timing in the network is shown in the two patterns of Fig. 12. The network is symmetric, the stimulus strengths are the same in both cases, and both cells 2 and 3 are firing regularily with $f_2=f_3=50$. In the left pattern, cells 2 and 3 are synchronized, and with $S_{24}+S_{34}=5$, cell 4 is also synchronized with $f_4=50$. In the right-hand pattern, cells 2 and 3 are desynchronized, so they fire out of phase. The effect corresponds to an S of 2.5 applied at the frequency $f_1 = 100$. The result is a chaotic pattern with $f_4 = 20$, so even if the two input frequencies are exactly the same in the two cases, the timing of the individual stimuli plays an important role for the result.

Taking into account the effect of the stimulus and the near-threshold behavior, which are inevitable ingredients in all nerve cells, a component of the neuronal network is revealed, where the exact timing of the incoming impulses and the previous behavior is crucial for its function.

Conclusion

When a nerve cell receives an external stimulus it has two options: It can decide to fire an impulse, or it can decide not to do so. In both cases, the decision influences future firing. Sometimes the stimulus and the threshold are so close to each other that the cell hesitates before the decision is taken. This hesitation also influences the future firing.

These are the general rules for the firing of a paced nerve cell and for the firing of the chaotic model. They may be put in different mathematical forms depending on the exact properties of a given nerve cell, but the fact remains that when the present and past firing influences the future firing in a non-trivial way, the behavior of even a single nerve cell becomes exceedingly complex. When many such nerve cells are combined in a network, a delicate interplay arises, where the information transmission and information processing melt together, and where the performance of the network continuously changes with the information it receives.

The key to the behavior - and thereby to the function of the network - is in the present model the Ca^{2+}-concentration just inside the cell membrane, because it controls the Ca^{2+}-dependent K^+-conductance, which in turn determines the firing threshold (the variable x_n in the model). The Ca^{2+}-concentration depends on many factors: Influx and efflux over the plasma membrane (Akaike et al., 1978; Ohnishi & Endo, 1981), intracellular buffering (Ahmed & Connor, 1988), diffusion (Gorman et al., 1984; Williams et al., 1985), and uptake into or release from intracellular stores (Dupont & Goldbeter, 1989; Dissing et al., 1990). All these factors are influenced by the cell metabolism, external and internal hormones, and the membrane potential, etc.

A way to describe the signal processing of even a simple neuronal network is therefore to start with the influence of the nerve impulses and the EPSP's on the firing, for example, as described by the chaotic model. If the changes in the metabolic and hormonal state of the cells are slow, they can then afterwards be introduced as variations in the model parameters. In this context the chaotic model can be used as a link between the cell state and the function of the network.

With fast changes in cell state, the separation of the two effects is impossible, and the changes become an integrating part of the behavior. As an example, hormonally driven intracellular oscillations in the Ca^{2+}-concentration (Dupont & Goldbeter, 1989) can interact with the oscillations driven by the impulse activity of the cell and produce complex impulse patterns, bursts, and even chaotic firing in a single cell.

References

AHMED, Z. & CONNOR, J. A. (1988). Calcium regulation by and buffer capacity of molluscan neurons during calcium transients. *Cell Calcium.* **9**, 57-69.

AKAIKE, N., LEE, K. S. & BROWN, A. M. (1978). The calcium current of *Helix* neuron. *J. Gen. Physiol.* **71**, 509-531.

BAK, P., BOHR, T. & JENSEN, M. H. (1985). Mode-locking and the transition to chaos in dissipative systems. *Physica Scripta.* **T9**, 50-58.

BENHAM, C. D., BOLTON, T. B., LANG, R. J. & TAKEWAKI, T. (1986). Calcium-activated potassium channels in single smooth muscle cells of rabbit jejunum and guinea-pig mesenteric artery. *J. Physiol.* **371**, 45-67.

BROWN, D. A. & GRIFFITH, W. H. (1983). Calcium-activated outward current in voltage-clamped hippocampal neurones of the guinea-pig. *J. Physiol.* **337**, 287-301.

COLDING-JØRGENSEN, M. (1977). Impulse dependent adaptation in *Helix pomatia* neurones: Effect of the impulse on the firing pattern. *Acta physiol. scand.* **101**, 369-381.

COLDING-JØRGENSEN, M. (1983). A model for the firing pattern of a paced nerve cell. *J. theor. Biol.* **101**, 541-568.

COLDING-JØRGENSEN, M. (1990). Fundamental properties of the action potential and repetitive activity in excitable membranes illustrated by a simple model. *J. theor. Biol.* **144**, 37-67.

COLDING-JØRGENSEN, M., MADSEN, H. Ø., BODHOLDT, B. & MOSEKILDE, E. (1990). A simple model for Ca^{2+}-dependent oscillations in excitable cells. *Proceedings of the 1990 European Simulation Multiconference,* 630-635.

DISSING, S., NAUNTOFTE, B. & STEN-KNUDSEN, O. (1990). Spatial distribution of intracellular, free Ca^{2+} in isolated parotid acini. *Pflügers Arch,* **417.** 1-12.

DUPONT, G. & GOLDBETER, A. (1989). Theoretical insight into the origin of signal-induced calcium oscillations. In *Cell to cell signalling: From experiments to theoretical models.* Ed. A. Goldbeter. Academic Press, London. 461-474.

GLASS, L., GUEVARA, M. R., SHRIER, A. & PEREZ, R. (1983). Bifurcation and chaos in a periodically stimulated cardiac oscillator. *Physica.* **7D**, 89-101.

GORMAN, A. L. F. & THOMAS, M. V. (1980). Potassium conductance and internal calcium accumulation in a molluscan neurone. *J. Physiol.* **308**, 287-313.

GORMAN, A. L. F., LEVY, S., NASI, E. & TILLOTSON, D. (1984). Intracellular calcium measured with calcium-sensitive micro-electrodes and arsenazo III in voltage-clamped *Aplysia* neurones. *J. Physiol.* **353**, 127-142.

HODGKIN, A. L. & HUXLEY, A. F. (1952). A quantitative description of membrane current and its application to conduction and excitation in nerve. *J. Physiol.* **117**, 500-544.

LUX, H. D. & HOFMEIER, G. (1982). Properties of a calcium- and voltage-activated potassium current in *Helix pomatia* neurons. *Pflügers Arch.* **394**, 61-69.

MATSUMOTO, G., TAKAHASHI, N. & HANYU, Y. (1987). Chaos, Phase Locking and Bifurcation in Normal Squid Axons. *Proceedings of the NATO Advanced Research Workshop on Chaos in Biological Systems, Dec. 1986.* Plenum Press, New York. 143-156.

OHNISHI S. T. & ENDO, M. (1981). The mechanism of gated calcium transport across biological membranes. Academic Press. New York.

RASMUSSEN, H. & BARRETT, P. Q. (1984). Calcium messenger system: An integrated view. *Physiol. Rev.* **64**, 938-984.

SCHUSTER, H. G. (1988). Deterministic Chaos. An introduction. VCH Verlagsgesellshaft mbH. Weinham, Germany.

TUCKWELL, H. C., WAN, F. Y. M. & WONG, Y. S. (1984). The interspike interval of a cable model neuron with white noise input. *Biol. Cybern.* **49**, 155-167.

WILLIAMS, D. A., FOGARTY, K. E., TSIEN, R. Y. & FAY, F. S. (1985). Calcium gradients in single smooth muscle cells revealed by the digital imaging microscope using Fura-2. *Nature.* **318**, 558-561.

ANALYSIS OF THE ADENINE NUCLEOTIDE POOL IN AN OSCILLATING

EXTRACT OF YEAST *SACCHAROMYCES UVARUM*

Joachim Das and Heinrich-Gustav Busse

Biochemisches Institut
Christian-Albrechts-Universität zu Kiel
Olshausenstrasse 40, D - 2300 Kiel

INTRODUCTION

The identification of the regulatory steps in a biochemical pathway
is greatly facilitated by the knowledge of the reaction steps which
proceed close to their chemical equilibrium. This is particularly
important in the case of the oscillating glycolysis. Here, we present
evidence which suggests that the adenine nucleotide pool is equilibrated
by the myokinase reaction during oscillatory glycolysis.

EXPERIMENTS

A cytoplasmic extract of yeast *S. uvarum* was prepared as described
in the literature (Das and Busse, 1985). The oscillation of the
glycolytic pathway was induced by the addition of 136 mM trehalose to
the extract. The optical density of NADH in the extract was recorded in a
dual wavelength photometer as the difference between the O.D. at 400 and
340 nm (see fig. 1 upper trace). After a few regular periods of
oscillations, 24 samples were drawn from the extract during one period
and analyzed by enzymatic methods (Estabrook et al., 1962; Jaworek et
al., 1974) for their concentrations of the adenine nucleotides (AMP,
ADP, ATP). The result is shown in the lower diagram of fig. 1.

RESULTS AND DISCUSSION

The sum Σ of the concentrations of the adenine nucleotides remains
within the experimental error constant during the oscillation (see fig.
1), i.e.:

$$\Sigma = [AMP] + [ADP] + [ATP] \qquad (1)$$

A phase diagram, in which the ratio of the concentrations of the adenine
nucleotides, relative to their total concentration Σ are drawn,
demonstrates their relationship (see fig. 2). Assuming that the
interconversion of the adenine nucleotides, a reaction catalyzed by the
enzyme myokinase (adenylatekinase):

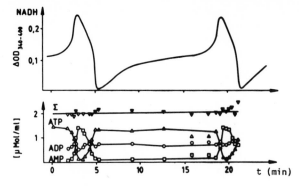

Fig. 1   Oscillatory time course of glycolytic metabolites.

Upper diagram: Difference between the optical density at 340 nm
and 400 nm in a 3 mm path of a flow cuvette, recorded as
reference to other data.

Lower diagram: Enzymatic determination of the concentration of
AMP, ADP and ATP in samples drawn at time intervals which were
chosen to display distinct changes.
Lines connect the points to improve the readability.
The constancy of the sum Σ of the adenine nucleotide
concentrations is demonstrated.

$$AMP + ATP \rightleftharpoons 2\ ADP$$

remains close to its equilibrium, the mass action law states:

$$[ADP]^2 = K \cdot [AMP] \cdot [ATP] \qquad (2)$$

where K is the equilibrium constant.

Since the sum Σ is constant during the experiment, it is related to
the nucleotide concentration by the equations:

$$M = [AMP]/\Sigma; \quad D = [ADP]/\Sigma \text{ and } T = [ATP]/\Sigma$$

Now, the equations (1) and (2) read:

$$M + D + T - 1 = 0 \qquad (1a)$$
$$D^2 - K \cdot M \cdot T = 0 \qquad (2a)$$

They display the following properties:

(a) Since M and T are exchangeable the resulting curves display an axis
of symmetry given by M = T (45° line).

(b) Equation (1a) is solved for M, D or T and the result is inserted into
(2a). This yieds a quadratic equation of the remaining variables. It
represents an ellipse because the plane (1a) with the cone (2a)
intersects in the threedimensional space (M, D, T).

(c) Special solutions:
for D = 0: Either (T = 0 and M = 1) or (M = 0 and T = 1).

180

For M = T: M = T = $1/(2 \pm \sqrt{K})$ and D = M $. \sqrt{K}$
and D reach its maximal value.

(d) The coordinate axes are tangents to the ellipses at (0,1) and (1,0)
for the M-T ellipse and at (0,0) for the other ellipses.

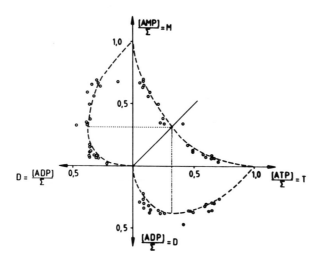

Fig. 2   Phase diagram of the adenine nucleotide concentrations normalized
to their sum Σ. The lines at 45° emphasize the mirror symmetry of
the curves. At the point at which the concentration of AMP equals
that of ATP, the concentration of ADP reaches its maximal value
(see dotted lines).

The experimental points in fig. 2 show these characteristic features. If a representative point is inserted into equation (2a) a value
of K $\approx$ 1.8 is obtained. It is within experimental error identical to the
value of the equilibrium constant 2.26 (Barman, 1969). Similar values are
obtained from a regression analysis or from the maximal value of D.

CONCLUSION

The concentration of the adenine nucleotide pool remains constant in
the oscillatory state of the glycolytic pathway in yeast and the
adenylate kinase reaction retains its equilibrium state.

REFERENCES

Barman, T.E., 1969, "Enzyme Handbook", Vol. I, Berlin-Heidelberg-
New York, pp 448.
Das, J. and Busse, H.-G., 1985, Long term oscillation in glycolysis,
J. Biochem., 97:719.
Estabrook, R.W. and Maitra, P.K., 1962, A fluorimetric method for
the quantitative microanalysis of adenine and pyridine
nucleotides. Anal. Biochem., 3:369.
Jaworek, D., Gruber, W., and Bergmeyer, H.U., 1974, Adenosin-5'-
triphosphat - Bestimmung mit 3-Phosphoglycerat-Kinase,
in: "Methoden der enzymatischen Analyse", H.U. Bergmeyer,
ed., Weinheim (3rd Ed.), Vol. 2, pp. 2147.

BOUNDARY OPERATOR AND DISTANCE MEASURE FOR THE CELL LINEAGE OF

CAENORHABDITIS ELEGANS AND FOR THE PATTERN IN FUSARIUM SOLANI

J. Das, E. Valkema*, and H.-G. Busse

Biochemisches Institut der Medizinischen Fakultät
Dept. Prof. Dr. B.H. Havsteen, and
* Inst. f. Informatik und Praktische Mathematik
Christian-Albrechts-Universität Kiel
Olshausenstr. 40, 2300 Kiel, FRG

SUMMARY

The dendogram of the cell lineage of the nematode C. elegans is
derived by a distance defined on a data matrix. Its elements are the
times needed to complete a period of revolution of the cell cycle or the
post-mitotic lifetime. The hierarchical clustering method of complete
linkage reproduces the experimental dendrogram. The construction of the
data matrix from a language and the relation to boundary operations is
discussed.

INTRODUCTION

The lineage of cell development has been established in detail for
the small nematode Caenorhabditis elegans (Deppe et al. 1978; Sulston et
al. 1983). This lineage is illustrated in a diagram as a binary tree in
which a branch signifies the division of a cell into its daughter cells.
The constancy of the number of individual cells in the adult nematode and
the invariance in the events which comprise the cell division allowed the
construction of the tree. The hermaphroditic adult consists of about 950
somatic cells and germline cells. It is generated by the cleavage of the
fertilized egg into 671 cells of which 113 cells die and by differentiation of the living cells into the functional units. After hatching,
the nematode passes through four "larval" stages without metamorphosis
before the adult stage is reached (Sulston et al. 1977; Kimble et al.
1979). During this period only certain postembryonic blast cells continue
to divide and some nuclei multiply endomitotically.

The particular properties of the cell lineage of this animal
species are likely to disclose general principles of embryology. In this
paper, it is demonstrated that the tree of the cell lineage can be
derived from a data matrix containing the periods of the individual cell
cycles which have been extracted from the experimental data (Sulston et
al. 1983). The matrix is considered as data to which a hierarchical
classification (Bock, 1974) may be applied. The original experimental
tree of the cell lineage is obtained by the complete linkage method of

hierarchical clustering under a specific distance measure which is
computed from the data matrix.

Classification may be important to developmental biology in general,
since other phenomena, like e.g. the aggregation of the amoeba
Dictyostelium discoideum to an organism (Gerisch et al. 1971), also may
be formulated by a distance measure.

Mathematical Preliminaries

A distance (or dissimilarity coefficient) is defined in mathematical terms as a function d into the non-negative real numbers with
the following properties (Bock 1974):

(I) $d(x,y) \geq 0$ (positive definite)
(II) $d(x,y) = d(y,x)$ (symmetrical)
(III) $d(x,x) = 0$ (equivalent)

where x and y are any objects in the set on which the function d is
defined. (Note that $d(x,y) = 0$ does not imply
$x = y$ e.g. grades form a distance, in which for two individuals x
and y having the same grade $d = 0$).

If in addition, for the distance $d(x,y)$, the inequality:

(IV) $d(x,y) \leq d(x,z) + d(z,y)$ (triangle inequality)

where x,y and z are any objects from a set on which d is defined
is valid, then the function $d(x,y)$ is said to be a metric distance.
Since $d(x,y) = 0$ does not imply $x = y$, it is not a metric in usual
mathematical terms. In many instances, metrical distances support their
interpretation as geometrical distances. The above demands on a function
are weak, therefore many functions d fulfill the laws (I) to (IV) and do
not assort the set of objects with much structure. However, these
functions are very useful in generating a preorder or classification on
the set of objects.

In an analysis of experiments, the objects usually are characterized by corresponding sets of data, which are represented in a
matrix of data (M_{ik}). A row of the matrix contains the data which belong
to one object and each column consists of data of the same type. The
distances between the objects can also be expressed in a symmetrical
matrix, the diagonal of which has zero entries.

Biological Preliminaries

The nematode *C. elegans* is a worm of approximately 1 mm length and
0.1 mm diameter as an adult. It can be cultured on bacteria on agar
plates in the laboratory (Brenner, 1974). The nematode releases
fertilized eggs of a length of 50 μm and a diameter of 30 μm. These eggs
are surrounded by a semi-rigid shell which does not change its shape
until the larvae hatch. The fertilized egg undergoes a series of nuclear
divisions with concomitant cell formation around the newly formed nuclei.
Hence, the fixed volume of the egg is successively divided into smaller
and smaller cells. When growth of the cellular network is nearly
complete, differentiational events become and stay dominant, until hatching. Since the shell of the egg is transparent, differential interference
contrast microscopy allows exact registration of the timing of divisions.
This series of divisions may be depicted as a tree (see Fig. 1) which
illustrates the embryonic cell lineage (Sulston et al. 1983). The
advantages of nematodes such as *C. elegans* for these experiments are
their small number of cells and the invariance of this tree for all

individuals of the species. As a consequence the number of cells in the just hatched animal is also invariant. Since postembryonic lineages are also stereotyped, this as well applies to the adult and four larval stages in which the nematode *C. elegans* by additional cell cleavage and differentiation approaches its adult stage. The review of Chitwood (1974) or the monograph edited by W.B. Wood (1988) may be consulted for more detailed information about nematode development.

CALCULATIONS

A Distance for Cell Lineages

A small subtree of the cell lineage of the nematode *Caenorhabditis elegans* (Sulston et al. 1983) is treated here in more detail to illustrate the method.

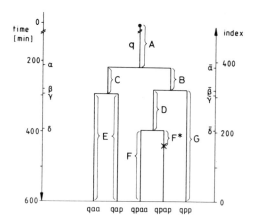

Fig. 1   A section of the tree of cell lineage of the nematode *C. elegans*. A vertical line represents the period in which a cell nucleus is visible in the egg. A nuclear division is given by a horizontal branch. B, C and D are lifetimes of single nuclei (i.e. time between nuclear divisions), whereas A is the sum of all lifetimes of the nuclei leading to cell q (including the lifetime of cell q). E,F,F* and G are the lifetimes of the corresponding cells at the bottom of the dendrogram. q may be substituted by ABalapppa to get the usual notation of cells in the nematode *C. elegans* (see: Sulston et al. 1983).

A similarity matrix S_{ik} may be constructed from a given lineage by defining the similarity between two objects as the time elapsed since the division of the last common ancestor nucleus. For the lineage given in Fig. 1, a matrix S_{ik} is obtained from the periods between the nuclear divisions (here denoted by A, B, C, D see Fig. 1), i.e. from the time schedule of the cell cycle events. Note, the similarity of a cell to itself is maximal, since the last common ancestor is the cell itself.

S_{ik}	qaa	qap	qpaa	qpap	pqq
qaa	τ	$\overline{W}$	$\overline{\alpha}$	$\overline{\alpha}$	$\overline{\alpha}$
qap		τ	$\overline{\alpha}$	$\overline{\alpha}$	$\overline{\alpha}$
qpaa			τ	$\overline{\delta}$	$\overline{\beta}$
qpap				τ	$\overline{\beta}$
qpp					τ

where:

$\overline{\alpha} = A; \quad \overline{\beta} = A + B; \quad \overline{W} = A + C; \quad \overline{\delta} = A + C + D; \quad \tau = 600$

(τ = time period elapsed since fertilization at which the cell population is examined)

The similarity matrix, S_{ik}, is transformed into a distance matrix D_{ik} by

$$D_{ik} = \tau \cdot E_{ik} - S_{ik}$$

where $E_{ik} = 1$ for all i and k. This transformation resets the time axis to an index scale with the origin at the bottom line in Fig. 1 (see right scale) and zeros all diagonal elements.

The matrix D_{ik} may be obtained from a data matrix M_{ik} of the objects (i.e. the cells), in which the elements are the lifetime of cells, e.g. the periods between the individual nuclear divisions:

D_{ik}	qaa	qap	qpaa	qpap	qpp
qaa	0	W	α	α	α
qap		0	α	α	α
qpaa			0	δ	β
qpap				0	β
qpp					0

M_{ik}	q	qp	qa	qpa	qaa	qap	qpaa	qpap	qpp
qaa	A	0	C	0	E	0	0	0	0
qap	A	0	C	0	0	E	0	0	0
qpaa	A	B	0	D	0	0	F	0	0
qpap	A	B	0	D	0	0	0	F*	0
qpp	A	B	0	0	0	0	0	0	G

The rows represent the terminal cells (Fig. 1) and the columns signify their corresponding terminal as well as intermediate cells.

The lifetimes are defined for the terminal cells as the time they have been alive , e.g. at the bottom of the diagram in Fig. 1, the pair of cells (qaa, qap) has a lifetime E. Sometimes, dead cells (e.g. cell qpap) are considered to be still living (i.e. F* = F) since this simplifies the application of algorithm.

The similarity matrix S_{ik} is obtained from the data matrix M_{ik} by:

$$S_{ik} = \sum_j (\sqrt{M_{ij}} \cdot \sqrt{M_{jk}})$$

The data matrix may be decomposed into a product:

$$M_{ik} = \begin{vmatrix} 1 & 0 & 1 & 0 & 1 & 0 & 0 & 0 & 0 \\ 1 & 0 & 1 & 0 & 0 & 1 & 0 & 0 & 0 \\ 1 & 1 & 0 & 1 & 0 & 0 & 1 & 0 & 0 \\ 1 & 1 & 0 & 1 & 0 & 0 & 0 & 1 & 0 \\ 1 & 1 & 0 & 0 & 0 & 0 & 0 & 0 & 1 \end{vmatrix} \times \begin{bmatrix} A \\ B \\ C \\ D \\ E \\ E \\ F \\ F* \\ G \end{bmatrix} \qquad (2)$$

of which the first matrix consists of the structural or connectivity information inherent to the tree structure, and the second is a column vector of the lifetimes of the cells which make up the tree in Fig. 1. If the rows and columns of the first matrix are labelled by the names of the corresponding cells (see Fig. 1), then the data entries in the matrix

	q	qp	qa	qpa	qaa	qap	qpaa	qpap	qpp
qaa	1	0	1	0	1	0	0	0	0
qap	1	0	1	0	0	1	0	0	0
qpaa	1	1	0	1	0	0	1	0	0
qpap	1	1	0	1	0	0	0	1	0
qpp	1	1	0	0	0	0	0	0	1

are given by the rule: if the name of the column is a substring of the name of the row then the entry is one, otherwise it is zero. Hence, given the words (names of all cells), the connectivity matrix can be constructed by the above law. All words constitute the language L of the nematode. The vector of lifetimes in equation (2) is actually a function f which assigns each word (nucleus) a positive real value (its lifetime T^{life}):

$$T^{life} = f(L).$$

Knowing the language L and the function f, the distance matrix D_{ik} may be constructed. For *C. elegans* the function is given as a table assigning the lifetime to each nucleus (word).

The words of a language have been determined by experimental observation of cell divisions in the eggs of the nematode (Sulston et al. 1983). This words of the language contain additional information on the spatial direction of the nuclear division defined as anterior (a) and posterior (p) nucleus. This spatial information is not contained in the connectivity matrix. Hence, instead of the letters a and p the names may be given by a capital and small letter. Such a language can be used to include e.g. the lifetimes or cell volumes in the name (in Fig. 1 e.g. qpaa may be replaced by AbDF).

The cell lineage tree is a binary tree, since each nucleus produces only two daughter nuclei. A computer language which is mainly based on binary trees is LISP. In LISP, the subtree of Fig. 1 is represented as a list σ = ((qaa.qap).(qpaa.qpap).qpp). The derivation (differentiation) of the list σ gives the structure of the tree. In this case, the terminal cells (qaa to qpp) constitute the elements of the list. The list σ is a linear representation of the tree (e.g. as information is coded on DNA) from which the words of the language L can be retrieved. The terminal cells (and depending on the representation the root cell) form the boundary of the tree (Desoer and Kuh, 1969) and hence of all names of cells in the lineage tree. From the language comprising the names of the terminal cells only the tree structure can be reconstructed.

Cell Lineage as a Dendrogram

Hierarchical clustering on a distance matrix may be used to classify individuals by a tree structure like a family tree in which the relationship between the ancestors is diagrammed. In Fig. 1, the ancestry of the cells is shown in a dendrogram. The dendrogram of Fig. 1 is constructed from the distance matrix D_{ik} by the method of hierarchical classification (Bock, 1974), which is named "complete linkage". In the method of complete linkage, one always combines (unites) the classes (sets) of individuals (cells) which have the minimal distance. However, the new distances of the combined (united) classes to other classes is obtained as the maximal distance of individuals between the two classes under consideration. The algorithm applied to the example of Fig. 1 shows that, of

$$\overline{\alpha} = 382 \quad \overline{\beta} = 319 \quad \overline{\gamma} = 305 \quad \overline{\delta} = 205$$

$\overline{\delta}$ is the smallest value. Thus, the cells qpaa and qpap will be grouped into a class giving the new distance matrix (3) obtained from the matrix (1).

D^2_{ik}	qaa	qap	(qpaa,qpap)	qpp
qaa	0	W	α	α
qap		0	α	α
(qpaa,qpap)			0	β
qpp				0

(3)

The values in the column of (qpaa, qpap) are the maximal values in a row for the column qpaa and qpap of matrix (1). The branching is indicated in the dendrogram at the level of δ (see Fig. 1). Now, W is the smallest value. Hence, qaa and qap form a new class. Then β has become the smallest value. Hence, (qpaa, qpap) and qpp should be combined and finally (qaa, qap) and (qpaa qpap, qpp) are grouped together. In this way, the dendrogram of the data matrix M_{ik} is constructed using the distance matrix D_{ik}.

Cell Volume as Baseline in the dendrogram

Sofar, the baseline of Fig. 1 has not been scaled. However, if the
volume of a cell is centered as an interval around its branch in the dendrogram then the intervals on the baseline give the division of total egg
volume into volumes of individual cells. Investigations of the cell cycle
show that as soon as cells reach a minimal volume they stop dividing.
Assuming, all terminal cells in the dendrogram have the same minimal
volume one may calculate volumes for the intermediate cells. Surprisingly
this calculated volumes agree roughly with the few experimental data
(Schierenberg, 1978) available. Such a dendrogram represents at each time
the division of the constant egg volume into the mosaic of volumes of
cells. Since the cell cycle needs a minimal time to complete a division
of a cell, branches in the dendrogram are longer than this minimal time
interval. Experimental data (Schierenberg, 1978) suggest that the minimal
time interval increases with decreasing volume of the cell until it moves
towards "infinity" for the minimal volume.

To get the mosaic of cells in the egg a similar procedure may be
applied in which a distance measure generates the mosaic. From the mosaic
of cells their volume for the dendrogram can be computed. Since in the
three dimensional space of the egg the procedure is of large scale here a
two dimensional example on a fungus is given. In Fig. 2 a two dimensional
cellular pattern is generated on the agar surface of a petri dish.

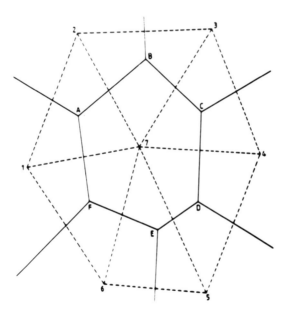

Fig. 2   Sketch of growing fungus *Fusarium solani*. At the center of the
         concentric ring pattern fungal growth was initiated by its
         spores. Each day a ring of the outward growing fungal mycelium
         is issued. If the mycelia moving outward from different center
         meet each other growth is stopped and a border line appear at the
         place of encounter. The border lines (----) form a cellular
         pattern within the petri dish of the experiment.

The boundaries which limits a cellular area may be derived from a distance measure and the structure of the pattern may be obtained from the matrix representation of boundary operators in a similar way as described above (Das and Busse, to be published). If the edge of the petri dish is the analog of the egg shell than the relation of the pattern and the mosaic of cells in the egg is obvious. The cellular network may be formulated by boundary operators. The spatial pattern of cells in the egg will not only supply the volume of the cells for the dendrogram but likely will reveal additional restriction on the language L of names of cells.

Order of the Cell Lineage

The arrangement of the cells at the bottom of the dendrogram (Fig. 1) is not fixed by the hierarchical cluster methods. However, an alphabetical order like in dictionaries may be used to order the cells by their names.

The above lexical procedure ranks the terminal cells in a specific order called total order. This order is displayed in the dendrogram of Fig. 1 but it cannot be derived from the distance matrix D_{ik}. The additional information comes from the axis of cell division in the developing nematode. The coordinate system of the organism defined by its anterior-posterior, left-right and dorsal-ventral axis (Sulston et al., 1983). The letters composing a word display the history in this coordinate system in that they abbreviate the axis of successive division.

DISCUSSION

The aim of this contribution is to show that distances relate cell lineage trees to data of individual cells and in particular to properties of their cell cycle. Conversely, a distance may be used to construct cell lineage trees from data of the individual cell cycles. Thus, as more data (e.g. duration time of the S or G phases) become available, they may be projected onto the dendrogram of the cell lineage and e.g. effects in timing of the cell lineage may be studied. Also the temperature dependence of the length of the cell cycle and its phases (e.g. S, G or M) may be reflected in the dendrograms in which the nuclear division as tree branch point has been replaced by a different criterion, e.g. a phase of the cell cycle.

Since algorithms and programs are available to compute the necessary transformations, this approach is especially useful to investigate models of biological development. In this paper, a part of the cell lineage of the nematode *C. elegans* is treated as an example. However, with a laboratory program package (NIH, 1984), the data matrix M_{ik} can be deduced from the entire cell lineage (Sulston et al., 1983) and the dendrogram of the total cell lineage can also be redrawn from data matrix M_{ik}. This modelling package of programs is well suited to study the influence of assumptions on the dendrogram.

The method of complete linkage does not take into account processes of reunion (or redistribution) of individual cells, e.g. the programmed cell death in the cell lineage. It is believed that the dead cell is phagocytized by its neighbor (e.g. by its sister cell). Since the dendrograms are with usual programs displayed, dead cells are depicted in

190

the dendrogram as if they had stayed alive. Thereby information about the time of death is omitted. It is a characteristic of methods of hierarchical clustering (e.g. a method of complete linkage) that as the time (or index) elapses, the classification is progressively refined. However, a different additional classification of the cells, e.g. in muscle cells, intestinal cells, cells which have died, etc. may provide an adequate description of differentiation. It seems to be more difficult but feasible perhaps to derive a reasonable distance measure which reproduces this classification, since spatial interactions and relations of the cells may likely determine the fate of a cell in the process of differentiation.

The method of complete linkage does not reproduce the complete ancestry tree of the cells of the nematode as it is given in Sulston et. al. (1983). The sequence of the cells at the leaves of the tree from the left to the right may be altered without affecting the complete linkage. The additional information needed is the direction of the separation of the dividing nuclei. The data matrix M_{ik} does not contain this information and therefore, the complete linkage method does not use it. The direction of the dividing nuclei may be represented in a similar way as entries in a matrix of the same morphology as M_{ik}, because nuclear division is an unique event in each cell cycle. The rank order of the cells is obtained from this matrix. It represents the original order of leaves of Sulston's dendrogram.

The example illustrate how distance measures may be used to describe developmental processes in biology. Here, the dendrogram of the cell linage is treated and transformed by the distances to the biologically relevant processes, the cell cycle and the nuclear division. However, there are many more processes in early embryogenesis which require a classification of biological material, e.g. into different cells. Also in such cases, distances might help to analyse the process. Since a distance is a mathematically simple construction, it may be more useful than many other more complex instruments for the description of biological processes. The applied distance measure does not uniquely reproduce the dendrogram. Equivalent methods of linkage may exist which reproduce the same dendrogram by a differently referenced distance measure. The above method has the advantage that its steps are simple and may be interpreted biologically, i.e. the data matrix is related to the events in cell cycles. Distances are often considered as a measure of similarity between objects. Here, similarity is not emphasized, since the biological significance of creating an order in the set of objects by classification seems more obvious.

The paper presents a general method to construct the cell lineage tree given the function which assigns to each cell its lifetime (or corresponding phase of its cell cycle). Cells are identified by names. Each name is a word of the language L of cell names. The method by which the data matrix is derived from words of the language and their corresponding lifetimes is independent of organism, as is the calculation of the distance matrix and the dendrogram from the data matrix. The specificity of the organism is inherent in the specific words (i.e. language) of cells which make up the tissue and in the function assigning lifetimes to each word of the language.

This function is here a table in which one column contains the cell names and the second the corresponding lifetime (e.g. the cell cycle data). Such a table may be used as a data base in calculation and facilitates computation with matrix calculus. Many symmetries within the tree can be used to reduce the actual size of the data base. A more systematic way of recognizing symmetries in the data should include the spatial relation of the cells. The spatial order of cells is already used in their names (e.g. anterior, posterior) to distinguish between daughter cells. Besides the ancestry, the language L provides also spatial information.

ACKNOWLEDGEMENT

For critical reading of the manuscript and many discussions of the topic we should like to thank Dr. B. Havsteen and J. Schumann. Dr. E. Schierenberg should be acknowledged for his support in the field of nematode embryology.

REMARK

The database for the dendrogram of *C. elegans* and programs are available with the authors upon request.

REFERENCES

Bock, H.H., 1974, "Automatische Klassifikation", Vandenhoek & Ruprecht, Göttingen.

Brenner, S., 1974, The genetics of *C. elegans*, Genetics, 77:71.

Chitwood, B.G., Chitwood, M.B., 1974,"Introduction to nematology", Univ. Park Press, Baltimore.

Deppe, U., Schierenberg, E., Cole, T., Krieg, C., Schmitt, D., Yoder, B., v. Ehrenstein, G., 1978, Cell lineage of the embryo of the nematode *C. elegans*, Proc. Nat. Acad. Sci. 75:376.

Desoer, C.A., Kuh, E.S., 1969,"Basic Circuit Theory", McGraw-Hill, New York.

Gerisch, G., 1971, Periodische Signale steuern die Musterbildung in Zellverbänden, Naturwiss., 58:430.

Kimble, J., Hirsh, D., 1979, The postembryonic cell lineages of the hermaphrodite and male gonads in *C. elegans*, Develop. Biol. 70:396.

NIH, 1984, (Modeling Laboratory, MLAB), Reference Manual, Div. Comp. Res., National Institutes of Health, Bethesda, Maryland.

Schierenberg, E., 1978, "Die Embryonalentwicklung des Nematoden *C. elegans* als Modell: computerunterstützte licht- und elektronenmikroskopische Untersuchungen am Wildtyp und an Genmutanten", Thesis Univ., Göttingen.

Sulston, J.E., Horvitz, H.R., 1977, Post-embryonic cell lineage of the nematode *C. elegans*, Develop. Biol. 56:110.

Sulston, J.E., Schierenberg, E., White, J.G., Thomson, J.N., 1983, The embryonic cell lineage of the nematode *C. elegans*, Develop. Biol. 100:64.

Wood, W.B., 1988, The nematode *C. elegans*, Editor, Cold Spring Harbor Lab., Monograph 17.

Section IV
Evolutionary Dynamics and Artificial Life

The mechanical world view will be swept away and replaced by the picture of a self-creating world.

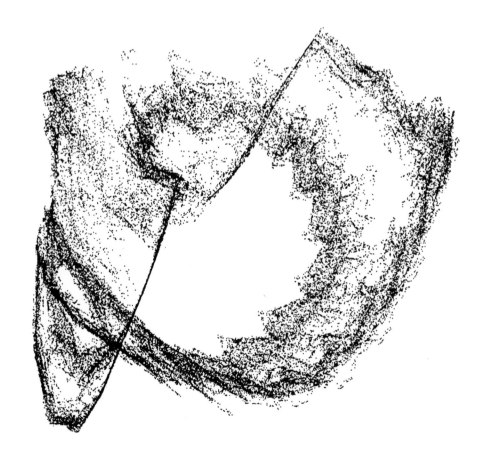

MUTATIONS AND SELECTION IN EVOLUTIONARY PROCESSES

Werner Ebeling

Humboldt University
Department of Physics
1040 Berlin
Federal Republic of Germany

ABSTRACT

The competition between species in a biological system is analyzed in terms of linear and nonlinear models. The significance of stochastic phenomena in evolutionary processes is emphasized. A qualitative model of evolution as a hopping process on the eigenvalue ladder is developed.

INTRODUCTION

The complexity of our world, including the system of biological species, the ecological communities and the human societies, may be viewed as the outcome of an evolutionary process that started some 20 billion years ago with the creation of the universe. As we understand it today, evolution is a historical, irreversible process which is based on a series of unstable transitions. There is no external program which controls this process, but it may be depicted as a long and practically unlimited chain of self-organizing cycles (Ebeling and Feistel 1982). In the biological realm, the cycle of self-organization has the following principal structure:

(i)   A stationary distribution around a master species (or a set of master species) is established.
(ii) A new species is introduced in one or a few exemplars.
(iii) The new species is valuated with respect to the established distribution. A "bad" new species is rejected; a "good" one is amplified.

Complexity, Chaos, and Biological Evolution, Edited by E. Mosekilde and
L. Mosekilde, Plenum Press, New York, 1991

(iv) Through an unstable transition, the system reaches a new stationary stage, and another cycle may begin.

In the next sections, the following problems are investigated: At first, general models of competition processes are developed. Then, a qualitative model of evolution as a hopping process on the eigenvalue is developed, and, finally, we study the process of infecting an established system with a few exemplars of a new species. The analysis is based on a stochastic model which considers the dynamics of the new species as a birth and death process.

GENERAL MODELS OF EVOLUTIONARY PROCESSES

The most simple model of an evolutionary process is the Fisher-Eigen model which considers a set of competing species $i = 1, 2, \ldots, s$ that have different growth rates

$$E_1, E_2, \ldots, E_i, \ldots, E_s. \tag{1}$$

The dynamics of the relative populations is given by the differential equations (Fisher 1930, Eigen and Schuster 1977)

$$\dot{x}_i = (E_i - k_o(t))\, x_i\, (t). \tag{2}$$

The decay rates are determined by the normalization condition

$$\Sigma\, x_i = 1 \tag{3}$$

which gives

$$k_o\, (t) = \Sigma\, E_i\, x_i = <E>. \tag{4}$$

The resulting set of dynamical equations

$$\dot{x}_i = (E_i - <E>)\, x_i \tag{5}$$

shows that species with values better than the social average $<E>$ will succeed in the competition, while the others will fail. At the end, only the species with the largest rate E_m will survive

$$E_m > E_i\,;\, i = 1, 2, \ldots, s\,;\, i = m \tag{6}$$

As already mentioned, this model is the simplest of all models of competition. It refers to an oversimplified case, and in some sense one can say even that the model reflects only pseudo-competition, since there is no real interaction between the species. In this respect, the following model developed in our earlier work (Ebeling and Feistel 1974, 1977, Ebeling 1976, Feistel and Ebeling 1989) is much more realistic.

We consider the case that the species compete for a common resource x_o, which flows with constant rate Φ_o into the system. The following dynamics is assumed

$$\dot{x}_o = \Phi_o - \Sigma\, k_i\, x_o\, x_i$$

$$\dot{x}_i = (k_i\, x_o - k'_i)\, x_i, \qquad i = 1, \ldots, s \tag{7}$$

Using stability analysis, one can now show that the winner of the competition, i.e. the master species m, is the species with the largest value of the relation of growth and decay rates k_i/k'_i. In other words, the master species has the property

$$k_m/k'_m > k_i/k'_i, \qquad i = 1, \ldots, s \tag{8}$$

More general situations of a competition for common resources were elaborated by Ebeling and Schmelzer (1979) and by Hofbauer and Sigmund (1984).

Let us discuss now briefly the class of stochastic models. Following earlier work (Ebeling and Feistel 1977, Ebeling and Sonntag 1986, Bruckner et al. 1986, Kristensen et al. 1990), we introduce first a rather general model which describes competition as well as several other basic processes such as mutation, exchange and imitation.

In order to have a sufficiently general basis for the description of competition processes, we need a stochastic birth, death and transition model for many interacting species. We assume that elements of different types (species) are present or potentially present in the system. The number of types may be very large or even infinite. Each self-replication or death process changes only the number of a single species

$$N_i \to N_i + 1 \quad \text{or} \quad N_i \to N_i - 1.$$

An exchange process, on the other hand, is accompanied by the change of two occupation numbers:

$$N_i \to N_i + 1 \quad \text{and} \quad N_j \to N_j - 1.$$

Higher-order processes in which more than two complementary occupation numbers change simultaneously will not be considered. In more highly developed systems, nonlinear growth rates may appear as a result of interactions between the various species. This gives rise to cooperative phenomena. To describe such nonlinear growth processes, either a deterministic or a stochastic approach may be applied.

However, the correct description of the initial phases of innovative instabilities is possible only on the basis of a stochastic model. This is because the innovation leading to a new field n is always a zero-to-one transition

$$N_n = 0 \rightarrow N_n = 1 \, ,$$

and such a birth process is strongly influenced by stochastic effects.

A generalization of the Fisher-Eigen equation that includes external sources and mutations reads:

$$\dot{x}_i = (A_i - D_i) \, x_i + \Sigma (A_{ij} \, x_j - A_{ji} \, x_i) + \Phi_i \, . \tag{9}$$

Here, x_i is the density of species i. A_i and D_i are the corresponding linear birth and death rate constants, and A_{ij} is the transition rate constant between field j and field i. Φ_i expresses the inflow of elements i into the system.

In order to describe higher-order effects and cross-catalytic phenomena, we generalize eq. (9) to include nonlinear contributions to the growth, decline and transition rates. This gives

$$\dot{x}_i = (A_i^{(0)} - D_i^{(0)}) \, x_i + (A_i^{(1)} - D_i^{(1)}) \, x_i^2 - D_i^{(2)} \, x_i^3$$

$$+ \Sigma \, (M_{ij} \, x_j + B_{ij} \, x_i \, x_j - C_{ij} \, x_i \, x_j)$$

$$+ \Sigma \, (A_{ij}^{(0)} + A_{ij}^{(1)} \, x_j) \, x_i - (A_{ji}^{(0)} + A_{ji}^{(1)} \, x_i) \, x_j + \Phi_i \tag{10}$$

The structure of the coupling between the species may be represented by graph theoretical concepts (Ebeling and Sonntag 1986, Rasmussen et al. 1989). These graphs (or the corresponding matrices) reflect the physics, biology, ecology or sociology of the particular problem. For the moment, we assume that the various coefficients and matrix elements are known or can be empirically determined. Let us note, however, that estimation of the coefficients of growth, decline or transition for particular problems can be very difficult (Bruckner et al. 1989).

In contrast to the previous models, the winner of the competition defined by (10) is not uniquely defined and depends in general on the initial conditions.

Another generalization of eq. (9) takes into account that the reproduction and death rates depend on the age of the individuals belonging to species i:

$$A_i = A_i \, (\tau) \, , \, D_i = D_i \, (\tau) \, .$$

Under the condition of constant overall number (3), the selection theory developed in our earlier work (Ebeling, Engel and Mazenko 1988, Feistel and Ebeling 1989) leads to the following eigenvalue problem:

$$\int_0^\infty d\tau \, A_i(\tau) \exp\left[-\int d\xi D_i(\xi) - \lambda\tau\right] = 1 \tag{11}$$

The winner of the competition is now the species with the maximal eigenvalue. As an example, let us consider the quasi-realistic rate functions:

$$A_i(\tau) = A_i^{m+1}\tau^m \exp(-a_i\tau)$$

$$D_i(\tau) = D_i \text{ if } \tau < \tau_0$$

$$D_i(\tau) = \infty \text{ if } \tau > \tau_0$$

Then the eigenvalue is estimated to be

$$\lambda = A_i - D_i - a_i \text{ if } a_i\tau >> 1$$

and in the general case follows

$$1 = A_i^{m+1}\left(-\frac{d}{da_i}\right)^m \left\{(a_i + D_i + \lambda_i)^{-1}\left[1 - \exp\{(-Ca_i + D_i + \lambda_i)\tau_0\}\right]\right\} \tag{12}$$

The winner of this competition is in general uniquely defined by the solution to (12). In most cases, early reproduction, i.e. small a, is of advantage in the competition.

In the stochastic description, instead of real numbers $x_i > 0$, integers $N_i > 0$ are used to represent the various types of elements. A given N_i means that the field of type i is occupied by a population of N_i representatives. The occupation numbers N_i are functions of time. In contrast to the smooth variation of $x_i(t)$, the dynamics of $N_i(t)$ is a discrete, stochastic hopping process. We may associate $x_i(t)$ with the ensemble averages of $N_i(t)$, i.e.

$$x_i(t) = <N_i(t)> \tag{13}$$

where the average is performed over a large number of identical stochastic systems. The complete set of occupation numbers $N_1, N_2, ..., N_s$ determines the state of the system at a given time. Because of the large number of potential types of elements in typical evolutionary systems, most of the occupation numbers are zero (Ebeling and Sonntag 1986). The probability that the system at time t is in a particular state may be described by the distribution function

$$P(N_1, N_2, .., N_i, ... N_s, t) .$$

We shall formulate now an equation of motion for this distribution function, considering the four fundamental processes of evolutionary behavior: self-reproduction, death (decline), transition between types, and input from external sources or spontaneous generation corresponding to (10). In self-reproduction, the species are assumed to produce elements of their own type, i.e., both identical self-reproduction and error reproduction occur. The failure rate is assumed to be controlled by mechanisms for error repression (quality control) and thus to remain small. For identical self-replication of an element of type i, the transition probability is assumed to be given by

$$W(N_i + 1 | N_i) = A_i^{(0)} N_i + A_i^{(1)} N_i N_j + B_{ij} N_j \tag{14}$$

where $A_i^{(0)}$ is the coefficient of linear self-reproduction, $A_i^{(1)}$ measures self-amplification (second-order self-reproduction), and B_{ij} measures sponsoring or catalytic assistance from other types of elements. Error reproduction of element i is assumed to be described by the linear relation

$$W(N_i + 1, N_j | N_i, N_j) = M_{ij} N_j \tag{15}$$

where the coefficient M_{ij} measures the probability that, through mutation, an element i is produced from an element j.

Elements belong in an active manner to their species only for a limited time. They may die or they may be forced to leave the system. This is expressed by the probability

$$W(N_i - 1 | N_i) = D_i^{(0)} N_i + D_i^{(1)} N_i^2 + D_i^{(2)} N_i^3 + C_{ij} N_i N_j \tag{16}$$

where $D_i^{(0)}$ measures the spontaneous death rate, $D_i^{(1)}$ and $D_i^{(2)}$ express nonlinear decay processes (self-inhibition), and C_{ij} measures the suppression of elements i by elements j.

The species are assumed to exchange elements. This especially may be connected with competition and selection and is assumed to be expressed by the probability

$$W(N_i + 1, N_j - 1 | N_i, N_j) = A_{ij}^{(0)} N_j + A_{ij}^{(1)} N_i N_j \tag{17}$$

Here, the coefficients $A_{ij}^{(0)}$ and $A_{ij}^{(1)}$ represent noncooperative and cooperative exchange, respectively. Finally, the change in occupation of a species i by inflow from the outside or by spontaneous generation is assumed to have a constant probability

$$W(N_i + 1 | N_i) = \Phi_i . \tag{18}$$

All coefficients in eqs. (14-18) have non-negative values. This guarantees that the system will not be absorbed by the empty state with all $N_i = 0$. At least one element of the inflow rate must be different from zero. Averaging the stochastic process over an ensemble of systems, eq. (11) follows from eqs. (14-18) in the limit of high occupation numbers. It should be stressed, however, that the deterministic equations remain true only in average.

Unlike the deterministic case, it is possible in the stochastic realm to pass over the barrier separating two different stable equilibria (Feistel and Ebeling 1989). The linear and nonlinear transition rates that occur in the deterministic as well as in the stochastic description give rise to a variety of complex dynamic behaviors, such as multistability, limit cycles and deterministic chaos (Rasmussen et al. 1989). The nonlinear growth terms describing self-amplification and self-inhibition typically lead to the existence of thresholds (or separatrices in state space). In the presence of such thresholds, the evolution of a species depends critically on the initial conditions, and multistability may arise (Mosekilde, Aracil and Allen 1988).

Self-amplification and self-inhibition are observed, for instance, in chemical and ecological systems as higher-order autocatalytic reactions and as self-obstruction. Interactions between species may result from cross-catalytic reactions in chemical systems, from mutual obstruction or promotion between populations in ecological systems, or from competition and imitation in social systems. The cross-catalytic interactions include sponsoring and suppression processes, modelled by the terms B_{ij} and C_{ij} where fields indirectly act on the reproduction rates of other fields.

EVOLUTION AS HOPPING ON THE LADDER OF EIGENVALUES

Most of the competition processes discussed so far are connected with the solution of linear eigenvalue equations. For example, let us come back to the generalized Fisher-Eigen equation which is closely connected with the linear problem

$$\left[(E_i - \sum_k A_{ik})\delta_{ij} + A_{ij}\right]x_i^{(n)} = \lambda_n x_i^{(n)} \tag{19}$$

The spectrum has in general the typical form shown in Figure 1. The high eigenvalues are well separated and correspond to a distribution of quasispecies which is well localized around certain master species. The lower eigenstates form a quasicontinuum which converges for $S \to \infty$ to a proper continuum. The corresponding eigenstates are extended and do not show a distribution peaked around one species. The solution may be written in the form

$$x_i(t) = \Sigma c_n \, exp(\lambda_n t - \langle E \rangle t) \, x_i^{(n)}. \tag{20}$$

Evolution corresponds to a hopping process on the ladder of eigenvalues. At the beginning, the average $\langle E \rangle$ will be located far left, i.e. in the region of the quasicontinuum; no species formation will be observed. After crossing the border to the localized eigenstates, the formation of well-separated quasispecies will be observed.

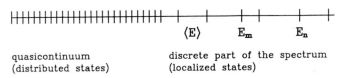

quasicontinuum   discrete part of the spectrum
(distributed states)   (localized states)

Figure 1. The ladder of eigenvalues.

Let us calculate now the average time which is necessary for the hopping from the eigenvalue λ_m corresponding to a master species which dominates at $t = 0$ to a new species n with the larger eigenvalue λ_n. Let us assume that the overlap at time $t = 0$ to the new species is

$$c_n/c_m = A_{nm} \, exp(- d(n,m)/l_n) \tag{21}$$

where $d(n,m)$ is the "distance" between the two species and l_n the localization length.
The transition will occur after a time

$$c_n \, exp(\lambda_n t) \simeq c_m \, exp(\lambda_m t) \tag{22}$$

$$t \simeq \frac{d(n,m)}{l_n(\lambda_n - \lambda_m)}$$

Assuming now that the species space has the quasidimension d and that the localization centers are uniformly distributed with the density $\rho(\lambda)$, we get in average

$$t(n,m) \simeq \frac{1}{l_n(\lambda_n - \lambda_m)} \left[\frac{V_d}{\rho(\lambda_n)} \right]^{\frac{1}{d}} \tag{23}$$

Here, V_d is the volume of the d-dimensional unit sphere. Since $t(n,m)$ has a pole at $n = m$ and another at $n \to \infty$, which arises from the disappearance of the eigenvalue density, there exists a minimum. As illustrated in Figure 2, this minimum corresponds to a most probable hop $\delta\lambda$, the so-called quantum of evolution and a well-defined time δt .

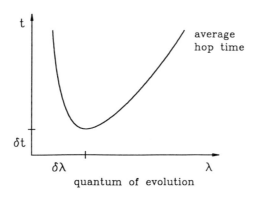

Figure 2. The average time necessary to make a finite step on the ladder of eigenvalues.

In this way, the velocity of evolution on the ladder of eigenvalues is defined as

$$V = \delta\lambda/\delta t \qquad\qquad (24)$$

The picture of evolution described here was first developed for a continuous model based on the Wrightian landscape and diffusion-like dynamics (Ebeling et al. 1984, Feistel and Ebeling 1989).

THE FATE OF NEW SPECIES

If a new species appears first by error reproduction or by exchange processes, in the framework of deterministic models only the value of E_n decides whether the system is stable or unstable with respect to the innovation. In case that E_n is larger than any of the E_i for the existing fields, the new field will outgrow the others, and the system is unstable with respect to the random generation of an element of field n. Asymptotically, the field with the largest selection value will be the "winner." Because of the unlimited possibilities for creation of new fields with higher selection values, the evolutionary process goes on, and the selection values in the system continue to increase. Such a situation is referred to as a simple selection process. With reference to the Darwinian principle of the survival of the fittest, the parameter E_i serves as a quantitative expression of the qualitative property of "fitness." The selection value concept thus includes a qualitative component which is related to the stability properties of the system.

When we consider processes that include exchange between the species i and j combined with nonlinear growth functions of self-reproduction and decline, calculation of the competition properties of an innovation becomes an extremely complicated mathematical problem which requires a detailed stability analysis of the deterministic problem (Prigogine, Nicolis and Babloyantz 1972). A new field is of higher selection value with respect to the existing occupation if the deterministic system is unstable with respect to a corresponding perturbation. In analogy to Eigen's concept, one can say that the stability properties represent the selection value of a new element appearing by a transition process, by inflow or by error reproduction.

We have already noted that cooperative processes, because of their property of mutual directedness (expressed by nonlinear terms), may significantly influence the result of selection processes. Further, we have shown that approval of a new element in the game of evolution is connected with a test of the deterministic system for stability. Let us assume that at a given time t the system consists of s different types in stationary equilibrium $x_1, x_2, \ldots, x_s$. Infecting the system with a new element N we get the linearized problem

$$\dot{x}_i = \Sigma \, a_{ij} \, x_j + a_{in} \, x_n \; , \quad i = 1, \ldots, s$$

$$\dot{x}_n = \Sigma \, a_{nj} \, x_j + a_{nn} \, x_n$$

This new problem has $(s + 1)$ eigenvalues $\lambda_1, \lambda_2 \ldots \lambda_s, \lambda_{s+1}$. If one of these eigenvalues has a positive real part, the system is unstable with respect to the new species. If the real part of one of the eigenvalues is zero, a more detailed analysis is necessary. However, if the real part of all $(s + 1)$ eigenvalues is negative, the system will definitely refuse the "infection" with the new element.

In this way, the question of selection value of a new element with respect to the already existing population becomes a well-posed problem. Mathematically, it reduces to an eigenvalues problem of rank $(s + 1)$ (Hofbauer and Sigmund 1984). In systems with nonlinear interactions of the type considered here, the selection process no longer depends only on the selection value of a new field but also on the existing configuration. We find multistability where the behavior of the system depends on the initial conditions in such a way that a new "better" field is a potential, but under the given conditions not a real possibility for the system. Such a situation (also called hyperselection) can describe the selection between two or more equivalent possibilities for the system (Eigen and Schuster 1977, 1978). On the other hand, the evolution in a system occupied by stable configurations like hypercycles is stopped forever (once-and-forever selection). This result of the deterministic analysis, which is clearly in contradiction to many observations of real systems, underlines the necessity of introducing a more complete stochastic description.

In the stochastic picture, the deterministic conclusions remain true only in average. By analyzing the survival probability of new mutants (which is analogous to a stability analysis in the deterministic case) and also by computer experiments, we can show that the deterministic order can be destroyed by fluctuations. Especially, the sharp border between unfavorable and favorable new mutants is broadened (Ebeling and Feistel 1982, Ebeling and Sonntag 1985, Allen and Ebeling 1983). Separatrices, which separate different parts in the state space from each other, can be crossed. In this way, new channels of evolution are opened by fluctuations. The fluctuations become more significant in the case of small occupation numbers, i.e. when a new species is created or when an existing species dies out. We shall give now several concrete results for the dynamics of new species.

Let us assume that a system is occupied by N master species with the reproduction rate E_m corresponding to a linear or quadratic growth law and that we infect this system with a few exemplars of a species having the rate E_n. We further assume that the total number of both species is constant $N = N_m + N_n$ and that the dynamics is described by

$$\dot{N}_k = E_k N_k^e - k_o N_k \;, \quad k = n, m; \quad e = 1, 2 \tag{25}$$

$$k_o = N^{-1}[E_n N_n^e + E_m N_m^e] \;. \tag{26}$$

Due to this deterministic dynamics, in the linear case $e = 1$ the new species will survive and outgrow the master if the relative advantage is larger than one

$$\alpha = E_n/E_m > 1 \tag{27}$$

In the case of quadratic growth $e = 2$, the fate of the new species depends on the initial conditions. The great interest in quadratic growth laws comes from the fact that the hypercycles of the Eigen-Schuster theory are of this type. If the initial number of infectants is low, i.e.

$$N_n(0) < \alpha (\alpha + 1) \tag{28}$$

then the new species has no chance to survive in a deterministic world described by eq. (25). In a stochastic dynamics, which is based on integer particle numbers

$$N_n = 0, 1, 2, ... ,$$

the picture changes completely. We here describe the growth of the new species by the master equation

$$(\partial/\partial t)P(N_n,t) = W^+(N_n - 1)\, P(N_n - 1) + W^-(N_n + 1)\, P(N_n + 1)$$
$$- [W^+(N_n) + W^-(N_n)]\, P(N_n) \tag{29}$$

where $P(N_n, t)$ is the probability to find N_n exemplars of the new species at the time t,

$$W^+(N_n) = E_n\, N_n^e\, (N - N_n) \tag{30a}$$

and

$$W^-(N_n) = E_m\, N_n\, (N - N_n)^e \tag{30b}$$

There are two absorber states $N_n = 0$ and $N_n = N$, and the final state will have the stationary distribution (Feistel and Ebeling 1989)

$$P(N_n, \infty) = \sigma\, \delta\, (N, N_n) + (1 - \sigma)\, \delta(0, N_n) \tag{31}$$

where σ is the probability of survival of the new. Due to the special structure of eq. (31), we need only one invariant of eq. (29), i.e. one quantity which satisfies

$$(d/dt) < h\,(N_n) > \; = 0 \tag{32}$$

in order to get σ. We find for the linear case $(e = 1)$

$$\sigma(\alpha, N) = \left[\frac{1 - \alpha^{-1}}{1 - \alpha^{-N}} \right]^{N_n(0)} \tag{33}$$

and in the nonlinear case $(e = 2)$

$$\sigma(\alpha, N) = \left[\frac{\alpha}{1 + \alpha} \right]^{N-1} \left[1 + \binom{N}{1}\left(\frac{1}{\alpha}\right) + \binom{N}{N_n(0) - 1}\left(\frac{1}{\alpha}\right)^{N_n(0) - 1} \right] \tag{34}$$

The most important result obtained so far is that the probability of survival is a rather smooth function which is continuously increasing with the relative advantage. There is no sharp difference between the better and the worse mutants except in the case $e = 1$, $N = \infty$. In a system with a linear growth law, any improvement (positive or negative) smaller than 10% has no significant effect on the survival. A finite population size improves in all cases the survival probability of the new species. In the case of quadratic growth laws $(e = 2)$, this effect is very essential since it guarantees the survival of mutants in hypercyclic systems with finite population size.

DISCUSSION

Evolution is a complex dynamical process which is connected with competition, mutation and selection processes. This process may be modelled as hopping on a ladder of eigenvalues. The behavior of new participants in the game plays a fundamental role for evolution. This leads to the basic importance of stochastic effects for evolutionary processes. The small initial number of mutants leads, in the case of linear growth, to a less definite selection than expected from the deterministic theory where the distinction between advantage and disadvantage and the decision about the fate of the new species is sharp. Due to stochastic range, we get a region of uncertainty which is about 10% of the relative advantage. The neutrality of selection with respect to a small advantage or disadvantage is presumably related to the empirical fact that species always exist in certain limits of variation and may also be related to the neutral drift phenomenon studied by Kimura. In the case of hyperbolic growth, stochastic effects destroy the once-and-forever selection predicted by the deterministic theory. Better mutants (hypercycles) can win the competition only if the reaction volume is sufficiently small. This may be the case, for instance, in coacervates, pores or absorbed layers. After winning the competition in a small volume, the new mutant may infect the whole system. In this way, the stochastic effects open new channels for the evolutionary process and deeply influence the results of mutation and selection in complex systems.

REFERENCES

Allen, P., Ebeling, W., Evolution and the Stochastic Description of Simple Ecosystems. BioSystems 16:113 (1983).

Bruckner, E., Ebeling, W., Scharnhorst, A., Stochastic Dynamics of Instabilities in Evolutionary Systems. System Dynamics Review 5:176 (1989).

Ebeling, W., Feistel, R., Physik der Selbstorganisation und Evolution. Akademie-Verlag, Berlin (1982, 1986).

Ebeling, W., Schmelzer, J., Koexistenz von Sorten in nichtlinearen autokatalytischen Parallelreaktionen. Z. phys. Chemie 261:31 (1980); studia biophysics 78:31 (1980); 80:53 (1980).

Ebeling, W., Sonntag, I., Stochastic Description of Evolutionary Processes in Underoccupied Systems. BioSystems 19:91 (1986).

Ebeling, W., Feistel, R., Stochastic Theory of Molecular Replication Processes, Ann. Physik 34:91 (1977).

Ebeling, W., Engel, A., Mazenko, V., Modelling of Selection Processes With Age-dependent Birth and Death Rates. BioSystems 19:213 (1986).

Eigen, M., Schuster, P., The hypercycle. Naturwissenschaften 64:541 (1977); 65:341 (1978).

Feistel, R., Ebeling, W., Evolution of Complex Systems. Verlag der Wissenschaften, Berlin (1989), Kluwer Academic Publ., Dordrecht (1989).

Fisher, R.A., The Genetical Theory of Natural Selection. Clarendon Press, Oxford (1930).

Haken, H., Advanced Synergetics. Springer-Verlag, Berlin (1983).

Hofbauer, J., Sigmund, K., Evolutionstheorie und dynamische Systeme. Parey-Verlag Hamburg (1984).

Kristensen, H., Risbo, L., Mosekilde, E., and Engelbrecht, J., Complex Dynamics in Bacterium-Phage Interactions, Proc. SCS European Simulation Multiconference, Erlangen-Nuremberg, West Germany, 636 (1990).

Mosekilde, E., Aracil, J., Allen, P.M., Instabilities and Chaos in Nonlinear Dynamical Systems. System Dynamics Review 4:14 (1988).

Prigogine, I., Nicolis, G., Babloyantz, A., Thermodynamics and Evolution. Physics Today 25:38 (1972).

Rasmussen, S., Mosekilde, E., and Engelbrecht, J., Time of Emergence and Dynamics of Cooperative Gene Networks, Proc. MIDIT Workshop on Structure, Coherence and Chaos, Manchester University Press, eds. P.L. Christiansen and B. Parmentier, 315 (1989).

CONSIDERATIONS OF STABILITY IN MODELS OF PRIMITIVE LIFE :

EFFECT OF ERRORS AND ERROR PROPAGATION

Clas Blomberg

Department of Theoretical Physics
Royal Institute of Technology
S-100 44 STOCKHOLM, Sweden

ABSTRACT

Models of simple self-replicating systems that represent a primitive form of life are investigated with the main aim of finding conditions for the occurrence of stable, self-sustained states. Such conditions are provided by requirements for 'food molecules', i.e. activated monomers, which build up the self-replicating polymers. For this, kinetic equations are used that explicitly contain both growth and degradation terms of all substances including the food molecules which are assumed to be created by some energy-driven process. The conditions for self-sustained states are essentially that the concentration of the food molecules must be large enough to allow growth of the polymers. In particular, we are interested in how these conditions change with the length of the polymers, with the complexity of the system and with the accuracy of the synthesis. For a system of independent self-replicating units, there do not seem to be any severe obstacles in reaching a stable state with errors taken into account even in a primitive system, provided the polymers are not too long. In such systems, a description of molecular evolution is relatively straightforward. On the other hand, in a system with cooperative units, the situation is more complex. A particular situation is studied with polymers built up by an adaptor (tRNA or a precursor of that). Here errors and the occurrence of error propagation are important features that should be properly understood in a description of the first cooperating forms of life.

INTRODUCTION

Mathematical models have been used in the past for the study of various aspects of the evolution of primitive life, and these have to a large extent been inspired by the work by Eigen and Schuster (Eigen, 1971, Eigen and Schuster, 1977). See also Schuster (1988) for a recent development of this approach. The main purpose is to study relatively simple features of possible self-replicating systems, which are driven by some (usually undefined) external energy source. Although such models necessarily are greatly simplified in some biological

Complexity, Chaos, and Biological Evolution, Edited by E. Mosekilde and
L. Mosekilde, Plenum Press, New York, 1991

details, they are important in order to emphasize certain features of evolution, to pinpoint relevant problems in general ideas of the origin of life and to show possibilities to answer arising questions.

The basic aim of this work is to provide a general framework for discussing questions of stability and the possibility to reach stable states of primitive self-replicating systems. The main problem is the following: The building up of polymers that are able to replicate themselves or to cooperate in an autocatalytic network is made by using activated monomers, provided by some steady influx of energy. For the self-replicating system to be able to grow and to reach a substantial size, it is necessary that the monomer concentration is above some threshold value, which leads to a requirement for the establishment of a self-replicating state. The condition may be more demanding for a complex system, a system consisting of large molecules, of many components or a system that makes a lot of errors, i.e. wasted products. The reason is that complex systems require a fairly high monomer concentration, which gives relevant constraints on the path of evolution.

Many models of early evolution do not take these effects into account. This is particularly so with models that use the so-called 'constant overall organisation' of Eigen and Schuster, where it is simply assumed that there is a constant organisation which will not decay (Eigen 1971). In other cases, one may look at growing systems, again without worrying about degradation and the effects of low monomer concentrations. We will here use a formalism that was used previously by our group primarily to consider questions of competition and coexistence in these models (Blomberg, von Heijne and Leimar 1981).

In particular, we will consider the effect of errors. All replicating systems can have errors. The processes of selection in biological systems are relatively well-studied (Blomberg 1983, Blomberg and Liljenström 1988). They are generally based upon differences in (free) energy for the binding of relevant molecules to some selection unit. These differences are limited by thermodynamic constraints. In present-day organisms, the accuracy in for instance DNA or protein synthesis is highly controlled and kept at a very high level where the systems certainly work satisfactorily. In the first primitive systems on their way to 'real life', probably no controlling mechanism existed, and the accuracy could only be moderately high. In such cases, errors might have a devastating influence. This could in particular be so in the case of error propagation, the positive feedback effect due to the fact that the synthesizing enzymes of a cell themselves are products of the processes they control. Thus, if there are errors, the selection enzymes become erroneous, make their task less efficient and lead to more errors, a situation that could blow up into an 'error catastrophe' and breakdown of the system. This was first discussed by Orgel (1963) and later studied through mathematical models by Hoffman (1974) and Kirkwood and coworkers (see Kirkwood, Holliday and Rosenberger 1984) who clearly demonstrated how the situation can be stabilized. Recently, the present author has studied this with a model that should be close to actual replicating processes of a cell (Blomberg 1990). We will here use a modified version of that model. What is seen in that model and what will be discussed here is that the most crucial step for the error propagation is the occurrence of an adaptor such as transferRNA in present-day cells that transfers the monomers that build up the enzymes.

We will here start with a discussion of the most simple processes with independent self-replicating units. Then, the model will be extended to the case with several cooperating units, what Eigen and Schuster refer to as hypercycles (Eigen and Schuster 1977). We will study a general, schematic scheme as well as an explicit scheme with adaptors. Finally, we generalize the models and take possible errors into account.

210

THE BASIC MODEL

We start with the simplest type of self-replicating system with one polymer, N, (perhaps RNA) which is built up by n activated monomers, M, (nucleoside triphosphate). The main reaction scheme is: $N + n \cdot M \rightarrow 2N$, which together with a constant creation rate of M as well as degradation reactions constitute the following kinetic equations which also were discussed by Blomberg et al. (1981):

$$\dot{N} = WMN - qN$$

$$\dot{M} = a - bM - nWMN \tag{1}$$

N represents the self-replicating macromolecule concentration and M the monomers. n is the number of monomers M in the polymer N. W is the self-replicating rate, and q the decay rate for the macromolecule. The monomer is assumed to be synthesized by the constant rate a which involves some energy turnover (e.g a photo-chemical process), and b is its decay rate. Without macromolecule production, the steady-state concentration of monomers is a/b. The length n should occur in the synthesis terms as shown in the formula. It could also occur implicitly in the synthesis rate or in the decay rate. The decay rate for a longer polymer should be higher than for a shorter, as there are more bonds that can break. The detailed reactions of polymerization can yield qualitative new features and a more complex dependence on length n. We will, however, not take up such questions here but postpone them for later work.

The analysis of the scheme (1) is quite simple. The macromolecules can grow if the concentration of monomers M is larger that q/w, otherwise it will decrease. There is a simple, stable stationary solution of (1) with:

$$M = \frac{q}{W} \quad \text{and} \quad N = \frac{aW - bq}{nqW} \tag{2}$$

Clearly, there is a condition for the existence of this solution:

$$aW > bq \tag{3}$$

This condition arises from the fact that the concentration of M that allows growth of the N must be lower than a/b, the concentration in the absence of macromolecules. The product of production rates should be larger than the product of decay rates. If (3) is not fulfilled, the polymer concentration N is zero.

The scheme (1) can be generalized in several ways. One way to make it more general is to include a term representing the synthesis of the macromolecule without self-replication:

$$\dot{N} = kM + WMN - qN$$

$$\dot{M} = a - bM - nkM - nWMN \tag{4}$$

The steady-state situation is given by the solution of a quadratic equation, and is more complicated than in the above case. One gets:

$$N = \frac{aW - qb - nkq + \sqrt{(aw - qb - nkq)^2 + 4naWkq}}{2nWq} \tag{5}$$

The main difference here compared to the previous case is that (5) always provides a finite N-concentration, while in (2), N should be exactly zero if (3) is not fulfilled. However, if (3) is not fulfilled and k is small compared to other rates, N in (5) will be small, and proportional to k. On the other hand, if (3) is fulfilled, N is close to the value of (2).

The advantage of this scheme is that it allows N to grow from the value zero, i.e. to explain the possible initial production of molecules that then are able to reproduce themselves.

We all the time assume that the monomer M is produced in a stationary non-equilibrium process, such as a photo-synthetic process. For this reason, it is not essential to consider detailed balance in the schemes. The decay rates b, and q are assumed to lead to products with low free energy, and the processs can be assumed to be irreversible. Back reactions by the enzyme N of the type $2N \rightarrow N + nM$ are neglected but can be included.

Coexistence

The possibility of coexistence of several species was discussed in Blomberg et al. (1981) in the framework of the scheme (1). The implications are quite simple. The production of another, competing species might be described by a production equation as the first one in (1):

$$\dot{N}_1 = w_1 N_1 M - q_1 N_1$$

The second equation of (1) for $\dot{M}$ should include a corresponding term. A stationary situation (with zero time derivative) would have a value of the monomer concentration equal to q_1/w_1. Unless this happens to be exactly the same as q/w in (2), N_1 can not coexist with the previous N. The condition for the stable macromolecule is quite simple: *only that one that allows the lowest value of the monomer concentration (i.e. the lowest value of the quotient q/w) will survive.*

The reason for this is quite clear in the model. The macromolecule N grows if the concentration of M is larger that q/k, otherwise it will decay. If a new species N_1 occurs, which can reduce this limit, M will decrease to a value below q/k, and N will decay. N_1 will survive since the corresponding steady state value of M, q_1/w_1, is lower that q/w.

Naturally, this is so only if the species are independent of each other. If they by some process help to build up each other, they may coexist. This is also so if the polymer by mistake also makes other, similar polymers. Mistakes are always possible, and these are the basis of spontaneous mutations. We will come back to this later.

Thus, in this model, evolution favours macromolecules that reduce the monomer concentration as much as possible. Apart from exceptional cases, a stable coexistence is not possible. (For the scheme (4) with a non-self-reproducing possibility, no macromolecules disappear completely. Most of these, however, have very low concentrations in (5), and it is not really meaningful to speak about coexistence in such cases as also in the case of low mutation rates.) *It becomes a general principle that the systems that can grow with lowest monomer concentration are the most stable ones.*

HYPERCYCLE TYPE OF SYSTEMS

The scheme (1) can be extended by including macromolecular cooperation, and we will do this in the following. Before that, however, let us consider a schematic model for a cooperating system with equations similar to the hypercycle equations of Eigen and Schuster (1977). (This model was also treated in Blomberg et al. (1981)). The basic idea is that there is a macromolecular cooperation which leads to a self-reproducing rate proportional to N^2. We will also keep a direct self-replicating possibility as in (1). (Non-self-replicating possibilities are neglected.) We write:

$$\dot{N} = kMN + WMN^2 - qN$$

$$\dot{M} = a - bM - nM(kN + WN^2)$$

$$(6)$$

The terms are similar to those in (1). The analysis is slightly more complicated with the following main results for a stationary solution:

$$M(k+WN) = q \qquad (7)$$

N is given by:

$$N = \frac{Wa-nkq \pm \sqrt{(Wa+nkq)^2 - 4nbq^2W}}{2nWb} \qquad (8)$$

Clearly, it must be required that:

$$Wa + nkq \geq 2q \cdot \sqrt{nbW} \qquad (9)$$

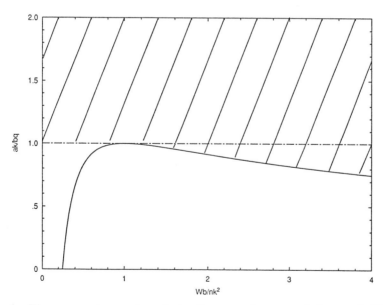

Figure 1. The parameter region that allows stationary solutions of (6) with non-zero N. The curved line represents the limit in (9), and the straight line ak = bq. The shadowed region is the allowed one.

If ak > bq (cf formula (3)), (9) is certainly fulfilled, and only the +sign provides a positive solution for N in (8). (Note that $(Wa+nkq)^2 - 4nbq^2W = (Wa-nkq)^2 + 4Wq(ak-bq)$). This solution is stable, and the result is qualitatively similar to that of equation (1).

On the other hand, if ak < bq, both signs of (8) provide fix-points of (7). Of these, the smaller one with a -sign corresponds to a saddle point, the higher one with a +sign corresponds to a stable state. (This also means the lowest value of M.) There is a separatrix going through the saddle point, and *N cannot grow to the solution (8) from an initial value below the separatrix.* This is analogous to the conclusion by Eigen and Schuster from a similar type of equation for a hypercycle.

One can here also investigate the possibility of coexistence. In this case, for certain parameter values, one gets a stationary solution corresponding to two coexistent species. Such solutions are, however, not stable. They yield stationary M-values that are larger than would be obtained for the pure species. Then, for a small deviation from that solution, at least one species can grow by decreasing the monomer concentration, and the other one decays.

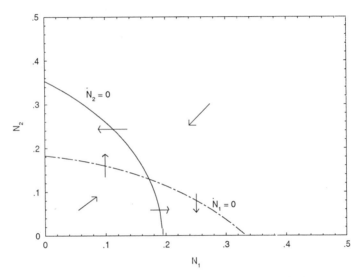

Figure 2. The flow pattern and stationary points (marked) for two competing hypercycles, described by equation (6). N_1 and N_2 represent the polymers, and the diagram shows the curves with the time derivatives of the respective N_i equal to zero and with M-values that make the time derivative of M equal to zero.
Values of the parameters are: (These are chosen arbitrarily) $ak_1/qb = ak_2/qb = 2$, $aW_1/k_1b = 3$, $aW_2/k_1b = 4$, $nq/a = 2$.

Two macromolecules and adaptors

An important example of a cooperating system is the adaptor system, where the syntheses of polymers are made via adaptor molecules, T, instead of directly using the monomer M. The adaptor is synthesized from the monomer via an enzyme N_p. There is also a set of basic reproduction molecules N_1, which build up the adaptor synthesizing enzymes as well as themselves by use of the adaptors and some coding instruction which we do not take into account here. In a present day cell, N_1 represents ribosomes, N_p, amino acyl tRNA synthetases, and the adaptors amino acylated tRNA. The scheme is suitable for describing the effect of errors and error propagation. It does not contain the translation of a genetic code nor the reproduction of code molecules. These can easily be included in the present framework, but for our present purpose, this scheme is sufficient. We have the following set of time evolution equations:

$$\dot{N}_0 = W_0 T N_0 - q_0 N_0$$
$$\dot{N}_p = W_p T N_0 - q_p N_p$$

$$\dot{T} = W_t M N_p - T(n_0 W_0 N_0 + n_p W_p N_0) - q_t T \tag{10}$$

$$\dot{M} = a - bM - W_t M N_p$$

The main notations are as before. The production and decay rates W and q can differ between the macromolecules, the lengths of which are n_0 and n_p respectively. T is the adaptor, produced from the monomer M by the enzyme N_p. As before, M is produced by a constant rate corresponding to some non-equilibrium factor and decays by the rate b.

The underlying scheme can be represented by:

$$N_0 + n_0 T \xrightarrow{W_0} 2N_0$$

$$N_0 + n_p T \xrightarrow{W_p} N_0 + N_p$$

$$M + N_p \xrightarrow{W_t} T + N_p$$

$$N_0 \xrightarrow{q_0}, \quad N_p \xrightarrow{q_p} \text{ decay products}$$

$$\xrightarrow{a} M \xrightarrow{b}, \quad T \xrightarrow{q_t} \text{ decay products}$$

The stationary solutions with all time derivatives equal to zero can be obtained in a straightforward way. One gets:

$$T = \frac{q_0}{W_0} \tag{11a}$$

$$N_p = \left(\frac{W_p q_0}{W_0 q_p}\right) N_0 \tag{11b}$$

$$N_0 = \frac{q_0 q_p q_t}{W_p W_t q_0 \cdot M - (n_0 W_0 + n_p W_p) q_0 q_p} \tag{11c}$$

which leads to a second degree equation for M:

$$M^2 - M \left[\frac{a}{b} - \frac{q_0 q_t}{b W_0} + \frac{q_p}{W_p W_t}(n_0 W_0 + n_p W_p) \right] + \frac{aq_p(n_0 W_0 + n_p W_p)}{b W_p W_t} = 0 \tag{12}$$

The condition to get real, positive solutions can be written in the form:

$$\sqrt{a} > \sqrt{\frac{q_0 q_t}{W_0}} + \sqrt{\frac{b q_p}{W_p W_t}} \; (n_0 W_0 + n_p W_p) \qquad (13)$$

The situation is qualitatively similar to that of the hypercycle equation (6) above. As there, only the lowest M-solution yields a stable situation. The condition (13) is of the same type as (9) and means that the production rates a, and the W's shall be large compared to the decay rates, b, and q's. Note that the length of the polymers occurs in (13) as well as in (9): *the stationary state may be less stable if the lengths are large, and other rates remain the same.*

As in the previous cases, this model does not allow two non-interacting systems to coexist in a stable, stationary situation apart from cases with exceptional coincidences of certain parameter values. Also, no new system can grow from low values of the molecule concentrations and replace an old system.

It is relatively straightforward to include an information carrier (nucleic acid) within this formalism which is coupled to the enzyme system with a 'polymerase enzyme' which would replicate the information carrier. The enzyme synthesizing molecules then read the information. With the aspects treated here, they are similar to the hypercycle type of equations. They, however, pose interesting problems about their evolution features which will be considered in a forthcoming paper.

EFFECT OF ERRORS

Now, we include the possibility of errors in our formalism. We will start from the type of system represented by equation (1), i.e. a simple, self-replicating system. We assume that for each inserted monomer, there is a probability p that it is of a proper kind. The error rate, which is used in some formulas below, is $E = 1-p$. Assume a situation that only polymers that have certain 'correct' monomers on m crucial places are able to self-replicate. m need not be equal to the total number of monomers, n. The time evolution of the molecules with correct monomers is:

$$\dot{N}_0 = W \cdot N_0 M \cdot p^m - q N_0 \qquad (14a)$$

The total polymer production (including all wrong ones) is:

$$\dot{N}_T = W \cdot N_0 M - q N_T \qquad (14b)$$

and as before, the equation for the monomers is:

$$\dot{M} = a - bM - nW \cdot N_0 M \qquad (14c)$$

The result for stationary solutions is similar to that of (1), given by (2). We get:

$$M = \frac{q}{W} \cdot p^{-m} \; ; \quad N_0 = \frac{a p^m - bq/W}{nq} \; ; \quad N_T = N_0 \cdot p^{-m} \qquad (15)$$

Here, we have the condition that:

$$a \cdot p^m > \frac{bq}{W} \qquad (16)$$

A relation of this kind, but derived from a slightly different aspect was discussed by Eigen and Schuster (1977).

Clearly, p^m must not be too small. Still, this may rather be a constraint on the lengths of the polymers than on the accuracy. p can well be 0.9 - 0.99 for an organic polymer (but probably not larger). Then m can be allowed to be of the order 10-100, which should allow reasonably large polymers also in a primitive system. However, m could not be allowed to be larger than so.

Active products of erroneous synthesis, superspecies

Clearly, mutations favour polymers which reproduce with few errors. It is of course possible that some of the products N₁ are capable of reproducing themselves with error rates and production rates that differ from those of N₀. This is a kind of 'mutation' situation, and the two molecule species will both occur together with all kind of faulty products. This provides what Eigen refers to as a 'super-species' (Eigen 1971). We consider that situation and assume that polymers that differ from the original ones by one unit (i.e. there is one error), but only these, are capable of self-reproduction (This is sufficient for providing the qualitative behaviour.) We get equations:

$$\dot{N}_0 = W_0 N_0 M \cdot p_0^m + m \cdot W_1 N_1 M \cdot p_1^{m-1}(1-p_1) - qN_0 \tag{17a}$$

$$\dot{N}_1 = m \cdot W_0 N_0 M \cdot p_0^{m-1}(1-p_0) + W_1 N_1 M \cdot p_1^m - qN_1 \tag{17b}$$

$$\dot{N}_T = W_0 N_0 M + W_1 N_1 M - qN_T \tag{17c}$$

$$\dot{M} = a - bM - nM[W_0 N_0 + W_1 N_1] \tag{17d}$$

Index '0' signifies the original polymers ('correct ones') and '1' those that differ fom N₀ by one unit. There are m of these, and we treat them as equivalent and neglect the activity of other polymers. As erroneous replication, N₁ will be formed by N₀, and the N₁ may make N₀-polymers. The decay rates are assumed to be the same for all polymers. As before, N_T is the total number of polymers, formed by N₀ and N₁.

The treatment of one main polymer, N₀, and m equivalent ones with one unit different from N₀ is a great simplification with no realistic foundation. However, what is important is to divide the replicating polymers in groups, one 'best polymer', represented by N₀, and those that are active but less efficient. The division into N₀ and N₁ is a well-defined grouping with easily interpretable results. It leads to the same qualitative behaviour as other groupings, which is the main motivation for the model.

The stationary solution of (17) is obtained in a straightforward way as in the previous cases. (17a) and (b) provide two relations for the quotient N₁/N₀, which give an equation for M:

$$\frac{N_1}{N_0} = \frac{q - W_0 p_0^m \cdot M}{mW_1 p_1^m \cdot (1-p_1)M} = \frac{mW_0 p_0^{m-1}(1-p_0)M}{q - W_1 p_1^m \cdot M} \tag{18}$$

This leads to a quadratic equation for M, and it is relatively easy to see that it always has two solutions, of which only the lowest is meaningful. (The largest solution yields negative values for the quotients of (18)). One sees that the solution for M is smaller than that given by (15) of the previous case with only inactive errors. The reason is clear from (17a): as there is some reproduction of N_0 from N_1, N_0 may grow with a smaller value of M than if the erroneous polymers were entirely inactive. If p_0 is sufficiently close to 1, the situation is stable and will not be destroyed by a low accuracy of N_1.

For relatively accurate systems when p_0 is close to one and $p_1 < p_0$, one can get a rough estimate of the solution of (18) with:

$$M \approx \frac{q}{W_0} \cdot p_0^{-m} \; ; \qquad \frac{N_0}{N_1} \approx \frac{m(1/p_0 - 1)}{1 - (p_1/p_0)^m}$$

The occurrence of active erroneous polymers (mutations) which leads to a coexisting family of polymers ('superspecies') can stabilize the situation compared to the model of (14). There is no 'error catastrophe' in this type of model.

(Note that there is some confusion on this point in the literature: when Eigen and Schuster (1977) speak about an error catastrophe, they refer to the situation above, represented by formula (16), which poses a constraint on the length of the polymer but is much less severe than the 'catastrophe' we shortly will discuss.)

Errors in adaptor systems

The possibility of error propagation is more serious in the adaptor system described by (10). This kind of problem was treated in a previous paper by the author (Blomberg 1990), and we here use a modified version of that in the present general framework with the following assumptions (compare formula (10)):

(1)   The enzyme producing polymer N_0 inserts correct monomers with probability p_0 (as in the above schemes). *Only correct versions of this are assumed to be active in synthesis.* (This is sufficient for the qualitative behaviour, and leads to relatively simple expressions, but it is not necessary for the treatment. The erroneous polymers of this kind play a less important role than the adaptors.)

(2)   As for the adaptor synthesizing enzymes N_P, those with one or two errors are considered to be active. The ones without any errors yield the best result, and the erroneous ones become less efficient. The most relevant feature is the fraction of wrong adaptors.

(3)   All adaptors, correct or wrong, are equally accepted by the enzyme synthesizing polymer. Because of this, correct monomers are inserted in the enzymes by a probability equal to Y, the fraction of correct adaptors.

We assume that the lengths of the two kinds of enzymes are n and m and get equations:

For the enzyme-synthesizing polymer:

$$\dot{N}_0 = W_0 (p_0 Y)^n \cdot N_0 \cdot T - q N_0 \tag{19a}$$

and for the adaptor synthesizing enzyme, made by N_0, the equation for correct polymers is:

$$\dot{N}_{p0} = W_0 (p_0 Y)^m \cdot N_0 \cdot T - q N_{p0}$$

And for those with one error:

$$\dot{N}_{p1} = W_0 (p_0 Y)^{m-1} (1 - p_0 Y) \cdot m \cdot N_0 \cdot T - q N_{p1} \tag{19b}$$

With two errors:

$$\dot{N}_{p2} = W_0(p_0Y)^{m-2}(1-p_0Y)^2 \cdot \frac{m(m-1)}{2} N_0 \cdot T - qN_{p2}$$

For simplicity, we assume all growth and decay rates to be the same.

The adaptor production is given by:

Total amount of adaptors:

$$\dot{T} = W_{p0}N_{p0}M + W_{p1}N_{p1}M + W_{p2}N_{p2}M - N_0(W_0n+W_1m)\cdot T - q_TT \qquad (19c)$$

and erroneous ones:

$$\dot{T}_1 = W_{p0}N_{p0}ME_{p0} + W_{p1}N_{p1}ME_{p1} + W_{p2}N_{p2}ME_{p2} - N_0(W_0n+W_1m)T_1 - q_TT_1$$

Finally, the equation for the monomers is:

$$\dot{M} = a - bM - (W_{p0}N_{p0} + W_{p1}N_{p1} + W_{p2}N_{p2})M \qquad (19d)$$

T_1 is the number of erroneous adaptors. The fraction of correct adaptors, Y, is given by:

$$Y = 1 - \frac{T_1}{T} \qquad (20)$$

E_{po}, E_{p1}, E_{p2} are the error rates of the adaptor synthesizing enzymes without errors, with one error and with two errors respectively.

For a stationary state, one gets the following relations from (18a) and(b):

The adaptor concentration is
$$T = \frac{q_0}{W_0}(p_0Y)^{-n}$$

Fractions are: $N_0/N_T = (p_0Y)^n$, $\quad N_{p0}/N_0 = (p_0Y)^{m-n}$

and

$$N_{p1}/N_{p0} = m\cdot(\frac{1}{p_0Y} - 1) \qquad N_{p2}/N_{p0} = \frac{m(m-1)}{2}(\frac{1}{p_0Y} - 1)^2$$

With (18c), this yields the following relation for Y:

$$\frac{T_1}{T} = 1 - Y = \frac{W_{p0}E_{p0} + W_{p1}m(\frac{1}{p_0Y} - 1)E_{p1} + W_{p2}\frac{m(m-1)}{2}(\frac{1}{p_0Y} - 1)^2E_{p2}}{W_{p0} + W_{p1}m(\frac{1}{p_0Y} - 1) + W_{p2}\frac{m(m-1)}{2}(\frac{1}{p_0Y} - 1)^2} \qquad (21)$$

This gives a third order equation for Y. To get solutions for the stationary state, the above relations are used together with the solution for Y. Then, equations (18c and d) provide a quadratic equation for N_o (or M), which will not always provide positive, non-zero solutions. This fact leads to conditions for solutions as are seen in the figures below. The solution depends to a high degree on the rate of the two-error-polymers, W_{p2} and on the length of the polymers. The absence of a solution can be interpreted as an 'error catastrophe': the system will not stabilize.

We show here results for the fraction of correct polymers, N_0/N_T, given by Y according to the formulas above as function of the relevant rate constant W_{p2}, or the length of the polymer. The stable solution vanishes drastically at certain parameter values as shown in the figures. Note the abruptness of the disappearance of the solution. (Of course this is due to the fact that N_0/N_T is given by Y for which solutions always exist in (21), while the existence of a solution depends on the further equation for N.)

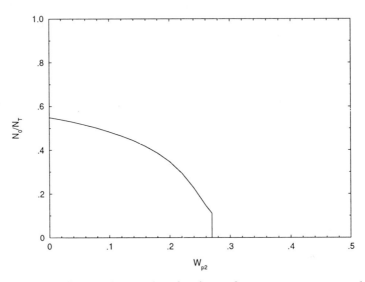

Figure 3. The figure shows the fraction of correct enzyme-producing polymers as function of W_{p2}, the rate constant of adaptor synthesizing enzyme with two errors in the model of the text. Other parameters are:
$E_0 = 0.02$, $E_1 = 0.1$, $E_2 = 0.5$, n=m=10, $W_1 = 0.5$, other rates = 1.

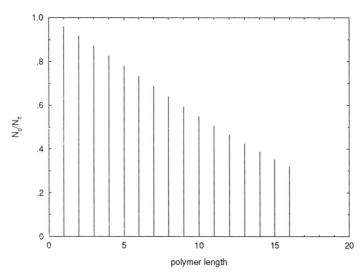

Figure 4. The fraction of correct enzymes as in figure 3 as funtion of the polymer length, n, which is assumed equal to m. $W_{p2} = 0$, other parameters are as in figure 3.

It is seen that the adaptor mechanism makes the situation more delicate than the case of independent replicators. In the latter case, a high accuracy of the 'correct' polymers guarantees stability. Here, it is also necessary that erroneous enzymes that may lack accuracy have a low activity. If this is not the case, the system becomes unstable. Also, the dependence on length is more delicate for this model.

CONCLUSIONS

What is seen here is that errors in simple, autocatalytic processes which well may have been networks of autocatalytic sets or just simple polymers with self-replicating capability can well work as a simple model of primitive life. With a crucial number of 10-20 relevant monomers in the molecules, errors or complexity should not pose any large problem, and the systems may easily stabilize.

The situation is more complex for cooperating systems of a hypercycle type. An adaptor system as described above is quite sensitive to errors and in order to stabilize, the rates must be tuned well and there should not be any 'moderately efficient' enzymes that do not provide a sufficiently high accuracy of the adaptors but still have an appreciable activity. With increasing length of the polymers, it should also have been important to improve the accuracy. Conclusions of this type seem to suggest that the step to a cooperating system was a great step in evolution which also should have involved some development of control mechanisms. There are evidently other features of a cooperating system which show difficulties at a too primitive stage. As will be discussed in another work, these systems probably need some kind of space organization in order to be meaningful, and thus also for that reason show a higher stage in evolution.

REFERENCES

Blomberg, C., 1983, Free energy cost and accuracy in branched selection processes of biosynthesis. Quart. Rev. Biophys., 16:415.

Blomberg, C., 1990, Modelling efficiency, error propagation and the effect of error-enhancing drugs in protein synthesis. Biomed. Biochem. Acta, 49:879.

Blomberg, C., von Heijne, G., and Leimar O., 1981, Competition, coexistence and irreversibility in models of early molecular evolution, in Origin of Life, Y.Wolman, ed., D.Reidel Publishing Company.

Blomberg, C., and Liljenström, H., 1988, Efficiency in Biosynthesis - Cellular synergetics, in Synergetics, Order and Chaos, M.G.Velarde ed., World Scientific.

Eigen, M. 1971, Selforganization of matter and the evolution of biological macromolecules. Naturwissenschaften, 58:465.

Eigen, M., and Schuster, P., 1977, The Hypercycle. A principle of natural self-organization. Part B. The abstract hypercycle. Naturwissenschaften, 65:7.

Hoffman, G.W., 1974, On the origin of the genetic code and the stability of the translational apparatus. J. Mol. Biol., 86:349.

Kirkwood, T.B.L., Holliday, R., and Rosenberger, R.F., 1984, Stability of the cellular translational process. Int. Rev. Cytol., 92:93.

Orgel, L. E., 1963, The maintenance af the accuracy of protein synthesis and its relevance to ageing. Proc. Natn. Acad. Sci USA, 49:517.

Schuster, P., 1988, Potential functions in molecular evolution, in Synergetics, Order and Chaos, M. G. Velarde, ed., World Scientific.

INFORMATION DYNAMICS OF SELF-PROGRAMMABLE MATTER

Carsten Knudsen,[†] Rasmus Feldberg,[†] and Steen Rasmussen[‡]

[†]Physics Laboratory III and
Center for Modelling,
Nonlinear Dynamics and
Irreversible Thermodynamics
Technical University of Denmark
DK-2800 Lyngby
Denmark

[‡]Center for Nonlinear Studies and
Complex Systems Group,
Theoretical Division
MS B258
Los Alamos National Laboratory
Los Alamos, New Mexico 87545
USA

ABSTRACT

Using the simple observation that programs are identical to data, programs alter data, and thus programs alter programs, we have constructed a self-programming system based on a parallel von Neumann architecture. This system has the same fundamental property as living systems have: the ability to evolve new properties. We demonstrate how this constructive dynamical system is able to develop complex cooperative structures with adaptive responses to external perturbations. The experiments with this system are discussed with special emphasis on the relation between information theoretical measures (entropy and mutual information functions) and on the emergence of complex functional properties. Decay and scaling of long-range correlations are studied by calculation of mutual information functions.

INTRODUCTION

A fundamental feature of living organisms is the ability to compute, or process, information. Information processing takes place over a wide scale of complexity, ranging from the simple processes by which an enzyme recognizes a particular substrate molecule, to complicated feedback regulations containing many different levels of information processing, to the extremely complex processes of the human brain.

An example of biological information processing in a feedback loop is provided by one of the negative feedback loops described by Sturis et al. (1991) in their model of oscillatory insulin release. The feedback control can here be divided into at least four different components: (i) an increased amount of glucose in the plasma stimulates insulin production in the pancreas and secretion of insulin into the plasma; (ii) from the plasma, insulin diffuses into the interstitial fluid; (iii) here insulin molecules attach to receptors on the surface of the cell; and (iv) insulin activated receptors enhance the uptake of glucose by the cells, which, of course, implies a decrease in glucose outside the cells.

This loop involves simple biochemical information processing such as the recognition of receptors by insulin molecules and the subsequent attachment of the molecules. In addition, there is a more complex information process involving active transport of glucose over the cell membrane facilitated by a cascade of conformational changes in the cell membranes protein molecules.

The most complex kind of biological information processing is probably the abstract and creative symbolic information processing in the human brain. Simple aspects of these processes are subjects of numerous investigations. In particular a variety of models of artificial neural networks have recently been proposed (see for example Palmer 1988, and Touretzky 1990).

The theoretical foundation of information processing in man-made machines can be described in terms of computation theory (Hopcroft and Ullman 1979). In computation theory, a number of different formalisms exist, of which the Turing machine (TM) for historical reasons is the most well-known. The Turing machine has been examined thoroughly by mathematicians and computer scientists because it is believed to be able to perform the most general type of computation, universal computation. This conjecture is known as the Church-Turing thesis (Hopcroft and Ullman 1979).

Besides Turing machines, several other systems have been shown to support universal computation, including cellular automata, the λ-calculus, Post systems, the hard billiard ball computer, general recursive functions, classifier systems, partial differential equations, von Neumann machines, and even C^∞ maps of the unit square onto itself. It has been shown that each of these formalisms is equivalent, since any one of them can simulate any other.

The information processing found in biological systems seems to be different in nature from that of a Turing machine. In fact, none of the above mentioned computational paradigms capture the full spectrum of biomolecular information processing. A fundamental property of computation in biomolecular systems arises from their ability to alter or program themselves. Self-programming occurs at all times and length scales in biomolecular systems. Although the above mentioned computational systems in principle can program themselves, this capacity has never been studied or used. It is known that any of the universal systems have the following properties: (I) the ability to store information, (II) the ability to communicate information, and (III) the ability to perform non-trivial information processing.

These abilities are also found in the living cell, although they are more difficult to classify. However, using the same scheme to discuss elements of biomolecular computation, we obtain:

(I) Storage and memory abilities: (1) single molecules, e.g. DNA and proteins, and (2) assembly/disassembly of supramolecular structures, e.g. the cytoskeleton and the cell membrane.

(II) Signal abilities: (1) diffusion (passive transport of materials, energy, and information) occurs everywhere in the cell, (2) active transport (non-specific (convection) and specific transport of materials, energy, and information) convection occurs in the cytoplasm and specific transport occurs for example over the cell membrane and along the microtubules, (3) conformational changes (transfer of energy and information), e.g. of dynein and kinesin in relation to cilia mobility, and (4) electromagnetic irradiation (transfer of energy and information), e.g. photochemical processes in chlorophyl.

(III) Transformation abilities: (1) chemical reactions - often using signal molecules as reactants to produce new signal molecules as products or using signals which act as catalysts or triggers, and (2) transcription of DNA to RNA and translation of RNA into protein molecules which fold up and act as constructive and regulatory units in the cell.

From this scheme it should be obvious that most fundamental biomolecular processes can be interpreted in terms of computation. These biomolecular processes are all coupled through a very complex network of functional interactions about which we only know certain details and in which the overall bauplan is still a mystery. The cell continuously programs and re-programs itself, and in multicellular organisms this self-programming also occurs at the organism level (recall the discussion of the feedback loop controlling insulin release).

Living systems can through a re-programming of some of their parts alter functional properties which are of vital importance for survival. Viewed over longer time scales this self-programming ability is also used to create new properties which are incorporated through the selection process of evolution. Since any computational universal system, in principle, is able to program itself, we shall modify one of them so that we can study self-programming as a phenomenon in a much simpler and more tractable system. We have chosen to modify the parallel von Neumann architecture. The modified von Neumann machine (MVNM) is easy to program since most modern digital computers are based on the von Neumann principle, and since the autonomous dynamics of such a system even at its lowest level (one single instruction) has a clear computational interpretation. We shall in the following focus on the emergence of new functional properties in MVNM's which most clearly reflect the evolutionary aspect of biocomputing.

SELF-PROGRAMMABLE MATTER

We can in general terms define self-programmable matter as a dynamical system of functional interacting elements, or compositions of elements, which through autonomous dynamics can develop new compositions of functionally active elements.

Such systems are characterized by an ability to construct novel elements within themselves. Thereby chemistry by definition becomes a particular kind of self-programmable matter. The physical properties (e.g. shape and charge) of the chemical species define the possible interactions with other molecules and thereby their functional properties. Chemical systems create new properties through recombination of molecules via chemical bonds. New combinations between existing molecules and combinations of new molecules with other molecules then define new functional properties. This defines a constructive or self-programming loop given by:

molecules → physical properties → functional properties → interactions → new molecules.

Description of Venus

As an example of a self-programming system, we have defined a modified von Neumann machine, called Venus. It consists of a one-dimensional memory array, called the core. This corresponds to the RAM (random access memory) in a modern digital computer. Each element of the core, a word, contains a machine code instruction. There are 10 different instructions, which are listed in Table 1. An instruction has three elements, an opcode and two fields, A and B. Each field consists of an addressing mode and a numeric field, the latter containing a non-negative integer. There are four different addressing modes, as shown in Table 2.

Many programs simultaneously execute instructions in the core. A monitor-like function always discovers whenever two or more instructions simultaneously try to obtain write access to the same core addresses. Thereby, write conflicts are resolved.

The model has several features which are not found in ordinary multi-tasking

Table 1

OPCODE	FUNCTION
DAT B	Non-executable statement. Terminate the process currently executing. Can be used for storing data.
MOV A,B	Copy the contents of A to B.
ADD A,B	Add the contents of A to the contents of B and save the result in B.
SUB A,B	Subtract the contents of A from B and save the result in B.
JMP A	Move the pointer to A.
JMZ A,B	If B equals zero, move the pointer to A.
JMN A,B	If B differs from zero, move the pointer to A.
DJN A,B	Decrement B, and if B differs from zero, move the pointer to A.
CMP A,B	If A differs from B, skip the next instruction, e.g. move the pointer two steps ahead instead of one.
SPL B	Create a new process at the address pointed to by B.

VNM's. One of the major differences is the presence of noise in our system. The task of an ordinary VNM is to perform very specific calculations, detailed via programs written by human subjects either in machine code or in a high-level language such as FORTRAN or C. In the presence of noise, most programs, e.g., ordinary differential equation solvers or bookkeeping programs, would crash or give more or less meaningless outputs. This is contrary to the computations in biological systems in which noise usually has a very limited effect. One notion of noise in Venus is built into the execution of the MOV instruction. When the MOV instruction (see Table 1) copies a word, something might go wrong, and the word written to the memory can be altered. This is the reason why ordinary programs have a hard time executing properly. Such routines rely on perfect copying of data. There is an additional source of noise that drives our system. Once in awhile a random pointer is appended to the execution queue. Since processes can terminate by executing the DAT instruction, we make sure the system is supplied with pointers via this stochastic mechanism. The mutation frequency is one per 10^3 copyings, and a pointer is appended to the execution queue approximately every twentieth generation. A mutation of a machine code program always yields a new legal program, as opposed to a change in a high-level language, which almost certainly will result in a syntax error.

The Venus system also incorporates a notion of computational resources. This prevents the simultaneous execution of too many processes both in total and within a limited spatial addressing area. The first limitation is expressed in terms of an execution queue of fixed length, which in all the simulations to be discussed were of size 220. The execution queue contains the addresses of the instructions to be executed. The second limitation is due to address-localized computational resources, which are measured in fractions of one execution. Each address in the core y is at any time t associated with a certain fraction $r(t,y)$ of one execution. This value is incremented by a fixed amount Δr at each generation. However, the value can never exceed some constant predefined fraction r_{max} of one execution. When the system executes a pointer from the execution queue, it looks to the addresses in the immediate neighborhood of the pointer and finds the sum of computational resources. If this sum exceeds one, then the instruction will be executed. If not, the pointer will disappear. The resource radius R_{res} defining the immediate neighborhood is three in all simulations.

Table 2

ADDRESSING MODE	EFFECTIVE OPERAND
# (immediate)	The effective operand is the value in the data field. Example: MOV #3,.. , has the effective operand 3.
$ (direct)	The effective operand is the word pointed to by the value in the data field. Example: MOV $2,.. , has the effective operand located two words towards increasing addresses.
@ (indirect)	The effective operand is found by looking at the data field pointed to by the actual data field, and then using the direct mode.
< (autodecrement indirect)	As indirect, only the value pointed to by the actual data field is decremented before being used.

Instructions are only allowed to communicate, e.g. to read and write data, locally. The allowed distance for read/write access is 800 in all simulations. However, in contrast with normal multi-tasking VNM's, all processes can overwrite anything in their neighborhood, as long as it does not occur simultaneously with other processes. This means that there is no notion of individual work space or, in biological terms, there are no predefined "proto-organisms" (cellularity).

Initialization of Venus

In all simulations, the system has been initialized randomly by a uniform distribution of opcodes, addressing modes, and data fields. Previous studies showed that the system can only evolve simple structures from a uniform distribution (Rasmussen et al. 1990). One can increase the complexity of the dynamics of the system greatly by supplying some kind of biasing of the initial core. In other words, we need to supply the system with a reactive potential, in the sense that the different machine code instructions need to be inhomogeneously distributed in the core to enhance many spontaneous computations. This potential is conveniently introduced by placing a self-replicating program in the randomly initialized core. This program has a replication cycle of 18 generations and will very quickly produce a considerable number of offspring, all of which will attempt to replicate unless they have been modified by noise or have been overwritten. As mentioned earlier, most programs designed to work in a noise-free environment very quickly crash by making erroneous copies. Another way in which these programs begin to malfunction is by copying on top of other offspring, which eventually happens since the core has a limited size, 3,584 addresses in the simulations to be discussed. By using our interactive graphics simulator, we have determined the lifetime of a well-functioning population of self-replicating programs to be around 200 generations. After this, no copies of the original programs are left. Only mutated versions with different functional properties are found. It is important to notice that the effect of the self-replicating program is a good mixing of instructions, and not a probing of the system with self-replicating properties. The last bit of this sophisticated self-replication is gone after 200 generations.

An alternative and conceptually more satisfying method for supplying the system with a reactive potential is to generate a random core by using a set of coupled Markov matrices. This approach is currently being investigated (Rasmussen et al. 1991).

Experimental Results

In the simulations with Venus, many different evolutionary paths have been observed. Typically, after extermination of well-functioning self-replicating programs, the system enters a phase in which massive copying of one or more instructions takes place. In the beginning, this is mainly caused by the copying loops of the former self-replicating programs. Typically such a partially malfunctioning loop will move copies of a single word out into the vicinity of the loop. As a result, large areas of the core will be filled with a single word, with unaltered opcode, amode, afield, and bmode, but possibly different bfields. This runaway process introduces a kind of sensitive dependence on initial conditions, as known from chaotic dynamical systems. However, after some generations the copying loops are destroyed, either by noise or by MOV instructions overwriting them. This signals the beginning of a new epoch for the system. From this point on, the dynamics of the core is governed by large coherent groups of single instructions of sizes ranging from one word to several hundred words. This is in strong contrast to the dynamics in the first phase of the evolution, in which relatively few instructions placed in the self-replicating programs were responsible for the dynamics.

Naturally, the further development of the core depends heavily on the distribution of opcodes at this point. The distribution of opcodes is determined by the instruction copying, and therefore from this point on we see different evolutionary paths. For instance, a core consisting mainly of SPL instructions will lead to an evolution involving large areas crowded with pointers, while a core containing mainly MOV instructions will lead to a very dynamical behavior concerning the contents of the core, but also to a core with a small concentration of pointers. In the following, we shall take a look at three different evolutionary paths that are often observed in simulations with Venus.

If the core contains relatively many jump instructions (e.g. JMP, JMZ, JMN, and DJN) with immediate amode, the pointers will get caught at these, because an afield with immediate addressing mode points to its own location in the core. In other words, a pointer meeting one of these instructions will keep jumping to the same instruction over and over again, until it is either killed by lack of computational resources, or the jump instruction is overwritten by noise or a MOV instruction. Of course, such a core will be quite static with respect to the distribution of different instructions, because most pointers will be trapped at the jump instructions. Therefore, possible MOV instructions will only rarely be executed. This kind of core will be referred to as a fixed point core. This state constitutes a quite trivial form of cooperation. The self-reinforcing mechanism is characterized by the special mixture of instructions. With the presence of noise in this core, a new pointer will occasionally activate a successive number of the MOV instructions, which, with a probability close to one, will repeatedly copy other MOV instructions or one of the jump instructions until the new pointer gets trapped at one of the jump instructions. Here, it may or may not survive, depending on how many pointers each jump instruction in the neighborhood can carry in terms of computational resources. Such a fixed point structure is very stable towards perturbations.

As indicated above, another common evolutionary path evolves from a core with a considerable number of SPL instructions. In such a core, a large pointer concentration in an area will imply the execution of many SPL instructions, leading to even more pointers, constituting a positive feedback loop. The pointer concentration in such areas will increase until lack of resources or the limited execution queue length puts an end to the growth. These SPL colonies have a more interesting cooperative dynamics than the fixed point cores. The SPL instruction colonies cooperate in the sense described above: Locally, they distribute pointers to their neighbors, which do the same, and globally, the colony occupies all available computational resources in terms of pointers. Sometimes we observe several addressing areas with this kind of behavior.

Above a certain size, the different colonies compete for pointers, and an intermittent behavior between the areas can be observed. However, because of the lack of MOV instructions, the contents of the core will be conserved.

A third and even more interesting evolutionary path is that of the MOV-SPL structure. If a core has a large number of MOV and SPL instructions, a very complex cooperative structure can emerge. The cooperation works in the following way: SPL instructions supply the structure with pointers, both to SPL and MOV instructions. This is somewhat similar to their function in SPL colonies, where they distribute pointers to themselves. We can interpret this as a supply of free energy in thermodynamic terms. The MOV part of the structure makes sure that both MOV and SPL instructions are copied, giving the structure the ability to move around in its environment and to locally explore the spatial resource in terms of addresses in memory. The copying is thereby also responsible for the growth of the structure. This of course means that the structure does not have a well-defined genotype in a contemporary biological sense.

It generally takes several thousands of generations from the extermination of copying loops before the cooperative MOV-SPL structures appear. Typically the system goes through several phases with relatively large concentrations of instructions other than SPL and MOV before reaching the MOV-SPL state. Usually a core with a MOV-SPL structure consists of up to 80 percent SPL instructions, with most other instructions being MOV instructions; however, the ratio of SPL to MOV instructions may change during the epoch.

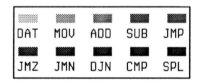

Figure 1. The shading used in core and cellular views.

The MOV-SPL structure is very stable, even though it is constantly subjected to perturbations, because the MOV instructions copy words continuously. However, the structure apparently does not change its basic functional properties. If we were to change the opcode of a word, the functional properties of this word would clearly be altered drastically. Since this happens all the time in the MOV-SPL structure, the system has found an area in rule space where it is stable towards such perturbations. Stability in Venus is an emergent property rather than an intrinsic element of the chemistry, because the elements of which the structures are composed are themselves very fragile.

In order to illustrate the micro- and macroscopic dynamics of the MOV-SPL structure, we have made two different kinds of projections of the high-dimensional

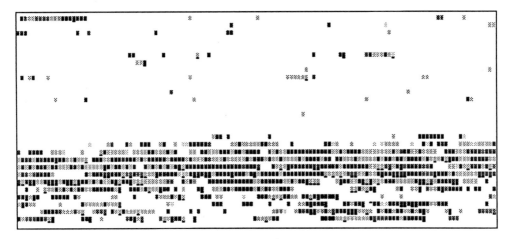

Figure 2. Core portrait of a MOV-SPL structure. The upper row shows the opcodes at addresses 0 through 127, the second row the addresses from 128 through 255, etc. Black underscores represent pointers. White squares indicate no recent references.

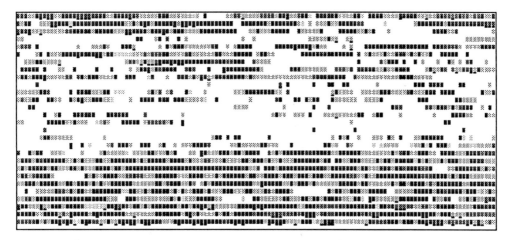

Figure 3. Core portrait of the structure shown in Figure 2, 100 generations later. Note how the activity has changed.

discrete phase space. The macroscopic dynamics is illustrated in Figures 2 and 3, in which each opcode in the core is represented by a small bar. The shade of each bar corresponds to the instruction type as explained in Figure 1. The figures show the opcodes in the core at two different times in the evolution. Note that in both of the core views, large parts are white. This merely means that these addresses have not been referenced for a while; there are no empty words in the core. In Figure 2, we see that the activity is restricted to higher addresses, whereas in Figure 3, approximately 100 generations later in the evolution, the structure has increased its domain. Note also that most words are occupied by either SPL- or MOV-instructions. The black underscores represent pointers.

Figure 4 shows a cellular automata-like view of part of the core. We have chosen 128 consecutive addresses within the MOV-SPL structure. The opcodes of the words in this addressing region are shown as horizontal lines at consecutive time points. In this figure, time increases downwards. The black underscores here also represent pointers. It is obvious that the microscopic dynamics is very irregular, although the macroscopic dynamics is preserved. We see how single words or sometimes consecutive words are overwritten. Also note that once in awhile an opcode different from MOV or SPL appears, caused by mutations. Groups of pointers suddenly appear or disappear. In biochemical terms, this structure has an irregular metabolism.

The evolutionary stories described here, and additional ones, were discussed in greater detail by Rasmussen et al. (1990 and 1991).

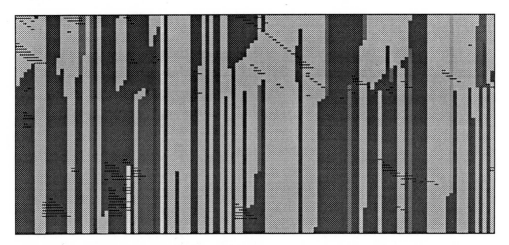

Figure 4. A cellular automata-like view of a part of the core.

INFORMATION DYNAMICS OF VENUS

The dynamics of self-programmable matter is generally quite complex, as described in the previous section, and at the same time simulations are computationally very time consuming. It is therefore preferable to have some quantitative measures that are easily computed and that signal changes in local or global dynamics. The calculation of such measures would enable us to characterize the system's behavior in terms of well-established quantities.

In order to make sampling frequent and simple, we have computed some simple information theoretic measures. The chosen measures contain both spatial and temporal ingredients. Such a combination of spatial and temporal measures is not necessarily ideal for all applications. It can be required, however, if the ingredients of the computational chemistries in question are such that there exists a preferred direction in space. To simplify things further, we shall here only consider the opcode, since the functional properties of a word are mainly determined by this element.

The simplest of the measures is the spatial entropy for templates of size one S_1. The spatial entropy is defined as the usual Shannon entropy, calculated from the probability distribution p_k defining the probability of finding an instruction with the opcode k at an arbitrarily chosen site in the core

$$S_1(t) = -\sum_{k=0}^{9} p_k \log_2(p_k).$$

Another quantity of interest is the mutual information M_1. This is defined in terms of the spatial entropies S_1 and S_2

$$M_1(t) = 2S_1(t) - S_2(t)$$

where

$$S_2(t) = -\sum_{k=0}^{9} \sum_{l=0}^{9} p_{kl} \log_2(p_{kl}).$$

The interdependence between two events A and B can be measured by the mutual information. The mutual information can therefore reveal the emergence of correlations between neighboring instructions and thereby the occurrence of new properties of interactions. Of course, too large correlations, such as for the pattern "101010101010...," are not desirable, since they simply indicate that everything is overly dependent. The appearance of mutual information of intermediate values is of more interest, since this could indicate that the system is able to perform non-trivial information processing (Langton 1990).

The spatial entropy for templates of size two S_2 is calculated from a probability distribution p_{kl} which, as we shall discuss shortly, also captures some very important temporal correlations that determine the system's potential functional properties. The probability p_{kl} is defined as the probability of finding an instruction with opcode k at an arbitrary address n, and an instruction with opcode l at address $n+1$. Note that this probability is not the same as the probability of finding opcodes k and l as neighbors. The reasons for the chosen definition are that pointers are incremented one word after each execution, and that the pointers therefore in general travel towards increasing addresses, with the obvious exception of the jump instructions. This means that changes in S_2 can signal a change in the typical sequential order of execution of two neighboring instructions. A change in the typical sequential execution order of instructions is of considerable interest, since a change in the potential functional properties can indicate that the system is in the process of changing its global dynamics through some self-organizing process. If the spatial entropy for templates of size one S_1 is approximately constant, the effect of a change in S_2 will immediately show up in the mutual information M_1.

Experimental Results

All experiments are performed with the parameters and initial conditions discussed in the previous section. During simulations, the above measures were calculated frequently. The sampling rate is the same in all computer experiments, namely one sampling every 10 generations. As can be seen from Figures 5 and 6, the monitored measures change reasonably with this frequency, i.e. there are apparently no sudden jumps in the measured quantities between samplings.

Figure 5 shows the entropy and the mutual information vs. time for a particular simulation. In Figure 6a we see the mutual information vs. entropy for the same simulation, and Figure 6b shows the mutual information vs. entropy for another simulation. In the following we shall describe these simulations in terms of their information dynamics.

It appears that the process starts with a drastic increase in the mutual information, while the entropy is almost unchanged. This is observed in both Figures 6a and 6b (the simulation starts in the lower right-hand corner of the plots) and is caused by offspring of the program initially placed in the core distributing their code. This process does not influence the opcode distribution very much, since the program has a distribution of opcodes fairly close to that of a randomized core. However, the spreading of this code influences the spatial correlations in the core and thereby the mutual information. After about 200 generations the self-replication of programs stops because of malfunction introduced by noise and programs writing on top of each other. Now the entropy starts falling while the mutual information remains in the vicinity of 0.6 bits. This is especially clear in Figure 6b but can also be observed in Figure 6a as a noisy plateau at the rightmost corner of the figure. The decreasing entropy in this phase is caused by partially malfunctioning copying loops spreading out a large number of a particular instruction.

For the simulations shown in Figures 5 and 6a, we see that after about 4,000 generations, the mutual information drops to a level of about 0.2 bits. At the same time the entropy drops to about 1.5 bits, and some oscillations in both the entropy and the mutual information are observed, giving rise to a "cloud-like" picture of the information dynamics in Figure 6a. In this part of the simulation, the core is mainly populated with MOV and SPL instructions and the pointer density is very high.

The epoch of the MOV-SPL structure continues until $t = 20,000$, where a large peak in the mutual information reflects a change in the functional properties of the system (recall the discussion on mutual information in the previous section). From this point on, the changes in the entropy as well as in the mutual information become less rapid and exhibit a stepwise character signalling yet another epoch of the system. In this phase the core mainly consists of MOV-, SPL-, and jump-instructions where most

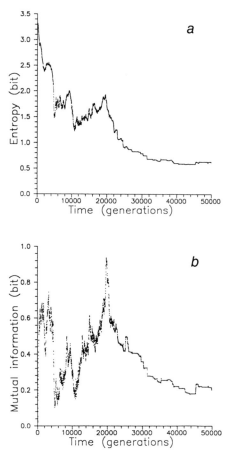

Figure 5. (a) The usual Shannon entropy of the opcode distribution versus time. (b) The mutual information versus time.

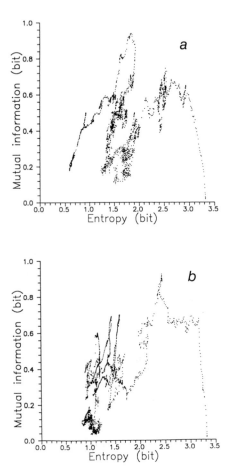

Figure 6. (a) shows the mutual information versus the usual Shannon entropy for the simulation shown in Figure 5. (b) shows mutual information versus entropy for another simulation.

of the pointers are trapped at jump instructions.

In the simulation shown in Figure 6b, the system leaves the plateau of partially malfunctioning copying loops at $t=1,100$. At this point there is a rise in the mutual information to a little above 0.9 bits followed by a drastic drop to a value of about 0.2 bits. After this we see the same cloud-like distribution of points in the figure as observed in Figure 6a. Note that during most of this phase we see rapid changes of the mutual information, while the changes in the entropy are more moderate. These changes indicate that some dramatic changes occur, while the opcode distribution is only affected slightly. This kind of behavior is likely to be caused by MOV instructions shifting the contents of the core as opposed to the partially malfunctioning copying loops multiplying single instructions into large areas of the core. These somewhat organized changes signal that the MOV instructions are activated many at a time, reflecting some kind of structure in the part of the core responsible for the dynamics. Finally, at $t=32,000$ we observe a drop in the mutual information to about 0.1 bits and the disappearance of the large oscillations in the mutual information. This phase of the simulation is observed in Figure 6b as a small point dense area just below the cloud-like point distribution. Compared to Figure 6a this final phase is somewhat more active, with small oscillations in the entropy and the mutual information. These oscillations are caused by MOV-instructions continuously re-organizing the contents of the core, however, in a less organized and dramatic way than in the previous phase of the simulation.

A rather coarse-grained resumé of the simple information dynamics would be that the system starts with a high degree of disorder, low complexity, and a high reactive potential in terms of the initial distribution of opcodes. The system then evolves, lowering the entropy towards intermediate values, while the complexity increases. Then the system wanders around in information space in a very complicated manner according to the information theoretic measures characterized by sudden changes in both order and complexity, where all major changes always are associated with changes in the functional properties. With a time horizon of 50,000 generations, most processes end up in one of two different dynamical states. One is best characterized as a frozen state, and the other may be characterized as recurrent or chaotic. The frozen state is a perturbed fixed point consisting of a variety of jump-instructions which have trapped the pointers. The fixed point dynamics is in this situation often perturbed by two factors. One factor is that some of the conditional jump-instructions periodically have their bfields counted down, allowing trapped pointers to escape. The other factor is the introduction of random pointers. The overall effect of both of these perturbations is small changes in the core reflected by small steps both in entropy and mutual information. This type of dynamics is seen in the simulation shown in Figure 6a. Another example of this type of dynamics is the collapse, where all pointers disappear. This also occurs quite often. The terminal state with recurrent dynamics is typically found in situations where a major part of the core is occupied by either MOV- or SPL-instructions. The MOV-dominated cores have many MOV-instructions executed at each generation changing the content of the core all of the time. This type of dynamics is seen in the simulation shown in Figure 6b. In SPL-dominated cores the composition of instructions remains virtually constant, whereas the pointer dynamics exhibits pronounced intermittent behavior. The SPL-dominated cores of course have maximal pointer density. The MOV-SPL structure often emerges as a transient structure active from generation 4,000 until generation 20,000. In rare situations the MOV-SPL structure can survive even after 100,000 generations. Since Venus is a universal system it of course supports the three main ingredients of computation (recall the discussion in the previous section): (i) the capacity to store information, (ii) the capacity to exchange or communicate information, and (iii) the ability to process information in a non-trivial way. Dynamical systems with these computational capabilities can apparently, in thermodynamical terms, be characterized as operating in the immediate vicinity of a phase transition (Langton 1990).

Since the dynamics of Venus changes the instruction contents of the core and thereby the rules governing the dynamics, the system often changes its computational capabilities during a simulation. An example is one of the terminal states, the frozen state (fixed point cores with pointers trapped at jump instructions), very efficient in storing information. However, the system has in this state, which in thermodynamical terms is equivalent to a solid state, lost its ability to communicate and process information. The chaotic or recurrent state (for example cores mainly populated with MOV instructions) exhibits a pronounced ability to exchange or communicate information but only a limited ability to perform non-trivial information processing. This state can be characterized as a fluid state in thermodynamical terms. The MOV-SPL structure is from a computational point of view more interesting. It is clear that the MOV-SPL structure does not have a well-defined genotype, and consequently it does not store information in the usual sense of data storage (recall (i)). However, it is also clear from the previous analysis that the macroscopic dynamics is preserved. The system therefore, through some complicated coding, stores its phenotype rather than its genotype. With respect to (ii) and (iii), it is fairly obvious that non-trivial computations are performed, and that information is being communicated by the MOV instructions, within the structure itself.

Scaling of Correlations

One can generalize the mutual information between neighboring words to investigate long-range correlations. When looking at long-range correlations, almost all temporal effects are removed. Changes in the functional properties/potential are certain to have taken place if long-range correlations should appear or disappear.

In the following, we shall discuss the simulation shown in Figure 5 from this point of view.

At time $t=5,000$, an interesting phenomenon is observed in the simulation concerning the spatial correlations. Here, the mutual information of opcode, amode, and bmode versus distance all decay smoothly. If we try to fit a power law to the decay, we can describe the decay by a scaling exponent α

$$M(n) \approx M(1)\, n^{-\alpha}.$$

What is particularly interesting here is that all three decays can be characterized by the same scaling exponent, $\alpha = 0.4$. One might suspect that this is always the case since the three elements of a word cannot be altered, but for example at $t = 1,000$, we find three different scaling exponents, ranging from 0.64 for the opcode to 1.1 for the amode. At later stages in the evolution, e.g. at time $t = 20,000$ and $t = 21,000$, the exponents differ by more than 30% from each other. In Figure 7, the decay of correlations for the opcodes is shown.

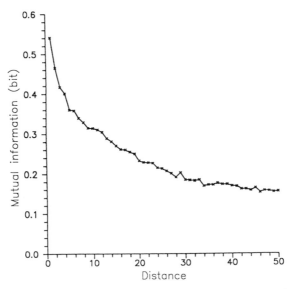

Figure 7. The figure shows the decay of correlations between opcodes as computed by the mutual information.

After approximately 20,000 generations, both the spatial entropy S_1 and the mutual information M_1 have maxima. Since both S_1 and M_1 change rapidly around this time, as can be seen in Figure 5, we expect to be able to find some indication in the long-range correlation. At time $t = 19,000$, we see from Figure 8 that there is a smooth decay of M_1 to values below 0.01 bit. The same is observed at time $t = 21,000$ (see Figure 8c), except that the value of the mutual information here slightly exceeds 0.01 bit. If we look at the correlations at time $t = 20,000$ (Figure 8b), several peaks are observed. At distances near 55 and 100, significantly greater values of the mutual information than in the surroundings can be seen, indicating long-range correlations. Since long-range correlations indicate some relationship between sites in the core separated with some distance, the emergence of such correlations signals that one or more MOV-instructions move the contents of the core that specific distance. Thus, the long-range correlations can be used to detect activity of one or more such MOV-instructions, enabling us to detect changes in the core that have only insignificant influence on the total distribution of opcodes, i.e. the entropy.

Yet another way of describing coherence in the core is by means of correlations defined by Markov matrices. In the case of opcode we can define a Markov matrix determining the probability p_{ij} of opcode i at address n is followed by opcode j at address $n+1$. Actually, such a matrix is intimately related to the spatial entropy S_2. Furthermore, as we mentioned earlier, such Markov matrices can be used for supplying the system with a reactive potential, instead of using a small self-replicating program, simply by generating the initial conditions on the basis of a set of low order Markov chains. We have calculated how the Markov chains determining the correlations between an opcode at address n and an opcode at address $n+1$ actually look at the sampling times corresponding to Figure 8. Figure 9 shows 3-D pictures of these Markov matrices. Note that the figure shows $\log(N_{ij})$, where N_{ij} is the number of occurrences of opcode j following opcode i. Note that p_{ij} can readily be calculated from N_{ij}.

In Figure 9a the strong correlations between subsequent MOV- and SPL-instructions indicate that the core is mainly populated with these instructions at $t = 19,000$. At $t = 20,000$ several smaller peaks appear indicating fluctuations introducing JMZ and DJN instructions. This signals that an instability is underway leading to an increased concentration of instructions other than the ones ruling the core earlier. Finally, at $t = 21,000$ a new set of high peaks appears revealing a strong presence of JMZ. The instability has thereby changed the macroscopic composition of the instructions. Note also that the small peaks indicating the emergence of DJN instructions have vanished. The DJN fluctuation has died out. The transition occurring in the system around $t = 20,000$ indicated by the information dynamics is thereby also reflected in the opcode correlation matrices. In addition these matrices indicate which instructions are involved in the transition and which new instructions emerge in the new epoch.

Obviously, the chosen measures are not sufficient to characterize the dynamics in all details. However, the proposed measures have the advantage of being very accessible from a computational point of view. Also, some of the measures are closely related to the functional properties of the system. Several other more complicated measures have been proposed, but they are either very complex to compute or cannot be computed at all. Some of these measures are discussed in Bennett (1988 and 1989).

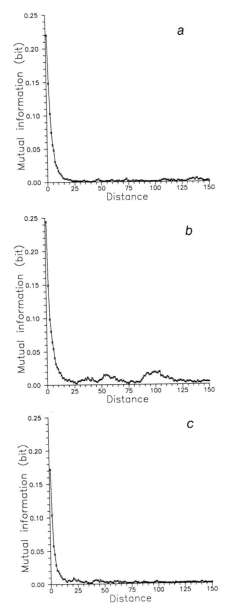

Figure 8. (a) The spatial mutual information vs. distance at $t = 19,000$. (b-c) as (a) except at $t = 20,000$ and $t = 21,000$. Note the peaks in (b) around distances of 55 and 100.

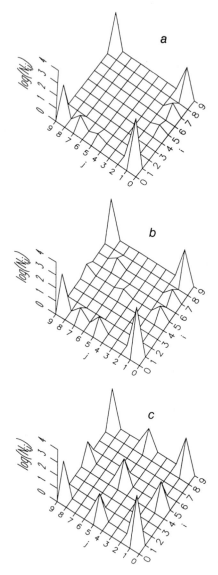

Figure 9. The occurrence of correlated opcodes at $t = 19,000$ (a), $t = 20,000$ (b), and $t = 21,000$ (c), respectively.

DISCUSSION

Of the information processing features associated with living systems, the ability of systems to alter themselves through an introduction of novel properties is what has made biological evolution possible. Novel properties are introduced through changes in the biomolecules that constitute the organisms. We refer to systems with these properties as being self-programmable. To investigate the self-programming property of living systems, we have designed a system far simpler and more tractable than contemporary biomolecular systems. Despite its simplicity, our system has the same fundamental constructive properties as contemporary biomolecular systems. We have seen how this parallel von Neumann based self-programmable system is able to successively develop novel functional properties and how complex cooperative structures spontaneously emerge in the system. Frozen accidents determine the different evolutionary paths in the system and thereby the particular details of the emerging structures. We have discussed cooperative properties of the MOV-SPL structure, of the SPL colonies, and of the jump cores.

These macroscopic cooperative structures are spontaneously generated in the computational system. They emerge in a similar way as macroscopic dissipative structures do, in, for instance, physico-chemical systems. A thermodynamical interpretation of the computational system yields an equivalence between the flux of free energy and the flux of computational resources (executions per iteration), and an equivalence between the microscopic degrees of freedom in the physico-chemical system and all the possible functional interactions in the computational system. A notable difference is, however, that our macroscopic computational structures change even with a constant pumping (executions per iteration). Due to our system's self-programming properties it does not stay in any fixed macroscopic pattern, as for instance a Raleigh-Benard convection or the chemical reaction waves in a Belousov-Zhabotinski reaction do. Our system has in addition the property biological systems also have: it can change itself and thereby undergo development.

How close the details of the processes and the details of the emerging cooperative structures are to evolutionary processes in biological systems and to the structures underlying contemporary living systems we cannot say. The detailed properties of biological evolution as well as the fundamental dynamics underlying life itself are yet unknown. However, our definition of constructive, or self-programmable, dynamical systems has allowed us to start a systematic investigation of truly evolutionary processes. We have freed our formal tools from any predefined evolutionary possibilities. Our system picks its own evolutionary route and constructs its own functional properties. This is in contrast to most formal approaches discussing evolutionary processes. A formal discussion of self-programmable systems is found in Rasmussen et al. (1991). Another constructive system based on the λ-calculus is discussed in Fontana (1991).

One of the major problems associated with self-programmable systems is their complexity. Since the functional properties of such systems depend on their dynamics, any characterization of dynamics as well as functional properties is difficult. We would of course like to be able to detect when novel functional properties are introduced and how such new properties are characterized. It turns out that the Shannon entropy, the mutual information, and Markov chains constructed from correlations, in particular when combined, can be used for that purpose.

The simultaneous calculation of S_1 and S_2 (or M_1) is an efficient way to determine when a complex system is in a quasi-steady state and when the system is in a transition. An alternative is to compare the states of the system at subsequent generations, which perhaps is a little faster. For large systems such as Venus, this requires a considerable amount of storage and retrieval of data, which can be rather time consuming. The calculation of the Shannon entropy S_1 alone is, of course, fast, but if the instruction set includes an instruction to exchange the contents of two words (which is an instruction actually found in most modern microprocessors), then the execution of a large number of exchange instructions would in general alter the functional properties drastically, while S_1 would remain constant. In this case calculation of the mutual information captures that something happens. Another example of the mutual information changing while the entropy is virtually constant occurs when the small self-replicating programs distribute their code into the core.

By also considering long-range correlations and the Markov matrices we obtain more detailed information about the dynamics of the system. The emergence of long-range correlations indicates changes in the local interdependence of the core. Similarities and differences in the scaling exponents describing the decays of the correlations of the different fields as a function of distance can be used to uncover details about the dynamics of the individual fields. Finally, the Markov matrices reflect instabilities and fluctuations in the opcode distribution, and they in particular signal which opcodes are involved in the transitions of the system.

Obviously, these measures are not sufficient to characterize the dynamics in all details, and they do not uncover all the interesting details of the emerging cooperative structures. However, the proposed measures have the advantage of being very accessible from a computational point of view. Several other measures have been proposed, but they are either very complex to compute or cannot be computed at all. Some of these measures are discussed in Bennett (1988 and 1989).

ACKNOWLEDGMENTS

We would like to thank Jeppe Sturis, Walter Fontana, Doyne Farmer, Chris Langton, and Erik Mosekilde for valuable discussions. Erik Mosekilde and Ellen Buchhave are acknowledged for arranging such an enjoyable workshop.

REFERENCES

Bennett, C. H., 1989, Dissipation, Information, Computational Complexity and the Definition of Organization, in: "Emerging Syntheses in Science," Pines, D., ed., Addison-Wesley, Reading.

Bennett, C. H., 1988, Computational Measures of Physical Complexity, in: "Lectures in the Sciences of Complexity I," Stein, D. L., ed., Addison-Wesley, Reading.

Fontana, W., 1991, Algorithmic Chemistry, in the proceedings of the Second Artificial Life Workshop, SFI Studies in the Sciences of Complexity, Farmer, J. D. et al., eds., Addison-Wesley (in press).

Hopcroft, J. E., and Ullman, J. D., 1979, "Introduction to Automata Theory, Languages, and Computation," Addison-Wesley, Reading.

Langton, C., 1990, Computation at the Edge of Chaos: Phase Transitions and Emergent Computation, Physica D 42, 12-34.

Palmer, R., 1988, Neural Nets, in: "Lectures in the Sciences of Complexity I," Stein, D. L., ed., Addison-Wesley, Reading.

Rasmussen, S., Knudsen, C., Feldberg, R., and Hindsholm, M., 1990, The Coreworld: Emergence and Evolution of Cooperative Structures in a Computational Chemistry, Physica D 42, 111-134.

Rasmussen, S., Knudsen, C., and Feldberg, R., 1991, Dynamics of Programmable Matter, in the proceedings of the Second Artificial Life Workshop, SFI Studies in the Sciences of Complexity, Farmer, J. D. et al., eds. Addison-Wesley (in press).

Sturis, J., Polonsky, K. S., Blackman, J. D., Knudsen, C., Mosekilde, E., and Van Cauter, E., 1991, Aspects of Oscillatory Insulin Secretion, these proceedings.

Touretzky, D., ed., 1990, Proceedings of the Neural Information Processing Conference, NIPS 2, Morgan Kaufman Publishers.

THE PROBLEM OF MEDIUM-INDEPENDENCE
IN ARTIFICIAL LIFE

Claus Emmeche
Institute of Computer and Systems Sciences
10, Julius Thomsens Plads
DK-1925 Frederiksberg C
Denmark

Abstract

One version of the Artificial Life research programme presumes, that one can separate the logical form of an organism from its material basis of construction, and that its capacity to live and reproduce is a property of the form, not the matter. This seems to contradict the notion of a cell within contemporary molecular biology, according to which "form" and "matter" do not represent separate realms. The information in a living cell is intimately bound to the properties of the material substrate. This condition may represent a restriction on the validity of formal theories of life.

Introduction

A new field of research called "Artificial Life" (henceforth referred to as AL) seems to be under establishment as a legitimate domain of scientific inquiry. Like "Artificial Intelligence," it brings together people from a lot of disciplines and provokes new questions and approaches to the study of complex phenomena. The purpose of this paper is to discuss a basic concept of this research programme and to look at the nature of the kinds of models that are claimed to realize genuine lifelike properties.

The strong version of Artificial Intelligence is based on the assumption that cognitive functions are computational and thus in principle independent of the material substrate supporting computational processes. In the same way, the proposed AL research programme seems to presume that one can separate the logical form of an organism from its material base, and that its "aliveness," its capacity to live and reproduce, is a property of the form, not the matter (Langton 1989). Therefore it is possible to synthesize life (genuine living behaviour) on the basis of computational principles. The claim that life, or lifelike behaviour is possible to realize in the computational medium, I will call the "strong version" of AL, as opposed to a weaker version that only claims to model various aspects of living behaviour by various kinds of computer simulation techniques. Strong version AL seems to contradict an immediate intuition of molecular biologists, that "form" and "matter" do not represent separate realms (at least on the intracellular level), and that information is intimately bound to the properties of the material substrate. In this paper, I will explicate this intuition by an example to give an idea of the problems facing computational attempts to synthesize life and give formal descriptions of living behaviour.

Artificial life as a step towards a theoretical biology

AL research may evoke a new dialogue between computer scientists and experimentalists about a set of related questions: Can we construct universal theories of life independent of the specific media that life on Earth is made of, as described by biochemistry and molecular biology? Are there any necessary relations between the material components and the formal processual structures that characterise living systems? Is the computational idea to synthesize life counterintuitive to biologists only because of a prejudice that experimental intervention in Nature is the only way to assess the structure of reality? In what sense may life be a computational, medium-independent phenomenon?

Part of the historical background of these questions is the way one has conceived of the relation between living and dead nature, and by implication, biology and physics. Traditionally, biology is seen as an empirical science concerned with pure local phenomena that are formed by natural selection, often too complex for detailed basic explanation. Physics is seen as the science of universal processes, ranging from the most minute particles to the evolution of cosmos. This picture is highly simplified. On the one hand, some types of physical processes only occur very rarely at highly specific circumstances, and are thus equally local. On the other hand, life, or lifelike processes, may be a much more global phenomenon than the picture of present biology can tell. Considering the size of the universe, we should expect on probabilistic grounds that forms of life have evolved on other planets, yet too far away to give access to empirical investigation (Papagiannis 1985). The specific earthly forms of life we know about are the result of a vast succession of historically frozen accidents constrained by some general principles of biological evolution and morphogenesis, which in turn depend on the mechanisms of heredity and biochemistry.

Unfortunately, neo- and post-darwinian evolutionary theory has been unable to give any satisfying account of the nature of developmental and evolutionary constraints (Webster and Goodwin 1982, Ho and Fox 1988). Thus, we cannot by the present theory of biology distinguish between possible and impossible forms of life (but for a small section of the biological possibility space very close to actual forms of life, such as the sets of lethal mutants). The genetic code, for example (specifying the transcription of DNA sequences into protein sequences), might have been differently composed. However, its presumed arbitrarity might not be due exclusively to historically frozen accidents and various external and (with respect to the living system) contingent causes; rather, some general biochemical constraints on possible forms of protein synthesis and regulation not yet understood may have acted lawfully in the process of creation of this specific code, disallowing the formation of other code tables (Crick 1968, Orgel 1968). Thus, theoretical biology could benefit from new approaches to its subject matter in order to make progress as to the general aspects of living systems.

Definition of artificial life

In this context, it is stimulating that Chris Langton proposes to characterise the field of AL as the study of "life as it could be" — so that other forms of life than those actually evolved on earth until present fall within the proper realm of bio-research.

The claim of empirical extension

The first point in Langton's definition of the subject of AL goes as this: "Artificial Life is the study of man-made systems that exhibit behaviors characteristic of natural living systems. It complements the traditional biological sciences concerned with the *analysis* of living organisms by attempting to *synthesize* lifelike behaviors within computers and other artificial media. By extending the empirical foundation upon which biology is based *beyond* the carbon-chain life that has evolved on Earth, Artificial Life can contribute to theoretical biology by locating *life-as-we-know-it* within the larger picture of *life-as-it-could-be*." (Langton 1989, p.1) This is a step towards a more comprehensive view of biology, although one can discuss if present biology is merely analytic. Biologists have recognized the synthetic organizational complexity of their objects ever since the word organization in the late 18th century was used by French and German naturalists to emphasise that the distinction between living, organized bodies and brute, inanimate ones was more essential than the earlier division of Nature into the kingdoms of animals, plants, and minerals. More important, one can discuss what kind of criteria can be used to evaluate theories and models of "life as it could be" in a non-trivial subset of possible worlds (a science of AL should be delimited from mere game construction and computer animation for science fiction movies). As anything is possible in pure imagination, AL has to take recourse to the earthly biology to see if a particular instance of an artificially constructed model of life has a plausible behaviour. Physically interesting models of reality should not represent violations of known physical laws. So biologically interesting models of some aspect of a living process should violate neither physical laws nor what is conceived to be possible living behaviour. The problem is, that the latter is ill-defined, and the computational paradigm of AL does not by itself provide a better description of the universal phenomena of biology.

Methodology

With respect the computational approach of AL and the crucial property of emergent behaviour in AL models, Langton rightly dissociates the AL research programme from classical research in Artificial Intelligence, where models are built "top-down" (general specifications of behaviour are recursively decomposed into simple algorithms), inference is sequential, and the global control of behaviour allows no emergence of really new patterns of behaviour. The computational paradigm of connectionism within cognitive science has an approach to modelling complex behaviour which is essentially the same as in AL, at least with respect to parallelism, "bottom-up" specification (recursive rules apply to local structures only) and emergence.

The claim of medium-independence of life

The third point in Langton's definition of AL is more troublesome. It is true that bottom-up *models* of emergent properties can be said to synthesize lifelike behaviours within a computer, but Langton intensifies this view and postulates that, in fact, by these means one can *realize* lifelike properties. Thus we can have *genuine life* in artificial systems (p.32, ibid.). For example, in a model of flocking behaviour of birds by Craig Reynolds, the individual simulated birds (or "boids") are not real birds, but their emergent behaviour in the model is for Langton as genuine and real as the behaviour of their natural counterparts. Thus Langton does not claim that computers themselves will be alive, but that the informational universes they can support can eventually be (p.39). According to Langton, the "artificial" in AL refers only to the component parts, not the emergent processes: "If the component parts are implemented correctly, the processes they support are *genuine* — every bit as genuine as the natural processes they imitate." (p.33)

The reason for this claim is a somewhat Platonic conception of life, according to which "...the dynamic processes that constitute life — in whatever material bases they might occur — must share certain universal features — features that allow us to recognize life by its dynamic *form* alone, without reference to its matter. This *general* phenomenon of life — life writ-large across all possible material substrates — is the true subject matter of biology." (p.2, ibid.) And thus "...the principal assumption made in Artificial Life is that the 'logical form' of an organism can be separated from its material basis of construction, and that 'aliveness' will be found to be a property of the former, not the latter." (p.11, ibid.)

Thus, while the simulated birds have no cohesive physical structure but only exist as information structures within a computer (as Langton admits), the phenomenon of flocking birds and the flocking of simulated birds should be two instances of the same phenomenon: *flocking*. To a hunter or an ornithologist, or even a bird, it may seem a little strange. To a logician or a computer scientist used to handling abstract symbolic structures, it is quite obvious. The problem of theoretical biology, I think, is to deal with the living material structures of organisms and their inherent "logic" at the same time.

But even if that can be done, our *models* of the logic of living systems are not necessarily an instances of the true logic inherent in the very systems themselves. Furthermore, the logic of life is a many level affair, spanning in time and space from molecular to ecological and evolutionary relationships. The physical/chemical causal processes within an organism are of a different kind (described by a different set of theories) than the processes within a computer running some programme. Their functions may be similar on some level of description, but the inherent logic of the processes, on the physical/chemical level (and probably on higher levels as well), is likely to be different.

Why "strong AL" is biologically counterintuitive

Why does strong AL seem to be so counterintuitive from a biological point of view? Hardly any biologists can disagree with Langton when he says that "Neither nucleotides nor amino acids nor any other carbon-chain molecule is alive — yet put them together in the right way, and the dynamic behavior that emerges out of their interactions is what we call life." (p.41) But we also learn that "Life is a property of *form*, not *matter*, a result of the organization of matter rather than something that inheres in the matter itself." (ibid.) Though this is a purely philosophical claim rather than a scientific proposition, it appears to be incompatible with an intuition nourished by the current paradigm of molecular biology. This intuition says that real life is both form and matter, and that the proper object of life science is to study both aspects and their dynamic interdependence.

All living organisms on Earth happen to be made up of cells. A cell is analyzed in terms of its materials, structures, and processes, and can be seen as a bag of chemicals (each of which has its own form), and as a complex self-organizing structure containing a sequence of digital non-complete self-description (Pattee 1977) and embodying a web of structural informational relationships in time and space (Løvtrup 1981, Alberts et al. 1989). In a material world, life can not be pure form. Molecular biology is often accused of only taking interest in the material components of the cell. The obvious answer is that, on the intracellular level, molecular biologists cannot separate form and matter, because the behaviour of the cell and its constituent molecules depends crucially on the *form* of the individual macromolecules (proteins, ribosomes, messenger-RNAs, etc.).

The "form," or biological information, of an organism is bound to the properties of the material substrate to such an extent that attempts such as von Neumann's to give "the logical form of the natural self-reproduction problem" (see Langton 1989, p.13) will encounter severe problems. This interdependence of form and matter can be illustrated by an example:

In the bacterium *Escherichia coli*, the synthesis of the aminoacid tryptophan from chorismatic acid occurs in three steps, each of which is catalysed by a specific enzyme. These enzymes are themselves synthesized from aminoacids by the ribosomes. This process of polymerizing aminoacids into protein chains is termed translation, because the sequence of the 20 different aminoacids in proteins is specified by a DNA sequence (gene) on the chromosome. In the case of the three tryptophan synthesis enzymes, the DNA regions specifying their aminoacid sequences are grouped together in a common control unit, the tryptophan operon. For the enzymes to be synthesized, the DNA regions have to be transcribed into mRNA chains. This is done by RNA polymerase which starts transcribing from the tryptophan promoter in the first part of the operon and terminates after all the genes have been transcribed. Since transcription and translation are very energy consuming processes, the regulation of these processes in the response to the need for the enzymes is highly advantageous to the cell. A sophisticated system for the control of transcription of the tryptophan operon is found in *E.coli*. The attenuation control system involves both the protein coding function and the physical nature of the RNA chain. The mechanism is thoroughly studied and shall only be briefly described here (see Landick and Yanofsky 1987).

Attenuation control is based on the tight coupling of transcription and translation in prokaryotes. The RNA leader region contains a ribosome binding site followed by a short gene encoding 13 aminoacids, a stop codon, and a non-coding region including the attenuator sequence. The latter consists of a transcription terminator, i.e., a short C-G-rich palindrome (that forms a "hairpin" secondary structure by base pairing) followed by eight U residues. The region can exist in alternative base-paired conformations (fig.1); only one of these allows the formation of the terminator. A substantial body of evidence has established a model according to which the tryptophan level determines the ability of the ribosome to proceed through the leader region, and this in turn controls the formation of mRNA secondary structure: a) When tryptophan is present, the ribosome can continue along the leader (synthesizing the leader peptide) to the stop codon between region 1 and 2. The ribosome now extends over mRNA region 2, preventing it from base pairing. Thus region 3 can pair with region 4, generating the terminator hairpin, that causes the RNA polymerase to terminate (it dissociates the RNA from the DNA so that no genes are transcribed). b) When the cells are starved for tryptophan, the ribosome stalls (because of deficiency of charged Trp-tRNA) when it encounters the two tryptophan codons (i.e., the RNA code specifying the insertion of tryptophan in the growing chain) in immediate succession within region 1 of the leader. By stalling, the ribosome sequesters region 1, so that regions 2 and 3 will base pair before region 4 has been transcribed by the polymerase. Thus, mRNA region 4 remains single-stranded, no termination hairpin is formed, and RNA polymerase will read through the attenuator and transcribe the remainder of the operon. In summary, the attenuation control responds directly to the need of the cell for tryptophan in protein synthesis.

The right timing of the events is essential. Evidence suggests that after initiation at the promoter, the RNA polymerase proceeds to a position (after the leader peptide sequence) where it pauses; this may be necessary to give time for ribosomes to bind to the ribosome binding site of the leader transcript before regions 3 and 4 are synthesized. The 1:2 mRNA secondary structure may function as the transcription pause signal.

The example shows several things. The "linguistic mode" of the cell (i.e., the instructions in the DNA) and the "dynamic mode" (the workings of the machinery) are so closely connected in the prokaryote cell that the "logic" that describes the behaviour of the cell is time-dependent and for some part implicitly represented in the machinery that reads the instructions (*pace* Pattee 1977). The argument of von Neu-

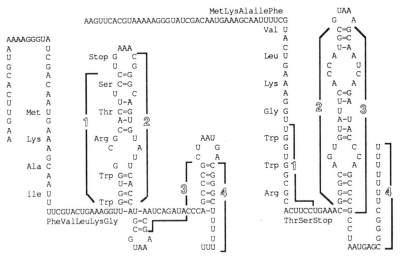

Fig.1. The alternative base-paired conformations of the mRNA *trp* leader region. The four regions that can base pair are shown. Region 1 contains the last five codons of the leader peptide. Region 4 and the last part of region 3 contain the attenuator sequence. Left: The 1:2 and 3:4 secondary structures. The pairing of region 3 and 4 generates the terminator (the 3:4 hairpin followed by the U residues). Right: The 2:3 secondary structure allows no formation of terminator hairpin. (After Lewin, 1983, and Landick and Yanofsky, 1987).

mann that it is possible to abstract the logical form of some feature of an organism's performance (such as self-reproduction) runs into difficulties when one attempts to "realize" this form in another medium. To describe logical aspects of biological systems in order to formalize them may be complicated, though possible (and if possible, it may only be trivial aspects of self-reproduction that are formalized, as argued by Kampis and Csányi (1987)). But attempts to realize these formal descriptions in a second medium may be much harder if the implementation of the formal description does not take into account the interdependence of form and matter at the cellular level. What is realized is our formal theory, not a duplication of the original living system.

One might object to the example above, arguing that it only shows that AL models must be adapted to another level of detail to encompass the mechanisms described. One could in principle make a cellular automaton model of the interactions of the attenuation control. But still, this would be a simulation; the model might be *formally* similar to the operon of the *E.coli* but would have no physical or causal similarity to the real system. It is often claimed (e.g., Burks 1975) that von Neumann's 1948 kinematic model of self-reproduction (the components of which are a factory, a duplicator, a controller, and a written instruction) was verified by the discovery of DNA structure and functioning (the analogous components being the ribosomes, the polymerases, the repressor + derepressor control molecules and the DNA). But the analogy is in no way complete, because the functions of the biological components are not separated in the real system and depend on the specific physical structure of the constituents. The dynamic information (Burks and Farmer 1984) stored in the 3D structure of DNA and the rest of the cell's components is not represented by the formalized model (a central instance of the symbol-matter problem described by Pattee (1989)).

The implicit functionalism in "strong AL"

The strong version of AL is in one respect very similar to the strong version of Artificial Intelligence, or the functionalistic stance within cognitive science. They both embrace the philosophical idea of medium-independence: the characteristics of life and mind are independent of their respective material substrates. Genuine living behaviour can be realized in the computer because *life* basically is (or belongs to) a class of complex behaviours that could haunt other media than the biochemical.

To a molecular biologist, functionalism may seem rather peculiar: a philosophical doctrine of "a person who believes that study of the functioning of a person or animal is all important and that it can be studied, by itself, in an abstract way without bothering about what sort of bits and pieces actually implements the functions under study." (Crick 1988) Functionalism in cognitive science has a background in psychology, developed in reaction to behaviourism (that did not allow psychologists to look into the black box of the brain), and in philosophy of mind, put forth as an alternative to an inconsistent materialistic theory of identity between mental states and neural states. Though one can distinguish between functional analysis as a research strategy (Cummings 1975), explanatory functionalism within psychology appealing to computation by representations within a "language of thought" (Fodor 1975), and metaphysical functionalism as a philosophical theory of mind (Putnam 1960, Block 1980), there are some main features shared by all forms of functionalism:

a) A more or less explicit notion of functional equivalence, where x is functionally equivalent with y, if x has capacities to contribute to the capacities of the whole in a similar (or the same) way as y.

b) A more or less strict reliance on the concept of a Turing Machine.

c) The assumption that the causal structures postulated to be identical with the mental states can be realized by a vast variety of physical systems.

ad.b) Mental states are often identified with Turing machine table states; and to give a true explanation of some psychological phenomenon is seen as something like providing a computer program for the mind — or some of its subroutines. One should therefore attempt to give a functional analysis of mental capacities broken down into their component mechanical processes. If these processes are algorithmic (which is often assumed without justification), then they will be Turing-computable as well.

ad.c) It is well known that a digital computer, in principle, can be of many different kinds of components; valves, transistors, chips, neurons, or jets of water. This multiple-realization argument may be true for any *formal* system, given the right interpretation of the structure that implements it. But three hurdles should be noted: 1. It does not guarantee that our formalization of specific systems — whether mental, biological, or physical — can catch all the essential factors that govern such a system. There might even be aspects of the system that are in principle unformalizable. For instance, the meaning of the symbols manipulated by a cognitive process is context-dependent, and the ultimate context of the human language is the natural and cultural world — that may be hard to formalize. 2. The construction (of any material kind) that implements the formal structure (a model of speech, for example) is still in need of our interpretation in order to give any meaning — a thing we might easily forget in the case of a system based on purely syntactical rules appearing to instantiate semantically meaningful behaviour. The semantics is not intrinsic to syntax but depends on our conscious interpretation of the system (Searle 1980). 3. The functioning of a construction implementing some formal structure may well be functionally equivalent to other implementations (or realizations) on one chosen level of description, while on another level it may show dissimilar properties that from a biological point of view may seriously effect its chances of survival in a realistic environment. This fact shows another problem with the property of functional equivalence: that it basically is a logical property; that it is level-dependent; and that it may not cope with "real life" situations where dependence on time-consumption and energetic efficiency on several levels of organization may be crucial for the proper functioning of a system.

Though some elements of functionalism are shared by the strong version of the AL programme, it does not follow that all problems facing functionalism in cognitive science will be the same in AL. I think, however, that two parallels can be drawn.

Some psychologists and phenomenalistic philosophers have objected to computational accounts of mind and cognition, arguing that *cognitive activity* is intimately related to a living human being (or animal) situated in a specific environment and cannot be abstracted from the sensuous, bodily actions of the organism without losing some crucial aspects of this activity, as, e.g., the view from inside (or the *umwelt*, i.e., the species-specific subjective universe (Uexküll 1926)). Only in theory is cognition guided by the formal rules of logic; in practice, it is subjected to a subject's specific bodily desires, feelings, material needs, interests, purposes, etc. Thus, one cannot separate cognition from volition and emotion, and these "psychical" properties are features of genuine biological processes. As the "psyche" of man or animal in this sense is medium-dependent, so is a living organism's teleonomic orientation and relation to its environment. Therefore, as we cannot have machines that "think" in the same way as humans or intelligent animals think, we cannot have machines that act and react, self-organize and reproduce (and sustain their "autopoiesis") in the same way as real organisms do. To generalize the concept of cognition to include machine as well as personal thinking leaves unanswered the question of the real nature of (the human type of) thinking. In the same way, though we could generalize the concept of *life* to include lifelike behaviour of machines, and postulate that wet organisms just instantiate some of the same abstract properties of reproduction, metabolism, irritability (or what might be selected as important features of an organism-machine), this would not reveal the specific constraints on the way life has evolved or could have evolved on Earth. And it does not tell us much new about life as it could be — not even in a silicon valley on a foreign planet. The processual characteristics of life will always be higher level phenomena constrained by specific lower level properties. The general phenomenon of emergence is probably a universal feature of life, but one must also look at the set of possible material substrates that can "support" emergence.

A second parallel between problems with functionalism in cognitive science and AL concerns the notion of computation and the relationship between the pattern generating properties of the physical functioning of a computer model and our specific interpretations of these patterns. Much research in what Haugeland (1985) dubs "Good Old Fashioned Artificial Intelligence" relies on *the formalists' motto*: "If you take care of the syntax, the semantics will take care of itself," i.e., if the system modelled is well formalized and the rules sufficiently strong, the automation of that system guarantees that any output when interpreted makes sense. However, this presupposes that such rules can be found, but many cognitive skills and topics (such as common sense and natural language use) resist formalization. Furthermore, the interpretation is still not intrinsic to the formal system itself but imputed by somebody ascribing meaning to the output symbols: semantics is not intrinsic to syntax. Although the computational paradigm of AL (and connectionism in cognitive science) is different, there seems to be a parallel *computationalists' motto* at stake: "If you take care of the computational setup, living behaviour will emerge by itself." Again, this presumes that the component units can be formalized appropriately and that "aliveness" exclusively is a property of a formal or computational system. But what if computation is not intrinsic to physical or biochemical systems? We normally conceive of computation as mathematical operations with numbers performed by man or manmade machines (interpreted by man). One may *talk* of the lac operon in digital-mechanical terms as a "chemical computer" and express one's amazement about DNA metaphorically, calling it "certainly the most sophisticated computer of which we are aware" (Burks and Farmer, 1984), but that does not by itself render the physical or biological world a computer or its processes computational. Anything that obeys physical laws can be simulated on a computer (with limitations on accuracy and speed), but that does not substantiate a computational viewpoint of physical processes.

One could argue that, in contrast to cognitive science that claims it possible to synthesize intelligence because the brain is information processing, AL is not committed to the view that life (e.g., a cell) is information processing; it may be, but that is not central to the possibility of AL. What is central is that the parallel, bottom-up computational approach allows the computer to support emergence of complex behaviour in the same way as a prebiotic chemical system allowed the self-organization of matter into living cells. However, that does not make the model an instance of the thing modelled. A cellular automaton (CA) model of some physical system such as weather, constructed by the same computational approach, is still not to be thought of as an (artificial) instance of weather, realizing the very causal phenomena of thunderstorms. To use a distinction of Kant in this context (as re-introduced by Sober, 1990), one recognises that the model system *follows a rule* (or a set of rules, namely the ones represented by the CA state transition function table), but the natural system acts *in accordance with a rule* (i.e., physical laws). The natural system does not consult representations in order to "update" its state but behaves *as if* it had consulted a set of rules. Thus, even if the possibility of strong AL is not committed to the view that life is information processing (that key features of life are governed by intrinsic representations within the living system), admitting this distinction should moderate the claims of computational realizability of life.

Conclusion

If AL or some other kind of "empirical mathematics" should have any bearing on the way biologists conceive of their subject matter, the question about the reality-status of the models will inevitably spring up — from both sides: What is the biological content of AL models, and what logical or computational lessons can be drawn from the biologists' empirical garden of model species? In biology, many theoretical generalizations have often been made on the basis of a small set of model species such as the fruit fly or *Eschericia coli*. (The question whether AL models are simulations of biological processes or essentially realize lifelike properties may be extended to the field of complex dynamics. Here, the relationship between mathematical properties (such as universality in the description of chaos) and measurable real behaviour of the systems described is not straightforward. It is not always clear whether real systems actually realize deterministic chaos or if their behaviour "simulates" instances of quasi-periodicity and noise.)

I am not really convinced that "real" artificial life is impossible; on the other hand, I'm far from persuaded that it's inevitable. I am dubious, because difficult questions about the nature of life and computation remain open. The following points are not meant to be conclusive, but to express the limits of my doubt about the AL research programme. There are reasons to believe that:

1. Life is not medium-independent, but shows an interdependence of form and matter.
2. Life may be realized in other media than the carbon-chain dominated as a result of a long, natural evolutionary process.
3. AL research may contribute to theoretical biology by:
   (i) simulating developmental and evolutionary phenomena of life on Earth,
   (ii) simulating life as it could have evolved in non-earthly environments given some set of realistic boundary conditions,
   (iii) providing new concepts and models of emergent phenomena belonging to a general set of complex systems of which biological systems (under particular kinds of descriptions) may be a subset.
4. AL may inspire attempts to realize life artificially in other media by *in vitro* experiments. Such prospects include the experimental approach of molecular biology and protobiology research. However, this is not yet the centre of interest in the present AL research programme.

Acknowledgments

I thank Mogens Kilstrup for his valuable comments on an earlier version of the manuscript.

References

Alberts, B., Bray, D., Lewis, J., Raff, M., Roberts, K., Watson, J.D., 1983, "Molecular biology of the cell," Garland, New York.

Block, N., 1980, What is functionalism?, pp. 171-184 *in*: "Readings in the Philosophy of Psychology, Vol.I," N.Block, ed., Methuen, London.

Burks, A.W., 1975, Logic, biology and automata — some historical reflections, *Int.J.Man-Machine Studies* 7: 297-312.

Burks, C., and Farmer, D., 1984, Towards modeling DNA sequences as automata, *Physica* 10D: 157-167.

Crick, F., 1968, The origin of the genetic code, *J.Mol.Biol.* 38: 367-379.

Crick, F., 1989, "What mad pursuit," Weidenfeld & Nicolson, London.

Cummings, R., 1975, Functional Analysis, *Journal of Philosophy* 72:741-764.

Fodor, J.A., 1975, "The Language of Thought," Crowell, New York.

Haugeland, J., 1985, "Artificial Intelligence: the very idea," MIT Press, Cambridge, Mass.

Ho, M.-W. & Fox, S., 1988, "Evolutionary processes and metaphors," Wiley, New York.

Kampis, G., and Csányi, V., 1987, Replication in abstract and natural systems, *BioSystems* 20: 143-152.

Landick, R., and Yanofsky, C., 1987, Transcription attenuation, pp.1276-1301 *in*: "Escherichia coli and Salmonella typhimurium. Cellular and molecular biology, Vol.2," F.C.Neidhardt, ed., American Society for Microbiology, Washington.

Langton, C.G., 1989, Artificial life, pp.1-47 *in*: "Artificial Life" (Santa Fe Institute Studies in the Sciences of Complexity, Vol. VI), Langton, ed., Addison-Wesley, Redwood City, California.

Lewin, B., 1983, "Genes," 2nd.ed., Wiley, New York.

Løvtrup, S., 1981, Introduction to evolutionary genetics, pp.139-144 *in*: "Evolution Today,"G.G.E.Scudder & J.L.Reveal, eds., Hunt Institute for Botanical Documentation, Pittsburgh, Pennsylvania.

Orgel, L., 1968, Evolution of the genetic apparatus, *J.Mol.Biol.* 38: 381-393.

Papagiannis, M.D., 1985, Recent progress and future plans on search for extraterrestrial intelligence, *Nature* 318: 135-140.

Pattee, H.H., 1977, Dynamic and linguistic modes of complex systems, *Int.J. General Systems* 3: 259-266.

Pattee, H.H., 1989, Simulations, realizations, and theories of life, pp. 63-77 *in*: "Artificial Life" (Santa Fe Institute Studies in the Sciences of Complexity, Vol. VI), Langton, ed., Addison-Wesley, Redwood City, Calif.

Putnam, H., 1960, Minds and machines, *in*: "Dimensions of Mind," S.Hook, ed., New York University Press, New York.

Searle, J., 1980, Minds, brains, and programs, *Behavioral and Brain Sciences* 3: 417-458.

Sober, E., 1990, Learning from functionalism — prospects for strong AL, talk delivered at The Second Artificial Life Workshop, Feb.5-9, 1990, Santa Fe, New Mexico (manus. version of Sept. 5th, 1990, unpubl.).

Uexküll, J. von, 1926, "Theoretical Biology," Kegan Paul, London.

Webster, G., and Goodwin, B.C., 1982, The origin of species: a structuralist approach, *J. Social Biol. Struct.* 5: 15-47.

AN OPTIMALITY APPROACH TO AGEING

Reimara Rossler[1], Peter E. Kloeden[2] and Otto E. Rossler[3]

[1]Medical Policlinic, University of Tubingen
7400 Tubingen, West Germany

[2]School of Mathematical and Physical Sciences
Murdoch University, Murdoch, Western Australia, 6150

[3]Institute for Physical and Theoretical Chemistry
University of Tubingen, 7400 Tubingen, West Germany

ABSTRACT

Medawar and Williams' partially predictive approach to ageing can be turned into a predictive theory. A recently found new ageing formula allows interpretation in terms of an optimality criterion: "Minimize competition with offspring." A fixed-point problem in function space results. If nature complies, two new open questions arise: (1) Identification of the evolutionarily stable mechanism which ensures selection for the new trait; and (2) identification of the physiological mechanism which enables statistical adherence to the new law on the level of the individual.

INTRODUCTION

Recently, a new empirical approach to ageing has been proposed. Two life expectancy curves, taken from human populations around the years 1900 and 1930, turned out to be fittable in their post-childhood portions by the same function if only the exogenous (age-independent) death rate was assumed to be different[1]. The new function is

$$x(t) = x(0) \exp \left\{ - bt - a \sum_{j=1}^{\infty} \frac{1}{(j+1)!} \ p^j \ (\{t-jr\}^+)^{j+1} \right\} . \qquad (1)$$

Here $x(t)/x(0)$ is the surviving fraction as a function of age t ; $j=1$ refers to the first filial generation and so forth; and the superscript $+$ to a parenthesis means that less-than-zero values are set equal to zero. The parameter b is the age-independent death rate; the parameter r is the pre-reproductive delay (from birth up to birth of the first offspring); and the parameter p is the average number of surviving offspring per individual per year, beginning after the r-th year.

This function contains as a special case the well-known, empirically

Complexity, Chaos, and Biological Evolution, Edited by E. Mosekilde and
L. Mosekilde, Plenum Press, New York, 1991

founded Gompertz-Makeham curve (cf.[2]),

$$x(t) = x(0) \exp \{ a\, p^{-1}\, (1 - e^{pt}) \} \ , \qquad\qquad (2)$$

if the fitting parameter a is put equal to b , see[1]. The Gompertz curve is known to provide only an inadequte fit to empirical data since the latter usually contain a "non-Gompertzian tail," cf.[3]. Moreover, the parameters entering the Gompertzian fit lacked an easy interpretation previously. On both counts, Eq.(1) represents an improvement. First, it is strictly greater than Eq.(2) - especially so in the tail region - , cf.[1]. Second, it draws on independently available biological parameters of the species in question.

Even though Eq.(1) still needs to be tested on further empirical data, both human and nonhuman, the question can be posed already why an equation like Eq.(1) may possibly be obeyed by nature.

DEDUCTIVE AGEING THEORY

The discovery of Eq.(1) was motivated and stimulated by the biological theory of Medawar[4-6] and Williams[7]. The basic idea in the M-W approach is the assumption that ageing somehow reflects the lawful decline of a quantity called "reproductive value" of the individual by R. A. Fischer (cf.[5]). The older an individual is, the more likely most of its genes have been planted safely in the population, so that its own survival becomes less and less important from the point of view of the survival of its genes.

The decline in reproductive value as a function of age leads to testable predictions. Two major predictions were put forward by M-W. First, unfavorable mutations whose effects become manifest only late in life (like certain cancer-generating genes, perhaps) should accumulate because they will be weeded out less efficiently than other detrimental mutations characterized by an earlier age of manifestation. Second, mutations that are ambivalent, generating a beneficial effect in the short run but causing unfavorable effects later in life (certain instances of primary hypertension come to mind here), ought to enjoy a selective advantage.

The very elegance and simplicity of the M-W approach suggests that an optimality approach[8] to ageing ought to be possible. This promise has, to our knowledge, not yet been consummated. The reason has to do with the fact that the notion of reproductive value is (at first sight) only a "unilateral" constraint. In the absence of a second contraint that pushes in the other direction, so to speak, a full-fledged optimality principle cannot be formulated. Moreover, even if a criterion with a well-defined optimum could be indicated, the two mechanisms suggested by M-W as underlying ageing would interfere with the very attainability of the new optimum whenever it lies markedly above the previously valid one. This follows from the "multigenic" origin of the ageing curve according to M-W. The effects generated by many different mutations would need to be "undone" on an almost individual basis by many new mutations, before a doubling of the life span could be achieved, for example[7]. Nature, on the other hand, is replete with closely related species which differ by a factor of two or more in their life spans (like chimpanzees and humans), cf.[9].

OUTLINE OF THE PHILOSOPHY WHICH LEAD TO EQUATION (1)

Equation (1) was derived on the basis of a single quantitative assumption: The likelihood to die from endogenous causes during a given infinitesimal time interval, as a function of age, was assumed to be strictly proportional to the cumulative number of offspring (in all filial generations) already born during that time interval. This assumption was motivated by the notion of reproductive value. However, it was at the same time also motivated by empiry. For it unexpectedly turned out that counting the cumulative number of offspring as such (without appending smaller and smaller weigths down the line since a grandchild contains fewer shared genes than a child) resulted in a better fit.

On the other hand, even this simple assumption is implemented in Eq.(1) in a not completely accurate fashion. Several idealizations enter for the sake of simplicity. Specifically, the offspring are assumed to be exempt from ageing themselves. This simplification can be considered as a zeroth-order approximation since offspring by definition is always markedly younger. In addition, the offspring was assumed to be exempt even from age-independent causes of death (except for a mean effect entering the fixed parameter p). It appears unlikely that a closed-form solution can still be found if these two idealizations are relaxed. On the other hand, there is no question that they can be easily be corrected for in the iterative, discrete-time algorithmic version to Eq.(1) which has also been indicated in[1].

A PROPOSED NEW PHILOSOPHY

The following optimality criterion is proposed to be applicable:

"Minimize competition with offspring!" (3)

This criterion makes intuitive sense since the more offspring are born, the more they are going to compete with the ancestor - and vice versa. Therefore a justification of the ad-hoc assumption of proportionality made above apperas to have been found. At the same time, a steady-state condition is assumed to be valid in the population in question as necessary. Only a fixed number of individuals are supposed to be supported by the carrying capacity of the ecological niche available to the population.

More important than compatibility with Eq.(1), however, is the question of whether criterion (3) is compatible with biological theory. In particular: Is there any way how Eq.(3) could be derived from biological first principles?

Unexpectedly, it can be shown that any other criterion is indeed less optimal - so that only (3) survives by default. To see this, assume that Eq.(3) had already been minimized and look at a perturbation. Specifically, assume that one particular "clone," deriving from a single individual, exists that differs from all others in one respect: Its members, starting with the ancestor, are longer-lived by a finite amount. (Think of a dominant gene mutation, for example). Then this clone will have an evolutionary advantage over all other clones in the population. It will therefore replace all others in the long run.

This replacement, however, is counterproductive. The advantage is entirely due to "intraspecific" (intrapopulation) selection and has nothing to do with greater fitness of the clone as far as competition with other populations living in the same niche is concerned. Hence the present trait actually decreases overall fitness. For suppose there exists another clone which is endowed with a genuine evolutionary advantage, under a condition of interspecific selection, but whose advantage happens to be numerically smaller than that conferred intraspecifically by the present trait. Then this other clone will be weeded out (competitively excluded) by Gause's[10] principle.

Once the longer-lived clone has taken over, so that the population has become homogeneous again as far as longevity is concerned, the above perturbation argument becomes applicable again; and so forth. Thus, a greater longevity - with attendant prolonged fertility in the simplest case - is <u>always</u> favored by intraspecific selection if all other traits are assumed to be equal. (If not - when there are late-acting detrimental mutations, for example, as assumed by M-W - , weeding the latter out indeed takes longer; therefore the present trend toward immortality is effectively slowed down. But as soon as genetic homogeneity has been reachieved, another increase in life-span becomes possible.)

Assume for the sake of the argument that the limiting case - complete absence of endogenous ageing - had been achieved. Then it is especially easy to see why this is a counterproductive situation. <u>Any</u> "excess longevity" is detrimental. For in the presence of immortality, there exists no bodily difference any more between individuals having a high and individuals having a low reproductive value. Hence the latter are allowed to compete fully with the former. The resulting situation can be characterized as "maximum competition with offspring."

This formulation reveals two facts. First, there indeed exists no difference in the present case between related and nonrelated (or less related) offspring. This is because for every "nonrelated" younger individual weeded out by an older one, another "related" younger individual will be weeded out by a fellow member of the age cohort of the older one (to the extent that related younger individuals exist for each). This <u>symmetry</u> - cross-inhibition induces an equal-sized self-inhibition - is a direct implication of the steady-state assumption made.

Second, the notion of reproductive value turns out to be strong enough to give rise to a full-fledged optimality criterion since competition between low-reproductive-value and high-reproductive-value individuals needs to be minimized.

In retrospect, this implication is obvious. For suppose that an older individual (whose genes have already been distributed in the population) eliminates through competition a younger individual. Then the latter's genes will be put at a selective disadvantage as we saw. However, this is not the only consequence. Even if there is <u>no</u> genetic difference, this death is counterproductive. For it is functionally equivalent to cannibalism on offspring. This result follows from the above equivalence principle between related and nonrelated youngsters valid under steady-state conditions. While all older individuals compete with all younger ones, the effect is the same as if each were competing only with its own offspring.

All resources which have gone into the weeded-out individual are lost in effect. This loss is   not   accompanied by any advantage to the species. It is as unnecessary as a loss through paedophagy would be. The unnecessary occurrence of those deaths puts the population in question at a selective disadvantage relative to another population in which such deaths occur less frequently. Only when medium-aged individuals are taken out through an endogenous mechanism can the effective narrowing of the niche (through "self-neutralization" of part of the population) be avoided. Thus, Eq.(3) is indeed respected on the level of competing populations.

STEPS TOWARD A MATHEMATICAL FORMULATION

The principle just described appears to be well-posed from a mathematical point of view. It is possible to describe from first principles a function (an ageing curve) which when respected in biology minimizes loss of fitness through competition with offspring.

Writing this function down nevertheless constitutes a difficult problem. The function sought corresponds to a fixed point in the space of all functions spanned by the relevant parameters. All parameters which enter into the production of offspring (p, r, b) are relevant. Moreover, in addition to those fixed parameters, also a "self-consistency constraint" needs to be obeyed. Any tentative change in the endogenous death rate entails the necessity to slightly readjust p if the steady state of the population is to be preserved (cf.[11]).

Therefore a second fixed point is bound to coexist with the one of interest. This trivial solution (runaway of both p and a) is not the solution sought. Thus, a problem with an unusual degree of mathematical difficulty appears to have been identified.

As a first step toward the final solution of this problem, an attempt at interactive numerical optimization will be justified. The discrete algorithm presented in[1] offers itself as a tool.[+] The hope is that in this way, it will be possible to arrive at an estimate of the constant of proportionality, a , which had to be put into Eq.(1) by hand. If all goes well, an "improved version" of Eq.(1) can be formulated which retains the latter's closed-form structure under the proviso that certain limiting ratios between some of the relevant parameters are not exceeded.

DISCUSSION

A "predictive" approach to ageing has been presented. The underlying philosophy goes back to Medawar and Williams. The mathematically defined notion "reproductive value as a function of age" is the main ingredient that enters. However, this notion is no longer defined genetically but in terms of the cumulative number of offspring (progeny) already brought into existence. This assumption, which was arrived at inductively,[1] can be justified if a steady-state condition is assumed to be valid as necessary.

The main new result is that under steady-state conditions, the notion of reproductive value is even stronger than previously realized. It constrains ageing not just in one direction but in two. Low-reproductive-value individuals are not only at a selective disadvantage (as shown by Medawar and Williams) but are also a liability to the population so that they need to be eliminated actively. The situation is not unfamiliar in a quite different, entirely rational human context. (Under a condition of competition for food, older couples may choose to "go onto the ice" as the Eskimo saying goes.) However, the same principle has apparently not been formulated before as an automatic control mechanism arising in a population-theoretical biological context.

While competition between individuals of differing age is as unavoidable as competition between individuals of the same age is, the former becomes selectively disadvantageous as soon as the younger individual represents an offspring (or is "steady-state equivalent" to an offspring). This type of competition therefore needs to be minimized since it reduces the fitness of the population in question.

Ageing, according to this view, reflects evolutionary economy. "Death

[+]Possibly, optimum ageing will occur when, under the harsh high-b conditions of the original niche, endogenous survivability is put equal to "unity minus number of offspring of the same sex likely to reach or have reached maturity." But this is only a speculation at the time being. (Cf.[12] for some on first glance compatible, extrapolated human data.)

is an artifice of life to have more life" is an old gnostic saying paraphrased by Goethe. Now, a quantitative recipe can be distilled form this qualitative insight it appears.

Still, mathematics is not biology. The mathematically prescribed and the biologically possible need not coincide. This fact holds true especially in the present case. For we have seen above (in the discussion following Eq. 3) that on a direct level - within the population - it is maximization of competition with offspring (immortality) that is selected for while on an indirect level - in between populations - , minimization of competetion with offspring is the optimality criterion. These two competing trends let is at first sight appear unlikely that the present, mathematically prescribed, ageing law will prove accessible to nature herself.

In this way, a second mathematical problem poses itself in effect. A mechanism needs to be sought which permits the new prescribed function to be optimal, not only indirectly under conditions of interspecific selection, but also directly under a condition of (mostly) intraspecific selection. Or, to use Maynard-Smith's[13] terminology, optimal ageing needs to be shown to be "evolutionarily stable."

So far, ageing appears to be in the same precarious position as sex (whose long-term advantages are beyond reproach but whose short-term stability is still a mystery). This analogy at the same time suggests a possible way out. If there exists a physiological machinery underlying ageing of a complexity which is comparable to that of the machinery of sexual reproduction, then this machinery may act as a buffer against the easy abolishability of ageing. For once such a machinery has arisen under exceptionally lucky evolutionary circumstances (with several weakly interacting populations competing in closely adjacent environments over an extended period of time, perhaps), it may from then on act as a "stabilizer." This is because on the level of direct competition, its maintenance will be less costly than its piecemeal destruction would be in view of the many links to vital functions that would need to be undone simultaneously.

This conjecture leads over to a second major open problem which will likewise become pressing only in case nature turns out to adhere to the abstractly prescribed law. The well-known life expectancy curve of a population then no longer reflects merely a property of a biological population; it in addition reflects a built-in propensity of the individual. Just as the statistics of many electrons are faithfully built into every individual electron - as when it is observed in a quantum interference experiment, for example - , so the ageing characteristics of the population would turn out to be the consequence of the faithful adherence to the same curve, as a rigidly prescribed expectation value to die from endogenous causes as a function of age, shown by the average (mutation-free) individual.

This prediction follows in a straightforward manner from the above perturbation analysis which started out from a single deviating individual and its built-in longevity. Therefore, "ageing" must exist on the level of the individual in a highly lawful way. While ageing (without quotes) on the level of the individual happens to be a well-recognized fact in society - being made use of by law-makers, for example - , the prediction of "ageing" (with quotes) on the level of the individual existing in nature comes almost as a surprise. This is because of the many secondary predictions that are implicit.

For suppose it is true that the individual is rigidly bound to a well-defined statistical ageing law - so that many identical individuals when put together still implement the requisite expectation value as a function of time (which seems to be the case, cf.[14]). Then a deterministic physiological machinery is bound to reside within each individual organism in order for this outcome to be guaranteed. More specifically, there must exist within each individual, (1) a highly accurate "clock" and (2) a system of "hands" that reaches every individual cell. In addition, the whole system must, (3) be easily "reprogrammable" through mutation in just a few loci.

More generally speaking, as soon as the possible existence of a lawful ageing process that is bound closely to time can no longer be rejected on a priori grounds, speculations as to the site and the nature of the new organ system implied become admissible and are seriously called for.[15]

To conclude, the "functional" approach to ageing initiated by Medawar is still not exhausted. The notion of reproductive value entails a new implication: Competition with progeny needs to be minimized. This principle implies the existence of a prescribed ageing law. This optimum function may or may not turn out to be obeyed by nature. Further refinements of a mathematical type are possible (for example, when seasonal breeding enters). More important in the long run, however, may be the fact that new empirical questions - of a demographic, physiological, and eventually therapeutic nature - can be formulated.

ACKNOWLEDGMENTS

R.R. thanks Helmut Engler for a discussion on the possibility of a lawful age-dependence of the sum of all incidences of cancer. P.E.K. thanks Keith Tognetti for discussions. R.R. and O.E.R. thank Norman Packard, Arvind Varma and Bob Rosen for discussions. O.E.R. would like to thank Erik Mosekilde for stimulation.

REFERENCES

1. P. E. Kloeden, O. E. Rossler and R. Rossler, A predictive model for life expectancy curves, BioSystems 23 (1990). In press.
2. B. Gompertz, On the nature of the function expressive of the law of human mortality, Phil. Trans. Roy. Soc. London I:513-585 (1825).
3. M. Witten, Information content of biological survival curves arising in aging experiments: Some further thoughts, in: "Evolution of Longevity in Animals, A Comparative Approach," A. D. Woodhead and K. H. Thompson eds,. 295-317, Plenum, New York (1987).
4. P. B. Medawar, Mod. Qt. 2:30 (1945).
5. P. B. Medawar, "An Unsolved Problem in Biology," H. K. Lewis, London (1952).
6. P. B. Medawar, The definition and measurement of senescence, in: "Ciba Foundation Colloquia on Ageing," Vol. 1, G. E. W. Hostenholme and M. P. Cameron, eds., 4-15, Churchill, London (1955); P. B. Medawar and J. S. Medawar, "The Life Sciences, Current Ideas in Biology," Harper and Row, New York (1967).
7. G. C. Williams, Pleitropy, natural selection and the evolution of senescence, Evolution 11:398-411 (1957).
8. R. Rosen, "Optimality Principles in Biology," Butterworth, London (1967).
9. M. R. Perrin, Alternative life-history styles of small mammals, in: "Alternative Life History Styles of Animals," M. N. Bruton, ed., 209-242, Kluwer Academic Publishers, Dordrecht (1989).
10. G. I. Gause, "The Struggle for Existence," William and Wilkins, Baltimore (1934).
11. H. R. Hirsch, Why should senescence evolve? An answer based on a simple demographic model, in: "Evolution of Longevity in Animals, A Comparative approach," A. D. Woodhead and K. H. Thompson, eds., 75-90, Plenum, New York (1987).

12. G. C. Williams and P. D. Taylor, Demographic consequences of natural selection, in: "Evolution of Longevity in Animals, A Comparative Approach," A. D. Woodhead and K. H. Thompson, eds., 235-245, Plenum, New York (1987).
13. J. Maynard-Smith, "Evolution and the Theory of Games," Cambridge University Press, Cambridge (1982).
14. R. L. Walford, "Maximum Life Span," ch. 7, W. W. Norton, New York (1983).
15. P. E. Kloeden, R. Rossler and O. E. Rossler, Does there exist a centralized clock for ageing?, Gerontology 37 (1991). In press.

MODELLING OF COMPLEX SYSTEMS BY SIMULATED ANNEALING

Bjarne Andresen

Physics Laboratory, University of Copenhagen
Universitetsparken 5, DK-2100 Copenhagen Ø
Denmark

1. Introduction

Simulated annealing is a global optimization procedure (Kirkpatrick et al. 1983) which exploits an analogy between combinatorial optimization problems and the statistical mechanics of physical systems. The analogy gives rise to an algorithm for finding near-optimal solutions to the given problem by simulating the cooling of the corresponding physical system. Just as Nature, under most conditions, manages to cool a macroscopic system into or very close to its ground state in a short period of time even though its number of degrees of freedom is of the order of Avogadro's number, so does simulated annealing rapidly find a good guess of the solution of the posed problem.

Even though the original class of problems under consideration (Kirkpatrick et al. 1983) was combinatorial optimization, excellent results have been obtained with seismic inversion (Jakobsen et al. 1987), pattern recognition (Hansen 1990), and neural networks (Hansen and Salamon 1990) as well, just to name a few. I am not aware of any applications to biological problems at the moment, but my expectation is that the method is just sitting there waiting to used. Because of its origin in statistical mechanics it is immediately applicable to the complex dynamics of biochemical pathways as well as artificial life. But also macroscopic, non-chemical biological systems can usually be cast in terms of a 'quality function' (the negative of a cost function) and rules for the dynamics which again can be fit into the annealing language

Complexity, Chaos, and Biological Evolution, Edited by E. Mosekilde and
L. Mosekilde, Plenum Press, New York, 1991

of the next sections. The virtue of simulated annealing is its efficiency as a general-purpose method for handling extremely complicated problems for which no direct solution method is known. It therefore holds great promise for the optimization of biological and other chaotic systems.

2. The Algorithm

The simulated annealing algorithm is based on the Monte Carlo simulation of physical systems. It requires the definition of a state space $\Omega = \{\omega\}$ with an associated cost function (physical analogue: energy) E: $\Omega \to$ R which is to be minimized in the optimization. At each point of the Monte Carlo random walk in the state space the system may make a jump to a neighbouring state; this set of neighbours, known as the move class $N(\omega)$, must of course also be specified. The only control parameter of the algorithm is the temperature T of the heat bath in which the corresponding physical system is immersed.

The random walk inherent in the Monte Carlo simulation is accomplished by the Metropolis algorithm (Metropolis et al. 1953) which states that:

(i) At each step t of the algorithm a neighbour ω' of the current state ω_t is selected at random from the move class $N(\omega_t)$ to become the candidate for the next state.

(ii) It actually becomes the next state only with probability

$$P_{accept} = \begin{array}{ll} 1 & \text{if } \Delta E \le 0 \\ e^{-\Delta E/T_t} & \text{if } \Delta E > 0 \end{array} , \qquad (2.1)$$

where $\Delta E = E(\omega') - E(\omega_t)$ is the increase in cost for the move. If this candidate is accepted, then $\omega_{t+1} = \omega_t$.

The only thing left to specify is the sequence of temperatures T_t appearing in the Boltzmann factor in P_{accept}, the so-called annealing schedule. Like in metallurgy, this cooling rate has a major influence on the final result. A quench is quick and dirty, often leaving the system stranded in metastable states high above the ground state/optimal solution. Slow annealing produces the best result but is computationally expensive.

This completes the formal definition of the simulated annealing algorithm, which in principle simply is repeated numerous times until a satisfactory result is obtained.

3. The Optimal Annealing Schedule

So far all suggested simulated annealing temperature paths (annealing schedules) have been of the a priori type and thus have not adjusted to the actual behaviour of the system as the annealing progresses. Examples of such schedules are

$$T(t) = a\,e^{-t/b} \tag{3.1}$$

$$T(t) = \frac{a}{b+t} \tag{3.2}$$

$$T(t) = \frac{a}{\ln(b+t)}. \tag{3.3}$$

The real annealing of physical systems often has rough parts where the surrounding temperature must be decreased slowly due to phase transitions or regions of large heat capacity or slow internal relaxation. The same behaviour is seen in the abstract systems, so annealing schedules which take such variations into account are preferable in order to keep computation time at a minimum for a given accuracy of the final result (Ruppeiner 1988).

At this point I would like to make the analogy between the abstract simulated annealing process and a real thermodynamic process even stronger. Specifically, if the correspondence with statistical mechanics implied in the simulated annealing procedure involving phase space and state energies is valid, then further results from thermodynamics will probably also carry over to simulated annealing.

Since asking a question (= one evaluation of the energy function) in information theoretic terms is equivalent to producing one bit of entropy, the computationally most efficient procedure will be the temperature schedule T(t) which overall produces minimum entropy. In the past we have derived various bounds and optimal paths for real thermodynamic systems using finite-time thermodynamics (Salamon and Berry 1983; Feldmann et al. 1985; Salamon et al. 1980b; Andresen 1983). The minimum-entropy production path which most readily generalizes to become the optimal simulated annealing schedule, is the one calculated with thermodynamic length [c.f. eq. (4.4) below]:

$$\frac{dT}{dt} = -\frac{vT}{\varepsilon\sqrt{C}} \qquad (3.4)$$

or equivalently

$$\frac{\langle E \rangle - E_{eq}(T)}{\sigma} = v. \qquad (3.5)$$

In these expressions v is the (constant) thermodynamic speed (see Sect. 4 below for further explanation), C and ε are the heat capacity and internal relaxation time of the system, respectively, $\langle E \rangle$ and σ the corresponding mean energy and standard deviation of its natural fluctuations, and finally $E_{eq}(T)$ is the internal energy the system would have if it were in equilibrium with its surroundings at temperature T. The physical interpretation of eq. (3.5) is that the environment should at all times be kept v standard deviations ahead of the system. Similarly eq. (3.4) indicates that the annealing should slow down where internal relaxation is slow and where large amounts of 'heat' has to be transferred out of the system (Salamon et al. 1988). In case C and ε do not vary with temperature, eq. (3.4) integrates to the standard schedule eq. (3.1). Figure 1 shows the successive decrease in energy for annealings on a graph partitioning problem following different annealing schedules.

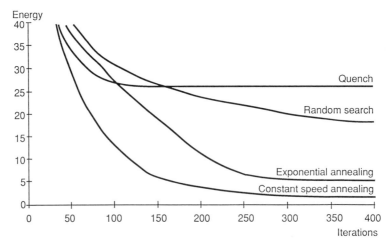

Figure 1. Energy for a graph partitioning problem as the annealing progresses, using different annealing schedules: quench (T=0), random search (T=∞), exponential (eq. (3.1)), constant speed (eq. (3.4) or (3.5)).

The extra temperature dependent variables of the constant thermodynamic speed schedule of course require additional computational effort. Since systems often change considerably in a few steps, ergodicity is not fulfilled, so the use of time averages to obtain $\langle E \rangle$, σ, C, and ε is usually not satisfactory. Instead we (Andresen et al. 1988) suggest to run an ensemble of systems in parallel, i.e. with the same annealing schedule, in the true spirit of the analogy to statistical mechanics. Then these variables can be obtained anytime as ensemble averages based on the system degeneracies $p_i = p(E_i)$:

$$Z(T) = \sum_i p_i \, \exp(-E_i/T) \tag{3.6}$$

$$E(T) = T^2 \frac{d \ln Z}{dt} \tag{3.7}$$

$$C(T) = \frac{dE}{dT} = \frac{\langle (\Delta E)^2 \rangle}{T^2} \tag{3.8}$$

$$\varepsilon(T) = \frac{-1}{\ln \lambda_2} \approx \frac{T^2 \, C(T)}{\sum_i p_i \sum_{j>i} (E_j - E_i)^2 \, P_{ji} \, \exp(-E_i/T)}, \tag{3.9}$$

where λ_2 is the second largest eigenvalue of the thermalized version of the transition probability matrix $\mathbf{P}$ among all the energy levels ($\lambda_1 = 1$ corresponds to equilibrium).

But where does one get the degeneracies p_i from? Actually (Andresen et al. 1988), information to calculate the temperature-independent (or infinite-temperature, if you prefer) transition probability matrix $\mathbf{P}$ can be accumulated during the annealing run by simply adding up in a matrix $\mathbf{Q}$ the number of attempted moves (not just the accepted ones) from level i to j as the calculation progresses. Normalization of $\mathbf{Q}$ yields a good estimate of $\mathbf{P}$,

$$P_{ji} = Q_{ji} \Big/ \sum_k Q_{ki}. \tag{3.10}$$

The degeneracies $\mathbf{p}$ are then the eigenvector of $\mathbf{P}$ corresponding to the eigenvalue 1.

This use of ensemble annealing is particularly well suited for implementation on present day parallel computers. A further analysis of its performance has been carried out by Ruppeiner, Pedersen, and Salamon (1990), and the trade-off between ensemble size and duration of annealing for fixed total computation cost has been addressed by Pedersen et al. (1990).

4. Thermodynamic length

Weinhold (1975, 1978) defined a metric on the abstract space of equilibrium states of a thermodynamic system, represented by all its extensive variables X_i: {U, S, V, N_i, ...}, as the second derivative matrix of one of these coordinates in terms of the others, e.g.

$$\mathbf{M}_U = \left\{ \frac{\partial^2 U}{\partial X_i \partial X_j} \right\}. \tag{4.1}$$

His purpose was simply to derive the classical Maxwell relations in a different way, but it occurred to us (Salamon et al. 1980a) that since this second derivative matrix does indeed satisfy all the requirements for being a metric on the entire equilibrium space, it was an invitation to calculate lengths along given paths, now called thermodynamic lengths. Later Salamon and Berry (1983) found a connection between the thermodynamic length along a process path and the (reversible) availability lost in the process. Specifically, if the system moves via states of local thermodynamic equilibrium from an initial equilibrium state i to a final equilibrium state f in time τ, then the dissipated availability $-\Delta A$ is bounded by the square of the distance (i.e. length of the shortest path) from i to f calculated with the metric $\mathbf{M}_U$, times ε/τ, where ε is a mean relaxation time of the system. If the process is endoreversible (Rubin 1979), i.e. if all irreversibilities are located in the couplings to the environment, then the bound can be strengthened to

$$-A \le \frac{L^2 \varepsilon}{\tau}, \tag{4.2}$$

where L is the length of the *traversed* path from i to f. For comparison, the bound from traditional thermodynamics is only

$$-A \le 0. \tag{4.3}$$

Equality in eq. (4.2) is achieved at constant thermodynamic speed $v = dL/dt$. The corresponding time path satisfies (Salamon and Berry 1983)

$$\frac{dT}{dt} = -\frac{vT}{\varepsilon\sqrt{C}} \tag{4.4}$$

which is the equation we used in the previous section, eq. (3.4).

An expression analogous to eq. (4.2) exists for the total entropy production during the process:

$$\Delta S^u \leq \frac{L^2 \varepsilon}{\tau}. \tag{4.5}$$

The length L is then calculated relative to the entropy metric

$$\mathbf{M}_S = -\left\{\frac{\partial^2 S}{\partial X_i \partial X_j}\right\} \tag{4.6}$$

which (when expressed in identical coordinates!) is related to $\mathbf{M}_U$ by (Salamon et al. 1984)

$$\mathbf{M}_U = -T\,\mathbf{M}_S, \tag{4.7}$$

where T is the environment temperature. In statistical mechanics, where entropy takes the form

$$S(\{p_i\}) = -\sum_i p_i \ln p_i, \tag{4.8}$$

the metric $\mathbf{M}_S$ is particularly simple, being the diagonal matrix (Feldmann et al. 1985)

$$\mathbf{M}_S = -\left\{\frac{1}{p_i}\right\}. \tag{4.9}$$

The same procedure of calculating metric bounds for dynamic systems has been applied to coding of messages (Flick et al. 1987) and to economics (Salamon et al. 1987).

5. Scaling in Simulated Annealing

The optimal temperature schedules eqs. (3.4) and (3.5) produce the lowest *average* energy for the allotted amount of computer time. However, it is quite likely that one random walker in the ensemble at some point in the simulation briefly visited a lower energy state which it later left again. Thus, since one is generally interested in the lowest possible energy, it will be worthwhile to maintain a record of the lowest energy ever visited by any walker. This extremal value has been called E_{VBSF} (Very Best So Far) (Jakobsen et al. 1987; Pedersen et al. 1990).

The analysis (Sibani et al. 1990) is at this point most readily carried out in terms of a random walk on the energy axis where one is interested in the mean first passage time for crossing the desired E_{VBSF}, or equivalently the waiting time probability $W_{E,E_0}(t)$, for the first crossing of E starting at E_0. A full knowledge of W(t) of course amounts to the exact solution of the problem, but an empirical ansatz of the form

$$W_{E,E_0}(t) = W\left(\frac{t}{t_0}\left[\frac{E}{E_0 - E}\right]^{1/\alpha}\right),$$

(5.1)

where W is any monotonically decreasing function with a value between 0 and 1, has worked very well. The constant α is a scaling factor in the range]0;1[, and t_0 is introduced in order to make the argument of W dimensionless. The origin of the ansatz is statistical mechanics on tree structures. If one imagines the energy surface of the problem in question as a real life landscape which is originally completely flooded, then slowly draining the water will close off one lake after another as the surrounding water falls below the enclosing saddle point of each lake. Each of these saddle points is then defined as a vertex in the tree. In this way the vertices indicate branching points between basins and thus ridges which must be climbed in order to move from one basin to another. Hence they are the important energy barriers which slow down the annealing.

Introducing the ansatz eq. (5.1) into the moment equations for the density of states (Sibani et al. 1990) leads to

$$\langle (E - E_g)^n \rangle \left(1 + \frac{t}{t_0}\right)^n = \text{constant},$$

(5.2)

where E_g is the (not yet encountered) ground state of the problem. During an annealing run a range of moments, including negative ones, are calculated. Through eq. (5.2) they put very strict limits on α as well as E_g and thus allows estimation of the ground state energy long before any member of the ensemble has been even close. Figure 2 shows an example from graph partitioning where it is obvious that the ground state energy estimate from scaling is way ahead of the lowest energy encountered from about 260 000 iterations onward, whereas E_{VBSF} improves only very slowly with time due to the low temperatures late in the annealing.

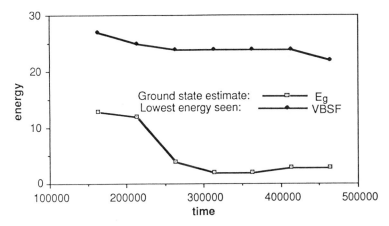

Figure 2. Comparison of the ground state energy estimate based on eq. (5.2) and the lowest energy ever visited during an annealing run.

6. Summary

Simulated annealing has proven to be a very useful general purpose optimization algorithm for extremely complicated problems, even with the fixed temperature schedules, eqs. (3.1) – (3.3). Elaborating the analogy to physical statistical mechanical systems with the introduction of optimality results from finite-time thermodynamics has improved the efficiency of the algorithm noticeably, and the use of ensembles of random walkers has made it self-adapting. Finally, inspiration from statistical mechanics on trees has produced an extrapolation procedure which dramatically reduces the computer time requirement for a desired quality of the result.

The method as described above mimics Nature as it equilibrates systems on an atomic or molecular scale (e.g. chemical reactions or crystal structures). Inspiration can also be lifted from Nature's way of developing the most favourable systems on the macroscopic biological scale: evolution. Several annealing procedures to replace the Metropolis algorithm of Sect. 2 have been developed based on evolution with its cloning of successful species and dying out of unsuccessful ones, notably by Ebeling and coworkers (Boseniuk and Ebeling 1988; Ebeling 1990) and by Schuster (Fontana et al. 1989). In my opinion the two lines of thought are complementary in the sense that they are advantageous for many trials and for a more limited set of tests, respectively.

7. Acknowledgments

Much of the work reported in this review has been done in collaboration with colleagues near and far. In particular I want to express my gratitude to profs. R. Stephen Berry, Peter Salamon, Paolo Sibani, and Jacob Mørch Pedersen.

References

Andresen, B. 1983. *Finite-time thermodynamics* (Physics Laboratory II, University of Copenhagen).

Andresen, B.; Hoffmann, K. H.; Mosegaard, K.; Nulton, J.; Pedersen, J. M.; Salamon, P. 1988. On lumped models for thermodynamic properties of simulated annealing problems, *J. Phys. (France)* **49**, 1485.

Boseniuk, T.; Ebeling, W. 1988a. Optimization of NP-complete problems by Boltzmann-Darwin strategies including life cycles, *Europhys. Lett.* **6**, 107.

Boseniuk, T.; Ebeling, W. 1988b. Evolution strategies in complex optimization: The travelling salesman problem, *Syst. Anal. Model. Simul.* **5**, 413.

Ebeling, W. 1990. Applications of evolutionary strategies, *Syst. Anal. Model. Simul.*. **7**, 3.

Feldmann, T.; Andresen, B.; Qi, A.; Salamon, P. 1985. Thermodynamic lengths and intrinsic time scales in molecular relaxation. *J. Chem. Phys.* **83**, 5849.

Flick, J. D.; Salamon, P.; Andresen, B. 1987. Metric bounds on losses in adaptive coding. *Info. Sci.* **42**, 239.

Fontana, W.; Schnabl, W.; Schuster, P. 1989. Physical aspects of evolutionary optimization and adaptation, *Phys. Rev. A* **40**, 3301.

Hansen, L. K. 1990. at *Symposium on Applied Statistics*, (UNI·C, Copenhagen); and in preparation.

Hansen, L. K.; Salamon, P. 1990. *Neural network ensembles*, preprint.

Jakobsen, M. O.; Mosegaard, K.; Pedersen, J. M. 1987. Global model optimization in reflection seismology by simulated annealing, in *The Mathematical Geophysics Fifth International Seminar on Model Optimization in Exploration Geophysics*, (Berlin).

Kirkpatrick, S.; Gelatt, C. D. Jr.; Vecchi, M. P. 1983. Optimization by simulated annealing, *Science* **220**, 671.

Metropolis, N.; Rosenbluth, A. W.; Rosenbluth, M. N.; Teller, A. H.; Teller, E. 1953. Equation of state calculations by fast computing machines, *J. Chem. Phys.* **21**, 1087.

Pedersen, J. M.; Mosegaard, K.; Jacobsen, M. O.; Salamon, P. 1990. Optimal degree of parallel implementation in optimization. *J. Phys. A*.

Rubin, M. H. 1979. Optimal configuration of a class of irreversible heat engines. I, *Phys. Rev. A* **19**, 1272.

Ruppeiner, G. 1988. *Nucl. Phys. B (Proc. Suppl.)* **5A**, 116.

Ruppeiner, G.; Pedersen, J. M.; Salamon, P. 1990. Ensemble approach to simulated annealing. *J. Phys. (France)*.

Salamon, P.; Andresen, B.; Gait, P. D.; Berry, R. S. 1980a. The significance of Weinhold's length, *J. Chem. Phys.* **73** 1001; erratum *ibid.* **73**, 5407E.

Salamon, P.; Berry, R. S. 1983. Thermodynamic length and dissipated availability, *Phys. Rev. Lett.* **51**, 1127.

Salamon, P.; Komlos, J.; Andresen, B.; Nulton J. 1987. A geometric view of welfare gains with non-instantaneous adjustment. *Math. Soc. Sci.* **13**, 153.

Salamon, P.; Nitzan, A.; Andresen, B.; Berry, R. S. 1980b. Minimum entropy production and the optimization of heat engines, *Phys. Rev. A* **21**, 2115.

Salamon, P.; Nulton, J.; Ihrig, E. 1984. On the relation between entropy and energy versions of thermodynamic length. *J. Chem. Phys.* **80**, 436.

Salamon, P.; Nulton, J.; Robinson, J.; Pedersen, J. M.; Ruppeiner, G.; Liao, L. 1988. Simulated annealing with constant thermodynamic speed. *Comp. Phys. Comm.* **49**, 423.

Sibani, P.; Pedersen, J. M.; Hoffmann, K. H.; Salamon, P. 1990. *Scaling concepts in simulated annealing*, preprint.

Weinhold, F. 1975. Metric geometry of equilibrium thermodynamics, *J. Chem. Phys.* **63**, 2479.

Weinhold, F. 1978. Geometrical aspects of equilibrium thermodynamics, in *Theoretical chemistry, advances and perspectives*, Vol. 3, edited by D. Henderson and H. Eyring (Academic Press, New York).

Section V
Biological Structures and Morphogenesis

Spontaneous form formation by self-organization in open thermodynamic systems can form the basis for a whole new discipline of theoretical study of developing organisms.

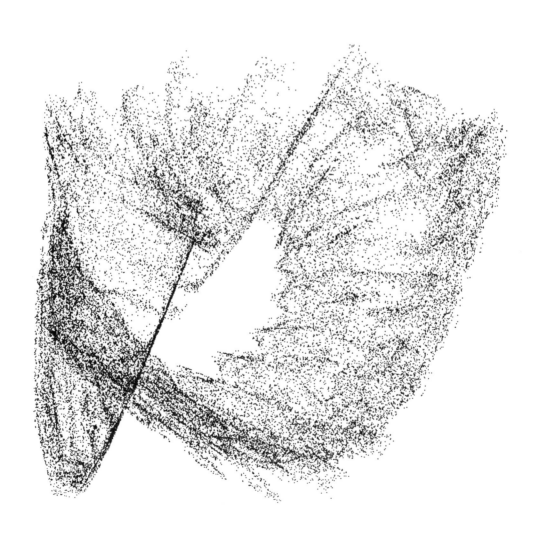

GENERIC DYNAMICS OF MORPHOGENESIS

B.C. Goodwin[†] and C. Brière[‡]

[†]Developmental Dynamics Research Group
Department of Biology
The Open University
Walton Hall
Milton Keynes MK7 6AA

[‡]Laboratoire de Biologie
Qualitative
CNRS ENSAT
Toulouse, France

ABSTRACT

Acetabularia is a single-celled alga that undergoes a characteristic pattern of morphogenesis to produce a giant cell of distinctive form. Because of its basic simplicity, this organism lends itself to experimental and theoretical studies of the components that make up the morphogenetic field, and their dynamic properties. A model of this field and a finite-element simulation of its behaviour are presented which show that spatial patterns generically similar to those observed in the alga arise naturally, suggesting that normal morphogenesis can be described as an attractor of a moving boundary process. The implications of this possibility in relation to morphogenesis in related species is considered.

INTRODUCTION

The giant unicellular green alga, *Acetabularia*, has been recognized for many years as an exceptionally favourable organism for the study of basic problems of cell and developmental biology, notably nucleo-cytoplasmic interactions and the molecular biology of cell differentiation (Brachet and Bonotto 1970, Puiseux-Dao 1970). It is now proving to be equally useful for studying the dynamic properties of the cytoskeleton and the cell wall that underlie the sequential transformations of cell shape observed during growth from a zygote, or in the process of regeneration of apical structures.

Complexity, Chaos, and Biological Evolution, Edited by E. Mosekilde and
L. Mosekilde, Plenum Press, New York, 1991

The basic characteristics of the life cycle of *Acetabularia acetabulum* (formerly *A. mediterranea*), whose habitat is the shallow waters around the shores of the Mediterranean, are shown in Fig. 1. Isogametes fuse to produce a roughly spherical zygote which breaks symmetry, producing a growing stalk and a branching rhizoid that anchors the alga to the substratum and houses the nucleus in one of its branches. When the stalk reaches a length of about 1 to 1.5 cm after several weeks of growth, a ring of small bumps arises around the tip, growing into a whorl of hairs or verticils that

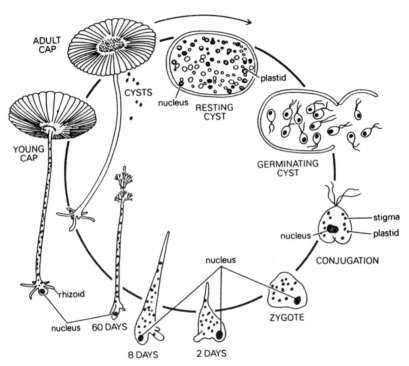

Figure 1. The life cycle of *Acetabularia acetabulum*.

branch successively as they grow. The tip renews its growth from the centre of the whorl and, after a few days during which the stalk grows several mm, another whorl is produced. This process repeats itself until a cap primordium is generated (Fig. 2), a structure with the same circular symmetry as a whorl but consisting of many rays joined together into a disc-shaped structure, the cap. This grows radially, the whorls drop off, resulting in the morphology of the adult: a rhizoid (still containing the nucleus), a stalk 3-5 cm in length and about 0.5 mm in diameter, and a cap whose diameter is approximately 0.5 cm: a giant differentiated cell (Fig. 3).

The alga is also capable of regenerating a cap after a cut through the stalk. It does so following the same sequence of shape changes as in normal growth. After the cut has healed with the production of a new cell wall, a tip emerges, grows, produces a series of whorls and then generates a cap. The regenerate with the rhizoid is, after the whorls have dropped off, indistinguishable from the original mature alga. In this chapter we discuss the dynamic processes responsible for the characteristic sequence of shape changes during normal development and regeneration.

Figure 2. Formation of a cap primordium, after three whorls.

CALCIUM-CYTOSKELETAL DYNAMICS

Our attention was drawn to the importance of calcium in morphogenesis by studies which showed that changing the concentration of this ion in the sea water in which the algae develop causes dramatic changes of morphology (Goodwin et al. 1983). These changes could be duplicated by adding to the sea water ions (Co^{2+} or La^{3+}) which block calcium channels, so the effects of reduced calcium are not simply on the mechanical properties of the cell wall but extend inside the cell. The wavelength of the whorl pattern can also be systematically altered by changing the calcium concentration in the medium (Harrison and Hillier 1985, Goodwin et al. 1987).

Figure 3. Morphology of the mature alga, showing rhizoid, stalk and cap. The whorls have dropped off.

Calcium is known to have significant effects on the mechanical state of the cytoskeleton (Kamiya 1981, Menzel and Elsner-Menzel 1989). It changes the viscosity and elastic modulus of the cytoplasm by influencing the state of polymerization of actin and tubulin, and by activating actomyosin contraction and enzymes such as gelsolin which cut actin filaments. Some of these influences are shown schematically in Fig. 4.

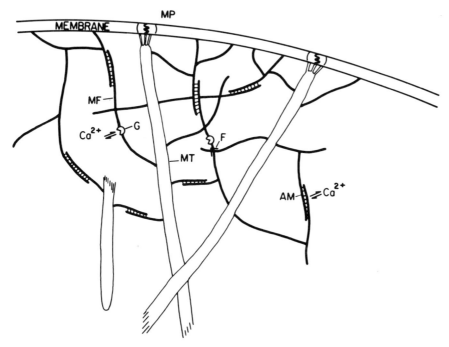

Figure 4. Schematic diagram showing structural compounds of the cytoskeleton. AM = actomycin, F = fimbrin, G = Gelsolin, MF = microfilament, MP = membrane protein, MT = microtubule.

In order for morphogenesis to occur it is necessary to have a medium in which spatial patterns are generated spontaneously. The most likely candidate for this in a cell is certainly the cytoplasm. It was established by Turing (1952) that coupled biochemical reactions combined with diffusion can produce spatially non-uniform patterns of reactants which he called morphogens. These reaction-diffusion systems have been extensively used by Meinhardt (1982) and by Murray (1989) to model morphogenetic processes in a variety of organisms, and they have been considered by Harrison and Hillier (1985) and by Goodwin et al. (1985; see also Murray (1989) in application to *Acetabularia* morphogenesis). However the morphogenetic effects of calcium and its influence on the cytoskeleton suggested that this system itself might play the role of primary pattern-generator. The first step in explaining this possibility was to derive equations that describe the mechanical properties of the cytoskeleton, the cytoplasmic regulation of calcium, and their interaction. This was done by Goodwin and Trainor (1985) and the coupled equations were shown to have the property of spontaneous bifurcation for particular ranges of the parameters. Within this range spatial patterns of cytosolic free calcium and mechanical strain in the cytoplasm develop from random perturbation of the system from a spatially uniform initial condition. The reason for this behaviour lies in certain basic properties of the calcium-cytoskeleton interaction, which will now be described.

Cytosolic free calcium is regulated in eukaryotic cells at concentrations of 100 nM or so by plasmalemma pumps, by sequestration mechanism involving the endoplasmic reticulum, vesicles or vacuoles, and by binding to cytoplasmic proteins and chelating agents such as calcitonin and calmodulin. Studies of actin gels have shown that as calcium rises above 100 nM it induces gel breakdown and solation by activation of enzymes such as gelsolin. At higher concentrations calcium initiates contraction of actomyosin filaments so that the cytoplasm becomes more resistant to deformation (Nossal 1988). At calcium concentrations above about $5\mu M$ depolymerization of filaments and microtubules, and the progressive action of gelsolin, have the consequence that the cytoplasm becomes progressively solated. A qualitative description of this behaviour in terms of changes in the elastic modulus of the cytoplasm as a function of calcium is shown in Fig. 5. This describes how calcium affects the mechanical state of the cytoplasm. We deduce that there is also a reciprocal action of the mechanical state of the cytoskeleton on free calcium concentration. It is assumed that strain or deformation of the cytoplasm results in release of calcium from the bound or sequestered state to free ions. Therefore regions that happen to have elevated strain will also have elevated free calcium levels. But increased free calcium causes gel breakdown and solation. This results locally in more strain (deformation) since the cytoplasm is assumed to be under tension, hence in further calcium release. The result is a positive feed-back loop in the regions of calcium concentration where the slope of the elastic modulus curve as a function of calcium is negative (see Fig. 5): a local, random increase of calcium above the steady state level initiates a run-away calcium release and increase of cytoplasmic strain. However, this is stabilized by the effects of diffusion which tends to reduce the calcium gradients and also by the opposing effects of calcium on actomyosin contraction, which increases the elastic modulus and so decreases the strain (region of positive slope, Fig. 5).

The argument also works in reverse: where calcium levels are decreased the strain is also reduced since the cytoplasm is more gel-like (higher elastic modulus) and so free calcium will be bound or sequestered, decreasing it still further. In terms of reaction-diffusion dynamics, calcium plays the role of short-range activator while mechanical strain is like the long-range inhibitor. The result of the interactions is that spatial patterns of calcium concentration and strain can arise spontaneously from initially uniform conditions when the equation parameters are in the bifurcation range. The model is qualitatively similar to that discussed by Oster and Odell (1983). A brief description of the calcium-cytoskeleton equations is given in section 1 of the appendix.

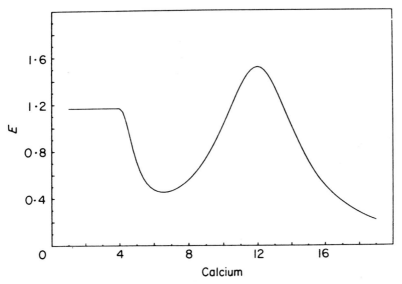

<u>Figure 5.</u> Variation of elastic modulus with free Ca^{2+} concentration (micromolar). Qualitative relations only.

CELL WALL DYNAMICS

The properties of the cytoplasm described above give it the characteristics of an excitable medium which can spontaneously generate spatial patterns, both stationary and dynamic – i.e., propagating waves. This is sufficient to initiate pattern; but morphogenesis involves changes of geometry. In the case of *Acetabularia* and many other developing organisms, morphogenesis is linked to growth, so the cell wall must undergo localized changes of shape together with elongation. The wall is described in the model as a purely elastic shell (about 2μ thick) whose state changes as a function of strain in the underlying cytoplasm, which is a thin shell about 10μ thick closely apposed to the wall, the plasmalemma separating them. This functional coupling is assumed to arise via strain-activated pumps in the plasmalemma that cause the wall to soften by excretion of protons or hydrolases. The large vacuole in the centre of the cell is an osmotic organelle, separated from the cytoplasmic shell by another membrane, the tonoplast. The vacuole exerts a pressure that is resisted by the wall. Patterns of strain in the cytoplasm are thus reflected in the elastic modulus of the wall, which undergoes elastic deformations as a result of the outward-directed osmotic pressure. A growth process was introduced into the model whereby new wall material was added wherever wall strain exceeded a threshold value so that elastic deformations led to plastic changes, in accordance with experimental evidence (Cleland 1971, Green et al. 1971). Growth of the cytoplasm was coupled to wall growth, while vacuolar pressure remained constant. The details of the growth algorithm are described in section 2 of the appendix, while the cytogel-wall coupling is presented in section 3. These relations are also described in Brière and Goodwin (1988).

SIMULATIONS OF *ACETABULARIA* MORPHOGENESIS

The calcium-cytogel and cell wall equations were used for a finite-element simulation of growth and morphogenesis. Parameters were adjusted so that the calcium-cytoskeleton equations were in the bifurcation range, making spatial patterns possible. The shape that developed depended on the characteristic wavelength of the pattern, and also on the parameters that describe wall growth. Simulations started with uniform initial conditions, on a dome representing a regenerating apex (Fig. 6). The first stage of pattern formation that typically occurred was the formation of a gradient of cytosolic free calcium that increased to a maximum at the apex. This is shown in Fig. 7. On the left is the outline of the shape of the tip along one of the longitudinal elements of the dome, from base to apex, while the other graphs show different variables of the cytogel as a function of distance from the base. Strain is measured both in the latitudinal (solid line) and longitudinal (dotted line) elements, showing the anistropy; similarly for the elastic modulus of wall elements (incorrectly labelled elasticity). All variables start off spatially uniform (flat) and spontaneously develop a

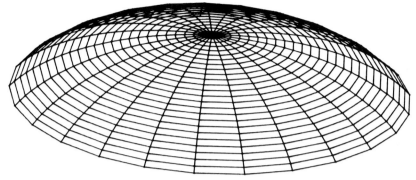

Figure 6. Computer simulation of regeneration in which the cytoplasm and the cell wall are described as shells made up of finite elements which obey equations describing their dynamics as mechanochemical or elastic media, respectively.

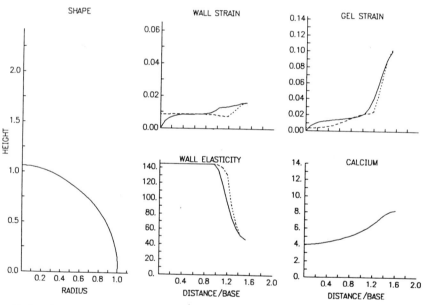

Figure 7. Section through the mesh describing the regenerating tip, showing shape, wall strain and elastic modulus, cytogel strain and free calcium concentration as a function of distance from base (origin, O, on the abscissa of the graphed variables, radius 1 in the curve showing shape). A gradient in calcium forms spontaneously, with a maximum at the tip.

pattern, calcium rising to a maximum at the tip, as does the gel strain. The result is that the wall softens in this region and there is an elastic deformation. A three dimensional view of this is shown in Fig. 8: a tip is produced.

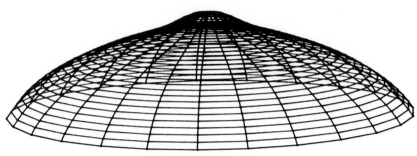

Figure 8. Tip initiation in the model, resulting from spontaneous symmetry-breaking of the cytogel-calcium dynamic, gradient formation with maxima of strain and calcium at the tip, and wall softening as a result of interaction between cytoplasm and the wall.

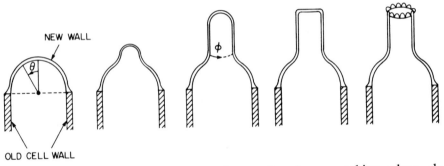

Figure 9. Representation of the morphogenetic changes taking place during regeneration of Acetabularia. (a) hemispherical dome (b) Tip formation (c) Growth (d) Tip flattening (e) Whorl formation.

This is the first stage of the regenerative process, shown schematically for the real alga in Fig. 9. A characteristic feature of whorl formation is the flattening of the conical tip just prior to the appearance of the ring of hair primardia that initiates a whorl. This was something we had never understood. The model gave us an explanation. As growth occurs at the tip, plastic changes of geometry following the elastic deformations, there is an interesting interaction between the shape generated and the dynamic behaviour of the calcium-cytogel system. After an initial stage of growth, the calcium gradient with the maximum at the tip becomes unstable and transforms into an annulus, the maximum level of calcium occurring away from the tip.

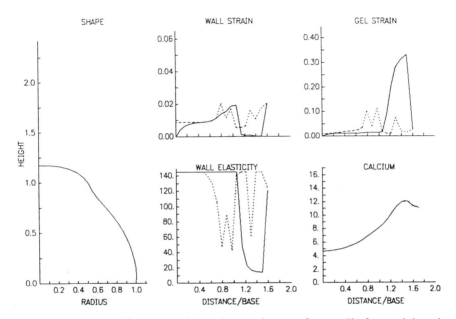

Figure 10. Later stage of regeneration: the maximum of cytosolic free calcium is now displaced from the tip, to the region of maximum curvature of the wall shape where the wall elastic modulus is reduced. In three dimensions, this defines an annulus.

The region of maximum cytogel strain also changes in a similar manner (Fig. 10). Wall softening is now greatest in this annular region, resulting also in maximal wall curvature proximal to the tip, with consequent flattening of the tip itself, where the elastic modulus of the wall is now larger than in the annular region. As the tip grows the amplitude of the annulus increases (Fig. 11).

A calcium annulus of this type was perturbed to see if it could spontaneously generate a pattern similar to that of a whorl. It did so, producing a ring of peaks of calcium that can be interpreted as the initiator of the whorl pattern. Unfortunately the finite-element program is not yet sufficiently robust to allow us to study the growth of such small elements, breaking the axial symmetry of the growing tip. This work is in progress. But the sequence of pattern changes observed, namely gradient and tip formation, elongation, annulus formation and tip flattening, and the bifurcation of an annular pattern to a ring of calcium peaks, provides a very natural dynamic explanation of a basic morphogenetic sequence. These observations suggest that we may be dealing here with an attractor in the moving boundary process with which we are dealing, a whorl being a natural form.

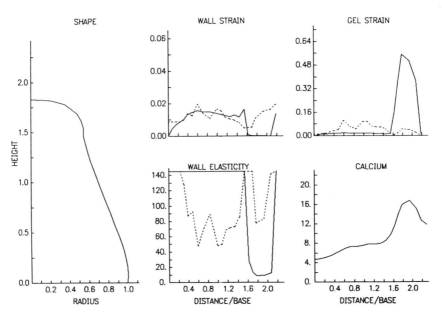

Figure 11. The pattern in the simulation corresponding to the state of tip flattening described in Fig. 2, with a well-defined annulus of calcium at the region of maximum curvature in the shape. This annulus then spontaneously breaks into a series of peaks, assumed to initiate the whorl pattern.

As the apex elongates the annulus itself becomes unstable and intermittently collapses back to a gradient with a maximum at the tip, an annulus then reforming and growing in amplitude. This occurs with a rather irregular frequency, suggestive of the somewhat irregular frequency of successive whorl initiations as the alga grows. So we seem to have here another dynamic aspect of the morphogenetic process that occurs spontaneously, without any parametric change. These aspects of the moving boundary process are again suggestive that the basic generative dynamic is robust and fundamental to this type of growth and morphogenesis. We shall return later to this issue in relation to higher plant phyllotaxis, where a multicellular meristem undergoing growth and morphogenesis commonly generates whorled patterns of leaves. Hairs in *Acetabularia* are basically simple leaves. They even fall off – in the adult. So they appear to serve no function. From a neo-Darwinian or functionalist perspective, this poses a problem: why are useless structures generated? The dynamic analysis presented suggests that whorls are produced because they arise naturally, almost inevitably, as an accompaniment of growth in this type of morphogenesis. We shall not pursue this aspect of the analysis here, which bears on evolutionary issues, but refer the interested reader to a more extended discussion elsewhere (Goodwin 1990).

Our simulations have so far failed to produce a cap. This requires another pattern of growth which we have approached, but never achieved, in which lateral growth exceeds longitudinal. The conditions for cap formation appear to require more anisotropy in the strain field than is currently in the model.

The results reported here are preliminary but very encouraging. What is surprising is how comparatively easy it was to find parameter values that result in a series of shape changes that simulate remarkably closely the morphogenetic sequences of normal growth. Tip formulation and elongation can be obtained equally readily starting from a sphere rather than a hemisphere, so the boundary conditions of the latter do not impose artificial constraints on the process. However, it is necessary to make a systematic study of the range in parameter space which results in the sequence observed, in order to characterise the robustness of the model. Although it is computationally complex, it is biologically extremely simple: cell wall, cytoplasm, and vacuole, modelled at their simplest. We may be looking at a basic eukaryotic morphogenerator.

VARIATIONS ON A THEME

Acetabularia belongs to a group of giant unicellular marine algae called the *Dasycladaceae*, all of which have basic morphological features in common. Some species are very similar to *Acetabularia acetabulum*, with slight variations of whorl and cap morphology. An example is *A.crenulata*, in which the rays of the cap are separated so that the overall form has spherical rather than circular symmetry, the rays emanating from the apical end of the stalk like radii from the centre of a sphere. Related species have no caps, the hairs of the whorls becoming large and bulbous and serving as gametangia, where the gametes are produced. These and other forms can be 'phenocopied' in *A.acetabulum* by environmental influences; i.e., this species can, without any change of genotype, be induced to take on phenotypes similar to those of other species simply by altering the environment. For example, Goodwin et al. (1983) found that changes in the concentration of calcium in the sea water caused characteristic changes of form. If the normal 10mM Ca^{2+} is reduced to 3mM, the algae fail to make caps so that the cells produced a series of whorls only. However, no gametes were produced under these conditions, so these cells are reproductively sterile. When the calcium is reduced to 2mM, the algae cease making whorls and the stalk simply continues to elongate. At 1mM, a bulbous tip is produced which stops growing. By altering the elastic modulus of the cell wall in the model, this bulbous form can be simulated, suggesting that there is a direct relationship between calcium in the medium and the mechanical properties of the wall, which is to be expected because of the role of calcium as a hardening agent in cell walls. Algae grown in the calcium range 2-4mM often have bulbous hairs in whorls and no caps, 'phenocopying' other species with this form. However, no systematic study of these variants by parametric modification of the model has yet been carried out.

CONCLUSION

The essential simplicity of *Acetabularia* morphology and its amenability to experimental study are attractive features for investigating growth and form. The model of morphogenesis described depends upon fundamental properties of the cytoskeleton in interaction with calcium, encouraging the view that this may be a primary pattern generator in the eukaryotes, as argued elsewhere (Goodwin 1989). What our studies have shown is that when the state of this excitable cytoplasm is coupled to the mechanical properties of the cell wall in ways that simulate experimentally-deduced processes of deformation and growth, the cell undergoes patterns of morphogenesis that reveal generic qualities of the normal sequence of shape changes. This encourages the view that there may be general dynamic principles underlying morphogenetic sequences in *Acetabularia* and its relatives in the taxonomic group *Dasycladaceae*. An extension of this idea is that the essential dynamic properties of the morphogenetic field described for *Acetabularia* may be applicable, with modification, to the higher plant meristem. What encourages this is the observation that the whorl is one of the basic forms of leaf phyllotaxis. The other two, distichous and spiral, can be seen as transformations of whorls, all then possibly arising from a similar generative dynamic in the meristem as a morphogenetic field. Each of these patterns may be a distinct morphogenetic attractor of the moving boundary process of the growing meristem. This view has been persuasively argued by Green (1987, 1989). Our own studies of asymmetric pattern initiation in the *Acetabularia* model support this. These speculations require further detailed investigation, which is in progress, but they point to the possibility of a remarkable unification of pattern-forming principles across a wide taxonomic range of organisms. Once again *Acetabularia* proves itself to be a model organism for studies which have very wide applications and implications.

APPENDIX

1. The Goodwin & Trainor model for calcium-cytogel interaction

1.1. Viscoelastic properties

The first equation of the Goodwin & Trainor model describes the viscoelastic displacements of the cytogel about an equilibrium state. The variables considered are the displacement ξ and the calcium concentration χ. From the displacement field, the strain tensor ε and the strain-rate tensor $\dot{\varepsilon}$ are derived.

The stress tensor σ depends on the strain and strain-rate tensors and of the mechanical properties of the cytogel. Applying Newton's second law to a very small unit element of cortical material, with a volume density ρ, gives

$$div\,\sigma + f = \rho\frac{d^2\xi}{dt^2} \tag{1.1}$$

where $div\,\sigma$ accounts for the elastic and viscous forces and f for the external forces.

Using a linear expansion of σ about the equilibrium we have

$$div\,\sigma = div\,\sigma_0 + div\,(S\cdot\varepsilon) + div\,(A\cdot\dot{\varepsilon})$$

Since σ depends on the calcium concentration $div\,\sigma_0 = \dfrac{\partial\sigma_0}{\partial\chi}\cdot\nabla\chi$ then

$$div\,\sigma = S\cdot\nabla\varepsilon + A\cdot\nabla\dot{\varepsilon} + \frac{\partial\sigma_0}{\partial\chi}\cdot\nabla\chi + \left[\frac{\partial S}{\partial\chi}\cdot\varepsilon + \frac{\partial A}{\partial\chi}\cdot\dot{\varepsilon}\right]\cdot\nabla\chi \tag{1.2}$$

For an isotropic material (which we assume for the cytogel), S can be expressed in terms of the Lamé coefficients λ and μ; similarly, in the theory of a simple liquid, A reduces to the shear viscosity ς and the bulk viscosity η. We finally get the Goodwin and Trainor equation for visco-elasticity

$$\rho\frac{d^2\xi}{dt^2} = \mu\nabla^2\xi + (\lambda+\mu)\nabla(\nabla\xi) + \eta\nabla^2\dot{\xi} + (\varsigma+\eta/3)\nabla(\nabla\cdot\dot{\xi}) - F\cdot\nabla\xi - R\xi + 2nd\ order\ terms \tag{1.3}$$

where $F = -\dfrac{\partial \sigma_0}{\partial \chi}$

F is a calcium dependent second order tensor and the external force term $R\xi$ stands, in a linear approximation, for the restoring forces due to structural components (e.g. microtubules) which resist local displacement of the gel.

1.2. Calcium kinetics

The second equation of the model attempts to describe the simplest aspect of calcium kinetics. A simple reaction with stoichiometry n is assumed between calcium ions and a macromolecule C. Assuming that the total concentrations of the binding macromolecules and of calcium are constant, a straightforward derivation leads to the kinetic equation

$$\frac{d\chi}{dt} = k_{-1}(K - \chi) - k_1(\beta + \chi)\chi^n \tag{1.4}$$

where χ represents the concentration of free calcium and β, K are constants.

In order to take account of stretching or compression effects on calcium release or calcium binding, the rate constant k_{-1} was assumed to be a function of strain (linear in first approximation). Finally the Goodwin and Trainor equation for calcium kinetic is

$$\frac{d\chi}{dt} = \left[a + a_{ij} \frac{\partial \xi_i}{\partial x_j} \right] (K - \chi) - k_1(\beta + \chi)\chi^n + D\nabla^2\chi \tag{1.5}$$

where a term for the diffusion of calcium, with diffusion coefficient D, has been added.

2. Growth model

2.1. Wall elongation

The cell wall may be considered as a viscoelastic-plastic material which behaves like a Bingham solid. When the tension (due to turgor pressure) is above a "yield threshold" the deformation, instead of being simply viscoelastic, becomes partially irreversible (plastic) and the cell grows. The mechanical process which is generally believed to be responsible for the plastic deformation of the cell wall is called "creep." It might correspond to displacements of the cellulose microfibrils of the wall, sliding relative to each other after rupture of load-bearing bonds.

From theoretical and experimental considerations Lockhart (1965) has derived a simple model relating the volumetric growth rate to the internal pressure (P). This model was then extended by Cosgrove (1985) to account for the effect of pressure variations:

$$\frac{1}{V}\frac{dV}{dt} = \phi[P-Y] + \frac{1}{E}\frac{dP}{dt} \tag{2.1}$$

where Y is the yield threshold, ϕ is the "extensibility coefficient" of the wall, E is the volumetric elastic modulus ($= VdP / dV$) and $[P - Y] = max\ (0, P - Y)$.

This model applies to the whole cell. If we consider now the level of the wall, it is interesting to note that the pressure corresponds to a mechanical stress and the volumetric growth rate is analogous to a strain rate. Therefore, the growth process described by the equation above could correspond, at the level of the wall material, to the following model:

$$E\frac{d\varepsilon}{dt} = \frac{1}{\tau}[\sigma-Y] + \frac{d\sigma}{dt} \tag{2.2}$$

where ε is the strain and σ is the stress.

It is worth noticing that this equation represents the strain-stress relationship in a linear viscoelastic fluid (Maxwell model).

When σ is above Y, then ε is a combination of the elastic strain and the growth strain Γ. Considering that, in (2.2), pressure variations are only related to the elastic strain, we may write

$$E\frac{d\Gamma}{dt} = \frac{1}{\tau}[\sigma-Y] \tag{2.3}$$

Let now ε be the elastic strain between the rest length L_0 and the actual length L of a small linear element. The growth rate $d\Gamma/dt$ is then equal to $\dfrac{1}{L_0}\dfrac{dL_0}{dt}$ and for a constant stress we may write from (2.3)

$$\frac{E}{L_0}\frac{dL_0}{dt} = \frac{1}{\tau}[E\varepsilon - Y]$$

where Hooke's law is applied to the elastic part of the strain.

Noting $s = \dfrac{Y}{E}$ we get finally

$$\frac{1}{L_0} \frac{dL_0}{dt} = \frac{1}{\tau} [\varepsilon - s] \qquad (2.4)$$

The relative growth rate is then expressed as a linear function of the strain in excess of a given threshold (s).

Equation (2.4) expresses the kind of growth which could correspond to the creep. But another type of growth should be also considered. When new layers of cell wall are deposited or when a softened wall is repaired, this occurs without change of shape, hence of the stress field. This reinforcement of the wall, which corresponds to an increase of the elasticity modulus E, implies also a decrease of the strain. Starting from Hooke's law at constant stress σ we have

$$\frac{d\varepsilon}{dt} = -\frac{\varepsilon}{E} \frac{dE}{dt} \qquad (2.5)$$

In the present case, the variation of ε is a result of a variation of L_0, L being constant.

Then, from $\varepsilon = \dfrac{L}{L_0} - 1$, we get

$$\frac{1}{L_0} \frac{dL_0}{dt} = \frac{\varepsilon}{1+\varepsilon} \frac{1}{E} \frac{dE}{dt} \qquad (2.6)$$

where only positive variations of E are considered (to account for wall synthesis only).

Combining equations (2.4) and (2.6), the relative growth rate of a linear element of cell wall can be expressed finally as:

$$\frac{1}{L_0} \frac{dL_0}{dt} = \frac{1}{\tau} [\varepsilon - s] + \frac{\varepsilon}{1+\varepsilon} \frac{1}{E} [\frac{dE}{dt}] \qquad (2.7)$$

2.2. Cortical growth

Wall elongation, and more specially wall creep, is necessarily followed or accompanied by the stretching of the cytogel. Considering that the cortex adheres closely to the cell wall, any deformation of the wall induces an analog deformation in the cytogel. In order to model cortical growth we assumed that the relative growth rate of a cortical element is proportional to the part of its strain due to wall elongation. That is

$$\frac{1}{l_0}\frac{dl_0}{dt} = c\left[\frac{L_0}{l_0} - 1\right] \tag{2.8}$$

where l_0 is the rest length of the cortical element associated to the wall element (L_0).

3. Control of wall elasticity

Among the various cortical variables which may have an effect on the rheological properties of the wall, we considered the cytogel strain field and the calcium concentration gradient. The cytogel strain, by mechanical triggering of proton pumps, might be responsible for acid induced wall loosening, hence for a decrease of the elasticity moduli. But acid induced wall loosening may also result from proton-calcium exchange due to calcium input through the plasma membrane. In the case of a calcium-facilitated calcium uptake, variations of wall elasticity would be related to variations of the cortical calcium concentration.

In both cases, a simple formulation of the variations of the Young modulus of elasticity E is to consider a balance between wall synthesis and wall loosening. In the simulations, we assumed that the wall synthesis rate was a decreasing function of the wall elasticity modulus (the more the wall is weakened, the more the synthesis is active) and the rate of wall degradation was a saturated function of the cytogel strain. The final expression used for the variations of the elastic modulus of the wall was then

$$\frac{\partial E(x,t)}{\partial t} = d(a - E) + \frac{b[\varepsilon - c]}{1 + f[\varepsilon - c]}E \tag{3.1}$$

In this equation, the term $[\varepsilon\text{-}c] = max(0, \varepsilon\text{-}c)$ means that wall degradation can occur only when the cytogel strain is high enough to activate the proton pumps and to undergo the wall acidification responsible for the rupture of specific bonds between microfibrils.

REFERENCES

Brachet, J. and Bonnotto, S. (1970), Biology of *Acetabularia*, Academic Press, London & New York.

Brière, C. and Goodwin, B.C. (1988), Geometry and dynamics of tip morphogenesis in *Acetabularia*. J. theoret. Biol. *131*, 461-475.

Cleland, R. (1971), Cell wall extension. Ann. Rev. Plant Physiol. *22*, 197-222.

Cosgrove, D.J. (1983), Cell wall yield properties of growing tissues. Evaluation by in vitro stress relaxation. Plant Physiol. *78*, 347-356.

Goodwin, B.C. (1989), Unicellular morphogenesis. In: Cell Shape: Determinants, Regulation and Regulatory Role (eds. W.D. Stein and F. Bronner), pp. 365-391.

Goodwin, B.C. (1990), Structuralism in Biology. In: Science Progress, Oxford (Blackwell) *74*, 227-244.

Goodwin, B.C., Brière, C. and O'Shea, P.S. (1987), Mechanisms underlying the formation of spatial structure in cells. In: "Spatial organization in eukaryotic microbes" (eds. R.K. Poole and A.P.J. Trinci), pp. 1-9.

Goodwin, B.C., Skelton, J.C. and Kirk-Bell, S.M. (1983), Control of regeneration and morphogenesis by divalent actions in *Acetabularia mediterranea*. Planta *157*, 1-7.

Goodwin, B.C. and Trainor, L.E.H. (1985), Morphogenese apicale et formation des verticilles chez *Acetabularia* sous l'action d'un champ de contraintes régulé par le calcium, IV Séminaire de l'Ecole de Biologie Théorique, pp. 3045-315, Editions due CNRS, Paris.

Green, P.B., Erickson, R.O. and Buggy, J. (1971), Metabolic and physical control of cell elongation rate. In vitro studies in *Nitella*. Plant Physiol. *47*, 423-430.

Green, P.B. (1987), Inheritance of pattern: analysis from phenotype to gene. Amer. Zool. *27*, 657-673.

Green, P.B. (1989), Shoot morphogenesis, vegetative through floral, from a biophysical perspective. In: "Plant Reproduction: From Floral Induction to Pollination" (eds. E. Lord and G. Barrier), Am. Soc. Plant Physiol. Symp. Series, Vol. 1, pp. 58-75.

Harrison, L.G. and Hillier, N.A. (1985), Quantitative control of *Acetabularia* morphogenesis by extracellular calcium: a test of kinetic theory. J. Theoret. Biol. *114*, 177-192.

Kamiya, N. (1981), Physical and chemical basis of cytoplasmic streaming. Am. Rev. Plant Physiol. *32*, 205-236.

Lockart, J.A. (1985), An analysis of irreversible plant cell elongation. J. Theor. Biol. *8*, 264-275.

Meinhardt, H. (1982), Models of Biological Pattern Formation, Academic Press, London.

Menzel, D. and Elsner-Menzel, C. (1989), Induction of actin-based contraction in the siphonous green alga *Acetabularia (Chlorophycea)* by locally restricted calcium influx. Bot. Acta *102*, 164-171.

Murray, J.D. (1989), Mathematical Biology, Springer-Verlag.

Nossal, R. (1988), On the elasticity of cytoskeletal networks. Biophys. J. *53*, 349-359.

Oster, G.F. and Odell, G.M. (1983), The mechanochemistry of cytogels. In: "Fronts, Interfaces and Patterns" (ed. A. Bishop), Amsterdam, North Holland, Elsevier Science Division.

Puiseux-Dao, S. (1970), *Acetabularia* and Cell Biology, Logos Press Ltd.

Turing, A.M. (1952), The Chemical basis of morphogenesis. Phil. Trans. R. Soc. B. *237*, 37-72.

MODELS OF BIOLOGICAL PATTERN FORMATION AND THEIR APPLICATION TO THE EARLY DEVELOPMENT OF DROSOPHILA

Hans Meinhardt

Max-Planck-Institut für Entwicklungsbiologie

Spemannstr. 35, D-74 Tübingen, FRG

Introduction

The complexity of a higher organism indicates that very many pattern forming reactions are at work that are coupled to each other in such a way that the final pattern can be generated with a high degree of reproducibility. Investigations of early development in *Drosophila* have provided us with much information about the molecular machinery on which development is based. About ten years ago, I proposed a model for pattern formation in early insect embryogenesis (Meinhardt, 1977). This model was based on a single morphogen gradient with a high point at the posterior pole of the egg. The gradient was assumed to be generated by short range autocatalysis and long range inhibition (Gierer and Meinhardt, 1972). This model was able to account for most of the experimental observations available at that time. More recently, this model of positional information has been complemented by a model for the hierarchical activation of gap-, pair rule and segment polarity genes (Meinhardt, 1985, 1986). In the meantime many additional genetic and molecular data have become available for *Drosophila*. Much of this new data supports the basic stipulations of these models, while some of it suggests modifications of these models. In this paper I will mention very briefly the basic ingredients of the models with reference to the *Drosophila* system and show of how these elements can be linked to obtain a reproducible pattern formation. This paper will be partially based on arguments previously put forward (Meinhardt, 1985,1986) and more recently updated (Meinhardt, 1988).

Generation of a primary pattern by autocatalysis and lateral inhibition

We proposed that primary embryonic pattern formation is accomplished by the coupling of a short ranging self-enhancing (autocatalytic) process with a long range inhibitory effect that acts antagonistically to the self-enhancement (Gierer and Meinhardt, 1972; Gierer, 1981; Meinhardt, 1982). A simple molecular realization of these processes would consist of an activator whose autocatalysis is antagonized by a long-ranging inhibition. The production rate of the activator (a) and inhibitor (h) can be formulated in the following set of coupled differential equations:

$$\frac{\partial a}{\partial t} = \frac{\rho a^2}{h(1 + \kappa a^2)} - \mu a + D_a \frac{\partial^2 a}{\partial x^2} + \rho_0 \qquad (1a)$$

Complexity, Chaos, and Biological Evolution, Edited by E. Mosekilde and
L. Mosekilde, Plenum Press, New York, 1991

$$\frac{\partial h}{\partial t} = \rho a^2 - \nu h + D_h \frac{\partial^2 h}{\partial x^2} + \rho_1 \qquad (1b)$$

where t is time, x is the spatial coordinate, D_a and D_h are the diffusion coefficients, μ and ν the decay rates of a and h. The source density ρ describes the ability of the cells to perform the autocatalysis. A small activator-independent activator production ρ_0 can initiate the system at low activator concentrations. A condition for stable pattern formation is $D_a \ll D_h$ and $\mu < \nu$. The patterns that can be generated by this type of interaction correspond with many patterns observed in living systems. In small or growing fields, monotonic gradients are formed that are appropriate to supply positional information (Wolpert, 1969). In fields large compared with the range of both substances, periodic patterns will result (Fig. 1). If the autocatalysis saturates at relatively low activator concentrations via the saturation term $1 + \kappa a^2$, stripe-like patterns will occur

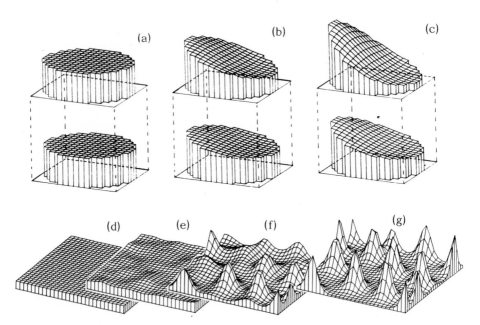

Fig. 1 (a-c) Stages in the formation of a stable gradient in a two-dimensional field of cells. Pattern formation is triggered by random fluctuations. If the range of the activator is of the order of the field size, only one marginal maximum can emerge since a maximum in the center would require space for two activator slopes. In the absence of strong polarizing influences, the pattern will orient itself along the longest extension of the field. The polar pattern can be maintained even after further substantial growth of the field. The gradient(s) can be used as positional information. (d-g) If the range of the activator and the inhibitor is smaller than the total field, many maxima are formed in a somewhat irregular arrangement. However maximum and minimum distance between the peaks is maintained.

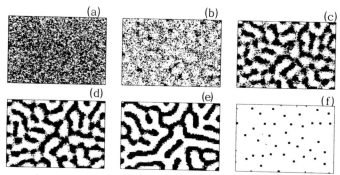

Fig. 2 (a-e) Stages in the formation of a stripe-like pattern. Assumed is an activator-inhibitor system and a saturation of autocatalysis at low activator concentration. This limits the competition among neighbouring cells. More cells remain activated, although at a lower level. Small diffusion of the activator leads to the tendency to form coherent activated regions. Stripes are the most stable pattern since activated cells have activated cells in the neighbourhood but, nevertheless, non- activated cells are nearby into which the inhibitor can escape. (f) For comparison, without saturation but otherwise the same parameters and initial conditions, a bristle-like pattern results. The maximum concentration is about ten times higher than in Fig. e. (from Meinhardt, 1988)

(Fig. 2, 3). These pattern are very frequent in many developmental situations. The development of the fruitfly *Drosophila* is discussed below as an example

An important feature of many developing systems is their ability to regenerate. The removal of some parts of an embryo are compensated by pattern regulation. This is a property of the activator-inhibitor system since, for instance, after removal of the activated region, the remnant inhibitor decays until a new maximum is formed via autocatalysis. A small activator-independent activator production ρ_0 can initiate the system at low activator concentrations. Conversely, a small baseline inhibitor production ρ_1 can cause that pattern formation is suppressed until an inducing signal is supplied that shift the activator concentration over a certain threshold. The further activator increase proceeds via autocatalysis.

In his pioneering paper, Turing (1952) showed that two substances of different diffusion rates can lead to pattern formation. He demonstrated this by analyzing interactions formulated as linear partial differential equations. These equations are not a realistic description of molecular mechanisms. For example he assumed, for one substance, that the number of molecules that disappears per time unit was independent of the number of molecules present. This can lead to negative concentrations, since molecules can disappear with a finite rate, even if no molecules are left. He realized that non-linear equations would be required, but that the number of possible interactions are very large. Our finding that self- enhancement and lateral inhibition are the driving forces for pattern formation, provides a handy tool to check whether an interaction has pattern forming capabilities and whether the resulting pattern can be stable. (Gierer and Meinhardt 1972, 1974).

The interaction between three substances given by Eq. 2 is an example that self-enhancement does not require a molecule with direct autocatalytic regulation. The autocatalysis can be a property of the system as a whole. For instance, if two substances, a and b exist, and a inhibits b, and *vice versa*, a small increase of a above an equilibrium leads to a stronger repression of the b production, and thus to a further increase of a,

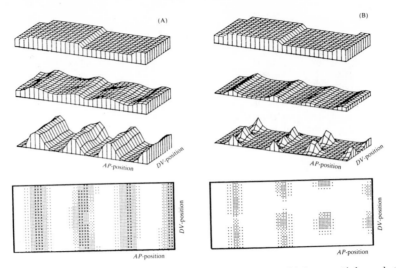

Fig. 3 Example for the initiation of a stripe-like pattern with a higher spatial resolution by a coarser pattern. Assumed is an activator-inhibitor system with saturation (see Fig. 2). The coaser pattern is assumed to have a modifiing influence on the subsequent pattern, shown is the ρ distribution (Eq. 1). In each subpicture the activating influence of the coarse pattern (top), an intermediate pattern and the final pattern is shown along the anterio-posterior and the dorso-ventral axis. At bottom, a separate simulation with different random fluctuations is provided in a top view. (a) Formation of stripes. The activation is favoured near the borders of the coarse pattern, since the inhibitor can escape into the neighbouring non-activated region, in contrast to the situation in the center. Two stripes are formed which coincide with their outer border with that of the triggering pattern. An additional stripe is inserted into the space between two activating regions. (b) If the saturation level is not reached due to a mutation, the stripes desintegrate into patches, a feature that is frequently observed in weak pair rule mutations.

and so on in the same way as if a would be autocatalytic. The same holds for b; a and b together form a switching system in which either a or b is high. Pattern formation requires that if, for instance, a has won the a-b competition in a particular region, b must win in the surroundings. This can be accomplished by a long ranging signal that interferes with the mutual competition of the two substances. A possible realization would be that the a molecules control the production of a substance h that, in turn, either inhibits the a, or promotes b production. In the example of Eq. 2, h undermines the repression of the b production by the a molecules.

$$\frac{\partial a}{\partial t} = \frac{\rho}{\kappa + b^2} - \mu_a a + D_a \frac{\partial^2 a}{\partial x^2} + \rho_a \tag{2a}$$

$$\frac{\partial b}{\partial t} = \frac{\rho}{\kappa + a^2/h^2} - \mu_b b + D_b \frac{\partial^2 b}{\partial x^2} + \rho_b \tag{2b}$$

$$\frac{\partial h}{\partial t} = \gamma a - \mu_h h + D_h \frac{\partial^2 h}{\partial x^2} \tag{2c}$$

In principle, the inhibitory action can also result from a control of the decay rate of the activator via the inhibitor as described by Eq. 3.

306

$$\frac{\partial a}{\partial t} = \rho a^2 - \mu h a + D_a \frac{\partial^2 a}{\partial x^2} \qquad (3a)$$

$$\frac{\partial h}{\partial t} = \rho a^2 - \nu h + D_h \frac{\partial^2 h}{\partial x^2} \qquad (3b)$$

However, such an interaction has severe disadvantages, since the range of the activator would change during the evolution of the pattern. Moreover, shortening of the lifetime of the activator can cause a transition from a stable to an oscillatory pattern.

Cell determination and region-specific gene activation

In higher organisms, the pattern generated by a reaction-diffusion mechanism is necessarily transient since, due to growth, the polar pattern cannot be maintained over the whole expanse of the growing organism. This requires that, at an appropriate stage, the cells make use of position- specific signals, i.e. that they become determined for a particular pathway by activating particular genes. Afterwards the cells maintain this determination whether or not the evoking signal is still present.

The simplest system with a long term memory would consist of a substance with a threshold behaviour. A possible mechanism would consist of a substance that feeds back on its own production rate in a nonlinear way that saturates at high concentrations. With appropriate parameters, only two stable states are possible. Eq. 4 provides an example.

$$\frac{\partial g}{\partial t} = \frac{cg^2}{(1 + \kappa g^2)} - \mu g + m \qquad (4)$$

If the signal m is above a certain threshold, only the high state is stable. The system remains at this state even if m becomes small later on (Meinhardt, 1976).

More important is the region-specific activation of several genes under the control of a morphogen gradient. The activation of a particular gene has many formal similarities to the formation of a pattern. In pattern formation, a particular substance is produced at a particular location while this production is suppressed at other locations. Correspondingly, determination requires the activation of a particular gene and the suppression of the other alternative genes of a set. Gene activation may thus be regarded as a pattern formation among a set of alternative genes.

In analogy to the activator-inhibitor system for pattern formation, one can formulate the following set of equations for activation of particular genes via its gene products g_i ($i = 1...n$, n is the number of alternative pathways) (Meinhardt, 1978, 1982).

$$\frac{\partial g_i}{\partial t} = \frac{c_i g_i^2}{r} - \mu g_i + m_i \qquad (5a)$$

$$\frac{\partial r}{\partial t} = \sum_i c_i g_i^2 - \nu r \qquad (5b)$$

Each gene product g_i regulates the transcription of the gene in an autocatalytic manner. Each active gene causes the synthesis of a common repressor r. The repressor acts on the transcription rate of each gene of the set. this has the consequence that within one cell only one gene can be active, since two active genes would compete with each other (Fig. 4). The last term in Eq. 5a describes a possible influence of the graded

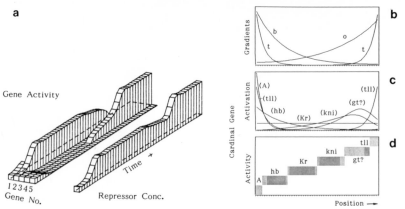

Fig. 4 Formation of cell states and space-dependent gene activation. (a) A set of genes (1,2...5) is assumed whose products feed back on the activation of the corresponding genes. In addition, all genes compete with each other by a repressor (Eq. 5a,b). This has the consequence that only one of the genes can be active within one cell. In this simulation, random fluctuations have been decisive that gene no. 2 initially wins this competition. By an external signal, gene no. 4 becomes activated. Full activation results from autoregulation and is accompanied by the repression of the previously active gene. (b) Simulation of gap-gene activation in *Drosophila* under control of maternally supplied positional information. (b) A multi-gradient system consisting of the posterior system *oskar-nanos (o)*, the anterior system *bicoid (b)*, and the symmetrical system *torso (t)*. (c) The actual activation of the genes. The activation of the most anterior cardinal region (called A) results from a cooperation of *t* and *b*. *b* without *t* causes *hunchback, (hb)* activation, *Krüppel (Kr)* has a baseline, position-independent activation such as proposed by Gaul and Jäckle (1987), the activation of *knirps (kni)* and *giant (gt)* results from the *o* gradient, the giant activation is assumed to require a dimer of the molecule forming the *o*-gradient and thus requires higher *o*-concentrations. Posterior-most structures (P) are formed if *t* is high but *b* is low; *t* is also assumed to inhibit competitively the *kni* and *gt* activation. (d) This mode of activation leads to sharply confined regions, the cardinal regions, in which the corresponding gap genes are active (after Meinhardt, 1988).

morphogen concentration m which provides the positional information. This could be, for instance, an activator or inhibitor gradient (Meinhardt, 1978). In Fig. 4, a simulation is shown in which several gradients of different slopes are involved in a combinatorial way. At different regions, different positional information dominate. Due to the competition via a common repressor (or via a direct negative influence of the alternative genes), a particular gene is active in groups of neighbouring cells. An abrupt transition from one activated gene to the subsequent one takes place between neighbouring cells despite the smooth distributions of the morphogenetic substances. Since each gene feeds back on its own activation, a gene remains active, independent of whether lateron the signals are present or not.

Generation of ordered sequences of structures by mutual activation of locally exclusive cell states

In the mechanism described above, the cells do not communicate directly with each other to obtain the correct determination. They measure the local concentration and behave accordingly in a slavish way. In other developmental systems, for example in the segments of the body or the legs of insects, a direct communication between the cells appears to be involved to obtain a particular determination. This requires some modifications of the gene-activation mechanism discussed above.

We have seen that self-activation of genes coupled with a mutual repression of the alternative genes leads to stable states of determination. If two (or more) such states not only exclude each other locally but also activate each other mutually over long ranges, these cell states stabilize each other in a symbiotic manner. Both cell states need each other in a close neighbourhood, while the local exclusiveness assures that the two states do not merge or overlap (Meinhardt and Gierer, 1980). Eq. 6 describes a possible interaction.

$$\frac{\partial g_i}{\partial t} = \frac{c_i g_i'^2}{r} - \mu g_i + D_g \frac{\partial^2 g_i}{\partial x^2} \tag{6a}$$

with

$$g_i' = g_i + \delta^- s_{i-1} + \delta^+ s_{i+1}$$

$$\frac{\partial s_i}{\partial t} = \gamma(g_i - s_i) D_s \frac{\partial^2 s_i}{\partial x^2} \tag{6b}$$

$$\frac{\partial r}{\partial t} = \sum_i c_i g_i'^2 - \nu r \tag{6c}$$

It is a feature of these interactions that they can also generate stripes. Stripes have necessarily a very long border between cells of different cell types. A particular cell type is close to cells of the other type. This facilitates the mutual support of the two cell states by the diffusible substances (s_i in Eq. 6). A stripe-like arrangement is therefore especially stable.

The mutual activation scheme can comprise several cell states. For example, state 1 activates state 2, state 2 activates state 3 and so on until state n. For the formation of the periodic structures of segments it is important that the system can be cyclic in that state n activates state 1. This system is able to generate self-regulating sequences. For instance, if cell types are juxtaposed which are usually not neighbours, the intervening sequences become intercalated, restoring in this way a normal neighbourhood of cell states. This need not necessarily be the natural sequence of structures. A biological system in which, after manipulation, a correct neighbourhood, but a non-natural pattern can be formed, is the intrasegmental pattern of insect legs (Bohn, 1970). As shown below, this scheme is supported by recent findings with segment polarity genes of *Drosophila*.

How are the legs and wings determined?

The primary pattern formation, such as that described above, must also control the position at which substructures such as legs, wings or antennae are to be formed. On the other hand, many experiments indicate that, after initiation, the pattern formation of these structures is a more or less autonomous, self-regulating process. The question is then, how can a particular group of cells be determined to form, e.g. a limb? Moreover, how is it achieved that a limb has a particular handedness (left or right) and a particular orientation with respect to the main axes of the developing embryo?

The determination of appendages clearly requires not only a primary determination along the antero-posterior (AP) axis but also a secondary pattern formation perpendicular to the first axis, along the dorsoventral (DV) axis. One could imagine that in this way a particular group of cells - specified like a particular field of a checkerboard - obtains the primary signal to produce, for instance, a leg, and that the further secondary pattern formation proceeds as described above for the primary pattern formation; namely by autocatalysis and lateral inhibition. However, an analysis of the experimental data indicate another mechanism. By the primary subdivision of the embryo, fields of cells with different determinations arise that are separated by sharp borders. Such a border can become the source region for a new morphogen if cells of two types cooperate to produce the new morphogen. The intersection of a border that arose by a subdivision along the AP axis with a border that results from a DV organization leads to unique pairwise intersections. The model accounts for many experimental observations in vertebrate and insect limbs (Meinhardt, 1980; 1982; 1983a; 1983b). It predicts that the structures are formed around preexisting borders. This proved to be correct for the initiation of imaginal disks in insects. The imaginal disks are separated into an anterior and a posterior compartment (Garcia-Bellido et al., 1973). The gene *engrailed* is responsible for the formation of the posterior compartment. However, *engrailed* belongs to the segment polarity group of genes (Nüsslein-Volhard and Wieschaus, 1980) and is expressed already during blastoderm stage (Kornberg et al., 1985), long before any imaginal disc could have been formed, in full agreement with the model proposed.

Segmentation of Drosophila as model system tho investigate the interaction of several coupled pattern forming systems

The formation of the (segmented) pattern is an important step in the development of higher organism. In this process, details of the primary anteroposterior pattern are laid down. Insect segmentation has been studied intensively to approach this problem experimentally. Saturation mutagenesis experiments affecting segment formation in Drosophila have shown that four major classes of phenotypes exist, that at least 25 genes are involved in segmentation and that new classes of mutant phenotypes are not to be expected (Nüsslein-Volhard, 1977; Nüsslein-Volhard and Wieschaus, 1980, Nüsslein-Volhard et al., 1982). The following phenotypes have been found (Fig. 5): (i) The (maternal) coordinate mutants which lead to a coordinated pattern rearrangement. For instance, the bicaudal (double abdomen) is of this type. (ii) The gap mutants in which a contiguous pattern deletion of about 7 segments occurs. (iii) The pair rule mutants in which structures around every second segment borders are lost, forming thus half the normal number of segments. (iv) The segment polarity mutants in which homologous pattern elements are lost in every segment and in which the remaining patterns show mirror-image duplications. The different classes of mutant phenotypes indicate a hierarchy of pattern forming events at four different levels. Many of the genes involved in segmentation have been already cloned (see Ingham, 1988)

The original model proposed for the segmentation of *Drosophila* (Meinhardt, 1985, 1986) was based on the assumption of the following four major steps. A gradient controls the overall antero-posterior pattern. Under its influence, four cardinal (and two marginal) regions are determined, defined by the regions in which the gap genes are active. The border between any two cardinal regions act as organizing region for the initiation of the first truly periodic pattern, the double segments. The double segment pattern results from the sevenfold repetition of four cell states. By an inductive process, these give rise to the periodic pattern of individual segments which results from 14 fold repetition of (at least) three cell states. According to this view, the formation of segments begins with a simple pattern - the gradient - and reaches the complex pattern by passing through a hierarchical series of intermediate patterns. Each pattern is more complex or has a higher spatial frequency than that by which it has been induced.

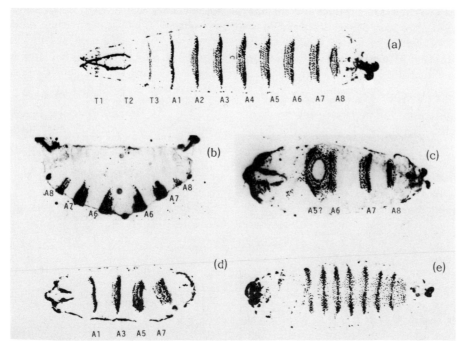

Fig. 5 Examples of the four classes of mutations affecting segmentation in *Drosophila* (Nüsslein-Volhard, 1977; Nüsslein-Volhard and Wieschaus, 1980). (a) Wild type embryo. (b) The coordinate mutant *bicaudal*: a mirror-symmetrical embryo with two abdomen. (c) *Krüppel*, a gap mutant: a single region of about 7 segments is missing (T1-A4/A5). (d) *hairy*, a pair rule mutant: Homologous pattern elements around every second segment border are deleted; the remaining denticle belts indicate the maintenance of the polarity. (e) *gooseberry*, a segment polarity mutant. Certain pattern elements are deleted in every segment, the remaining pattern of each segment is duplicated. The polarity of the segments is lost. (Negatives of dark field photographs, kindly provided by Christiane Nüsslein-Volhard, after Meinhardt, 1986).

Many features of this model have turned out to be correct, such as: the gap genes are expressed in a region half as large as gaps which are formed if the gene product is absent due to a mutation (Knipple et al. 1985), that the pair rule genes are initiated by specific combinations of two gap gene products (Pankratz et al. 1990), the expression of a gap gene in a large region on its own does not promote the formation of the corresponding segments, and the segmentation that results from a discrete sequence of cell states which mutually stabilize each other (Martinez-Arias et al, 1988).

On the other hand, several findings have been unexpected on the basis of the model, such as: gaps are asymmetrically located in relation to the region of activity of the corresponding gap genes in the wild type (Knipple et al., 1985), and the pair rule genes *even-skipped (eve)* and *fushi tarazu (ftz)*, despite the fact that they have complementary regions of activity, show little mutual interference (Carroll and Scott, 1986; Harding et al., 1986); while genes which show partial overlap in their region of activity, such as *hairy (h)* and *ftz* have a strong but hierarchical influence on each other (Howard and Ingham, 1986). Unexpected also was the finding that the positional information employed by the embryo consists of three positional information systems which are to a large degree independent of each other (Nüsslein-Volhard et al., 1987).

The maternal positional information system and the activation of the gap genes

The results of investigation of early insect development by classical methods can be summarized as follows (for review see Sander, 1976): (i) At the anterior pole, a general instability exists such that posterior structures often develop instead. The result is a symmetrical double abdomen embryo (see Fig. 5b). This malformation can be induced in different species by different experimental manipulations, such as UV- irradiation (Kalthoff and Sander, 1968), centrifugation (Yajima, 1960), temporary ligation (van der Meer, 1984) or by mutagenesis (Bull, 1966, Nüsslein-Volhard, 1977). (ii) In eggs of the leaf hopper *Euscelis*, a shift of posterior pole material can lead to ectopic abdominal structures at the new position of the posterior pole material (Sander, 1960), accompanied by a shift of the fate map, such that a normal embryo can be formed at a more anterior position. (iii) A ligation of an insect egg before the blastoderm is reached frequently leads to a gap in the sequence of segments formed. The earlier the operation the larger the gap will be (Sander, 1976).

Many of the experiments listed above can be accounted for in a quantitative way by the assumption of a single gradient generated by an activator-inhibitor mechanism discussed above (Meinhardt, 1977). The inhibitor, spreading out from the posterior pole, would have two functions: to provide the positional information necessary for segmentation and to localize the autocatalytic center there. According to this view, the region of high activator and/or inhibitor concentration is the signal to produce and abdomen. The double abdomen malformation result if a second autocatalytic center is initiated at the anterior pole, for instance, by a temporary reduction of inhibition. The gap phenomenon after ligation is thought to result from an accumulation of the morphogen at the posterior side of the ligation and a decrease at the anterior side due to the interruption of diffusion. For the explanation of the shift experiments with posterior pole material in *Euscelis*, it was assumed that the activator is shifted with the posterior pole material. New high points of gradients can emerge at the new location of the posterior pole material in a self-regulatory manner followed by subsequent restoration of the gradient.

The shortcomings of such a mechanism were also clearly recognized. If a gradient is generated by a local source at one end and decays everywhere else, the slope is very shallow at the pole opposite to the source region and thus inappropriate as positional information in that region. In addition, size-regulation is bad. As long as the other parameters, such as source strength, decay and diffusion rate remain constant, low gradient levels are poorly regulated if the total field, i.e. the egg length, becomes reduced. For many insect embryos, this does not create a problem since the embryo proper is formed anyway only in a posterior portion of the egg. For instance, in the *Euscelis* egg mentioned above, only the posterior half of the blastoderm is used to form the embryo. A complete embryo is formed even if most of the anterior egg half is removed by ligation (Sander, 1959). However, *Drosophila* fate mapping has shown that almost all of the blastoderm is used to form the embryo (Lohs- Schardin et al., 1979), and here a change of the egg length in some mutants nevertheless leads to complete embryos (Nüsslein-Volhard, 1977). Both properties proved to be difficult to integrate into a single gradient model.

As mentioned, in *Drosophila* at least three positional information exist (Nüsslein-Volhard et al., 1987). Elsewhere, I have provided a detailed computer simulation for such a system (Meinhardt, 1988; Fig. 3).

The mechanism of how the gap gene transcription becomes so sharply confined is not yet completely clear. So far, no direct autocatalyc regulation has been demonstrated. However, the *hunchback* protein seems to have a binding site on its own gene (Treisman and Desplan, 1989). Some gap genes show strong repression of other gap genes, for instance *knirps* on *hunchback* (Pankratz et al., 1990). If such repression is mutual, this can lead to a switch-like behaviour. However, for example, *Krüppel* increases the transcription of the *knirps* activation without any influence on the size of the region in which *knirps* is transcribed (Pankratz et al, 1989). The gap genes are transcribed only for a short period of time. No long term memory is required for the gap gene activation. Therefore, a very non-linear activation of gap genes by the maternal positional information may be sufficient. Indeed, several binding sites of different affinity have been reported for the *bicoid* protein to the *hunchback* gene (Driever and Nüsslein-Volhard, 1989).

The fourfold subdivision of double segments and the activation of the pair rule genes by the borders of gap gene activity

A mutation of a pair rule gene shows periodic pattern deletions around every second segment border. Either posterior parts of the even numbered, and anterior parts of the odd numbered segments, are deleted or vice versa. Thus, mutant phenotypes show half the normal number of segments and those segments that remain usually have normal polarity (see Fig 5d). These mutations suggest a transient formation of a periodic pattern with a repeat length corresponding to two segments.

The explanation of these phenotypes requires the assumption of at least four cell states per double segment for the following reason. If there were an alternation of only two states, even (E) and odd (O) (i.e. a sequence ...OEOE...) the loss of any one state would lead to the loss of the periodic character (e.g. OOOOO). In a three state model, the loss of any one state would lead to a symmetrical pattern. Thus, we must assume (at least) four cell states - to be called 1, 2, 3 and 4 - as the basic building blocks of a double segment. After the loss of one of the four elements, the remaining three elements still possess an unambiguous polarity, in agreement with the pair-rule phenotypes. For instance, if cell state 2 is lost, the (polar) sequence ...134134... remains.

In-situ hybridization experiments have shown that the pair-rule gene *fushi tarazu (ftz)* is transcribed in 7 regularly spaced bands at the early blastoderm stage. Each band and interspace has the width of one future segment (Hafen et al., 1984). Later in development, the regions of *ftz* activity shrink while interspaces extend (Martinez-Arias and Lawrence, 1985). The transcription of *hairy (h)* takes place in similar bands and combined *h* and *ftz* transcriptions have shown that the anteriormost *h* and the posteriormost *ftz* band are isolated, while between these extremes 6 bands exist in which *h* and *ftz* transcription partially overlap (Ingham, 1985).

These findings can be integrated into the following scheme. The first pattern of pair rule gene activity consists of two sequences, each has two members, ..1313.. (corresponding to *hairy* and *runt*) and ..2424.. (corresponding to *even-skipped* and *ftz*). Both are out of register. These two sequences give rise to the double segment pattern proper, the 1234 sequence (Fig. 6).

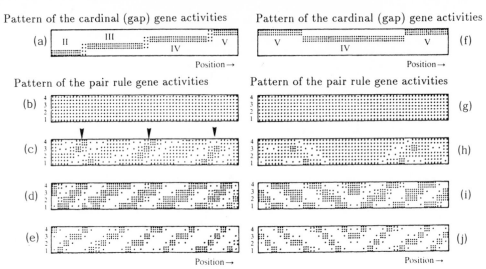

Pattern of the cardinal (gap) gene activities   Pattern of the cardinal (gap) gene activities

Pattern of the pair rule gene activities   Pattern of the pair rule gene activities

Fig. 6 Stages in the activation of the pair-rule genes by the gap genes. (a) Normal pattern of the gap genes (II-V, see Fig. 4). (b-c) The borders between cardinal genes activate the pair-rule genes (1-4, corresponding to *hairy*, *even-skipped*, *runt* and *fushi tarazu*). For instance, the II/III border is assumed to activates a 341-sequence. These borders of gap gene expression are the first regions of increased pair rule gene activities (arrows)(c). The polarity residing in the sequence of cardinal genes is transmitted. (d) If state 1 and 3 compete via the same repressor and this repressor is different from the repressor by which state 2 and 4 compete, the 1 and 3 region as well as the 2 and 4 region are complementary. The phase shift is achieved only by the mode of ..131.. and ..242.. induction at the borders of the cardinal regions. If the 131 and 242 sequence compete via a common repressor (or if both repressors have a physical similarity), both sequences merge. The extension of an active region shrinks in such a way that in a particular cell only one of the 1,2,3 or 4 genes would be active. Such merging would stabilize the phase shift between both sequences. (f) If the pattern of the cardinal genes is symmetric (*bicaudal* phenotype, Fig. 5) the resulting 1234-pattern (g-j) is mirror- symmetric too. The polarity resident in the cardinal borders is transferred to the pattern of pair-rule genes (after Meinhardt, 1986).

To get a predictable pattern of the pair rule genes, their activation at particular location in relation to the gap gene pattern is required. The phenotypes of the gap mutations provide some hints of how this takes place. A straightforward explanation for the overlap of the gaps would be that it is not the cardinal region in which the gap genes are expressed, but rather the borders between two cardinal regions that control the formation of the double segment pattern. Imagine a series of cardinal regions, I,II,III,IV.... (corresponding to *...hunchback, Krüppel, knirps...*). If, for instance, the gap gene II is lost, the I/II and the II/III border would be lost. In other words, the gap is expected to comprise half of the I-region, the II-regions and half of the III-region. Thus, the gap is expected to be twice as large as the region in which the gap gene is transcribed. This prediction (Meinhardt, 1985) has been found to be true. In-situ hybridisation experiments of the *Kr*-gene (Knipple et al., 1985) show a band of *Kr* activity which is 3 - 4 segments wide. This is much smaller than the gap of 7 segments caused by a loss of *Kr*. The reason why the extension of a cardinal region possibly corresponds to a non-integer multiple of a segment length (3.5) may be as follows. This facilitates the initiation of the ..131.. and ..242.. sequence with a half segment phase shift since the anterior border of a cardinal region would be in register with one sequence (e.g. ..131..), the posterior border in register with the other (..242..).

A further support for the stipulation that a cardinal region on its own is insufficient to initiate the pattern of pair rule genes but that borders with neigbouring regions are required, can be derived from experiments of Gaul and Jäckle. (1987). They found that removal of the maternal positional information leads to a large expansion of the *Krüppel* region. However, no pair rule pattern is initiated in such constructs.

In several gap mutations, some of the remaining segments close to the border of the gap show a polarity reversal. The *Kr*-mutation (Fig. 5c) is an example. It is reasonable to assume that (maternally determined) gradients remain unchanged whether a (zygotic) gap mutation is present or not. Therefore, these occasionally occurring polarity reversals suggest that the polarity of the segments is not under the control of the (unchanged) maternal positional information, but results normally from the sequence of cardinal regions. If a cardinal region is missing, this transmission of polarity may fail. According to the model, one reason for a polarity reversal could be an incorrect merging of the 131-242 sequence, so that instead of $...1_2 3_4 1_2...$ the sequence $_2 1_4 3_2 1_4...$ is formed which has reversed polarity. The transmission of polarity is presumably also the reason why the borders between the cardinal regions, and not the cardinal regions themselves, organize the double segment pattern. The asymmetry of a border between two regions of gene expression contains polarity information, while just a region of gene activity does not.

For the transmission of polarity, a particular cardinal border must activate at least the genes coding for two neighbouring regions. In the simulation Fig. 6 it has been assumed that, for example, the II/III border (corresponding to the *hb/Kr* border) activates the pair rule genes 3, 4 and 1 . Cell state 3 has to appear at the anterior side, 1 at the posterior side of the border and 4 has to be centered over the border. This is achieved by an activation of gene 4 if II and III molecules are present in equally high amounts, 3 is activated by two molecules of II and one molecule of III and so on. In this way, the polarity is transmitted from the sequence of gap genes to the double segment pattern. The initiation of the pair rule pattern around the gap gene borders is clearly visible in the simulation of Fig. 6 at intermediate steps (arrows). The pattern sharpens by the selforganizing property at the pair rule genes. A symmetrical pattern of gap gene activity (as given in the double abdomen malformation) must cause double segments of reverted polarities in the anterior half since there, both cardinal borders have opposite polarities (V/IV vs. IV/V). The simulations Fig. 6f-j show that the model has this property. I found that particular scheme for the activation of the pair rule genes by the gap genes so inelegant that I have described it only in the appendix with the details of the computer simulations. In the meantime, it has turned out for the *hairy* as well as for the *eve* gene that indeed stripe-specific promotors exist (Howard et al.1988, Goto et al, 1989, Harding et al, 1989). For the 6th and 7th *hairy*- stripe, the mode of activation by the gap genes as been investigated in details (Pankratz et al., 1990).

According to the model proposed, the activation of pair rule genes proceeds in two steps. On the one hand there is the activating influence of genes from the next higher hierarchical level, in this case the gap genes. On the other hand there is the assumption that the pair rule genes have by itself pattern forming capabilities such that the pattern becomes refined in a self-organizing way. From the model, I expected that not every stripe is determined by the gap gene overlap, but that only every second stripe is fixed by the gap genes in a scaffold-like manner. The remaining stripes are formed due to the internal pattern forming capability of the pair rule system. However, at least for the *ftz* gene, seven weaker stripes appear even if the "zebra" region of the *ftz*- promotor is lost (Hiromi and Gehring, 1987). It is not yet clear whether each stripe is determined individually and the sharpening is a cell-local process (as proposed in the model of Edgar et al., 1989) or whether cell- communication and selforganization takes place as assumed in simulation Fig. 6 or in the model of Lacalli (1990).

The (at least) threefold subdivision of segments and the activation of the segment polarity genes by the pair rule genes

Segmentation is a periodic pattern. The early subdivision of the thoracic structures into two compartments (Garcia-Bellido et al., 1973) has suggested that segmentation results from a repetition of two cell states. However, in a periodic pattern of two cell states ...APAPAP..., neither the position of the segment borders (/) nor the polarity of the segments would be determined: it could be .../AP/AP/AP/... or ...A/PA/PA/P... For this reason, I have proposed that segmentation results from the repetition of (at least) three cell states, to be called S, A and P (Meinhardt, 1982, 1984). The A and P regions would correspond to the well-known anterior and posterior compartments. The third cell state S is assumed to occupy the anterior-most portion of each segment and forms mainly the larval denticle belts. A sequence of three (or more) structures always has a defined polarity. According to this model, a segment border is formed whenever P and S cells are juxtaposed (...P/SAP/SAP/S...). The juxtaposition between A and P is used in a similar manner to induce new structures: the AP border is a prerequisite to form legs and wings (Meinhardt, 1980, 1983b).

The loss of any one of the three states would lead to an alternating pattern of the remaining two. Three basic types, ...SASASA..., ...S/P/S/P/S/P... and ...APAP... are expected. All three patterns are symmetric since each cell state would have the same neighbour on both sides. This loss of polarity fits well with the phenotypes of the segment polarity mutations mentioned above which form, in contrast to the phenotypes of the pair rule genes, a symmetrical pattern.

Direct support for the proposal that a third cell state separates the P and the A cells of two adjacent segments can be derived from an experiment of Bohn (1974) with cockroaches. He removed epidermal cells close to the anterior metathoracic segment border. After wound healing, he observed the formation of a supernumerary leg. In terms of the model, this corresponds to the removal of a S region $(../SA|P/_SA|P/... = .../SA|P|A|P/-$ (the region removed is written as a subscript, the leg initiation sites by |). This leads to a new intersection of an AP border and a DV border which creates the prerequisite for leg formation (Meinhardt, 1984).

More recent experiments have suggested that segmentation result from the repetition of four cell states (Martinez-Arias et al., 1988), to be called ..D/ABCD/ABCD/A.. The formation of the symmetrical pattern caused by the mutation mentioned above could be based on the insertion of an additional cell state at an ectopic position in order to produce a normal neighbourhood. For example, after a mutation of a gene that is required for the structure B, instead of a sequence ../ACD/ACD/.. a sequence ..A/D/ACA/D/AC.. will be formed, which is obviously symmetric (see Martinez-Arias et al., 1988). However, it could be that the fourfold subdivision of a segment is a secondary event. Genes responsible for or that mark the extreme ends of a parasegment, *engrailed* and *wingless* (Baker, 1987) are activated very early. Another early expressed gene is *patch* (Nakano et al., 1989, Hooper and Scott, 1989). It becomes transcribed in all cells in which *engrailed* is not expressed. Therefore, the initiation of a segment may start with three cell states, two that mark the border and the third that keeps the space inbetween. The function of *patch* could be to allow an ordered intercalation of the remaining structures, for instance by suppression of ectopic *engrailed* transcription.

Pattern of the pair rule gene activities

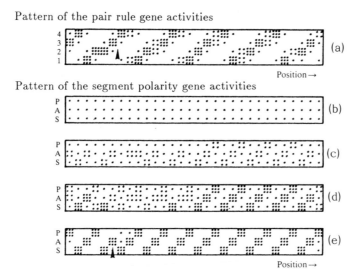

Pattern of the segment polarity gene activities

Fig. 7   Activation of the segment polarity genes (cell states S, A and P) by the pair rule gene pattern (1234-pattern, see Fig. 6) (a) The pattern of pair- rule gene activity. (b-e) Stages in the activation of the SAP pattern. The S, the A and the P cell states are activated at two different positions of the 1234 pattern, A at 1 and 3, S at 2 and 4 and P at the 1/2 and the 3/4 border. The result is a doubling of the spatial frequency such as required for the double segment-single segment transition. If not enough cells carry a particular activation in the double segment pattern the resulting single segment pattern may contain a locally symmetric pattern (arrow heads).

Each double segment has to direct the formation of two normal segments. In terms of the model, each ..1234.. sequence has to induce two ..SAP.. sequences. A possible mode would be as follows: State 1 induces state A, state 2 induces S and 1 and 2 together induce P. In a similar way, states 3 and 4 induce the second APS-sequence. A computer simulation made under this assumption is shown in Fig. 7. This model provides a straightforward explanation of the main pattern alterations found in the pair rule mutants. For example, if state 2 is lost due to a mutation (..341$_2$34..), the P and the S region of every other segment would be lost (...AP/SA$_{P/S}$AP/S... = P/SAAP/S). S is lost because it is directly induced by the state 2, and a P is lost because it is induced by the cooperation of state 1 and 2. Since S is assumed to form essentially the denticle belt and A and P the naked region, we expect a normal sized denticle belt (S) separated by a large naked region (AAP). This is the phenotype of a weak allele of *even skipped* (Fig. 5d). An analogous phenotype, shifted by one segment, is expected if state 4 is lost.

If state 1 (or 3) is lost a deletion of the states A and P - i.e. the naked region - of every second segment will result (..34$_1$234.. =..AP/S$_{AP/S}$SAP/S.. = ..P/SSAP/S..). The expected phenotype is a broad denticle belt (SS) of approximately the same size as the remaining naked region (AP). This corresponds to the phenotype of the mutation *hairy* or *paired*. In agreement with the model, the weaker allele *paired2* reveals that the loss of pattern elements starts in the naked region, and that pairs of two denticle belts fuse (Nüsslein-Volhard and Wieschaus, 1980).

The model describes correctly essential features of the pair rule mutants. Two frames of deletions exist for both even and odd numbered segments. According to the

317

model, a loss of state 1 or of state 2 leads to a deletion of the same segment borders (A1/A2, A3/A4...) but with different pattern deletions surrounding that borders, either ../S_{AP}/S.. or ..$SA_{P/S}$AP. The alternative segment borders (..T3/A1,A2/A3...) are deleted if state 3 or 4 is lost and again, two different frames of deletion exist. The borders of pattern deletion do not coincide with segment borders. According to the model, primarily only two-thirds of a segment is removed when one of the four cell states is missing. Thus, the resulting pattern cannot span from one segment border to the next. A deletion $P_{/SAP}$/S = P/S should not occur. Since the pattern around the segment border results from a cooperative interaction of P and S cells, the loss of the one part, for instance the P state in a SA_P/S deletion, should lead to a complete loss of a border. The isolated formation of the pattern on only one side of the border should not occur.

The transition from double to single segments requires that a particular cell state belonging to the single segment sequence is turned on at two positions in the double segment pattern. For instance, the P-state is activated by the 1/2 and by the 3/4 border. These two P-state activations need not be completely equivalent. This may be the reason why the mutation *engrailed* is differently expressed in every second segment (Nüsslein- Volhard and Wieschaus, 1980) and the even-numbered stripes of *engrailed* activity appear earlier than the odd-numbered stripes (Kornberg et al.,1985)

In conclusion, the assumption of an activation of the segment polarity genes by essentially four pair rule genes describes correctly that (i) mutations of pair rule genes lead to half the normal number of segments due to the loss of every other segment border, (ii) the remaining segments still have normal polarity, (iii) pattern deletions never coincide with segment borders, and (iv) two major frames of pattern deletion exist for both, the even-numbered and the odd-numbered segments.

Comparison with other models

A number of models has been proposed for steps in the *Drosophila* development. In the model of Nagorcka (1989), an universal prepattern has been assumed. Its formation depends on the density of nuclei (energids). With increasing density of nuclei, the spatial frequency increases too. No attempt is made in this model to integrate the role of the known genes nor to provide an explanation e.g. for the particular phenotypes. The finding of Sullivan (1987) that the pattern of *fushi tarazu* stripes is independent of the density of nuclei argues against the basic stipulation.

Goodwin and Kauffman (1990) have proposed a model that is based on the assumption that all four systems, maternal, gap, pair rule and segment polarity genes are pattern forming systems and that the genes of each level have an influence on the bifurcation parameter that are critical for the subordinated level. The appearance of a second *hunchback* domain at the posterior pole is therefore regarded as due to a shrinking wave length. In the model I propose, the activation of the gap genes is based solely on the maternally supplied positional information and the mutual interaction of the gap genes. I regard the fact that the posterior *hunchback* domain mentioned above is under control of a separate promotor (Tautz et al., 1987) as an argument for the positional information scheme. It could well be that the gap gene system emerges as genuine pattern forming system in evolution and that their control by positional information is a later evolutionary addition to make the early development as fast as we observe it in *Drosophila*.

Lacalli (1990) and Lacalli et al. (1988) have proposed a model for stripe formation that is very close to the model I have proposed. They have shown that an activator-depletion mechanism can have the tendency to form stripes. This is the case if a substrate that becomes depleted during autocatalysis has an independent modus of decay or removal. I regard this as an alternative mechanism for limiting the autocatalysis. As shown above, this generates stripes.

An entirely different model has been proposed by Edgar et al. (1989) for the activation of pair rule genes under gap gene influence. In this model, the activation of the pair rule genes follows exclusively a positional information scheme, i.e. no long-range communication is involved among the pair rule genes. Their interaction are assumed to be strictly local. Several arguments support this idea. So far, no gene product is known for the pair rule level that could be involved in the communication between the nuclei or the cells (such as the EGF-like *wingless* gene product at the segment polarity level, see Baker, 1988; Rijsewijk et al., 1987).

Conclusion: Essential elements and open problems

The details of the interactions between the components have certainly to be modified with the more experimental data that become available. However, the following features of the model I regard as essential and to be maintained in a revised version.

(i) The genes belonging to a higher hierarchical level activate the genes of the subsequent level with a particular combinatorial code. In order to generates a more complex pattern from a simpler pattern, a gene belonging to the lower hierarchical level has to become activated at several different positions in respect to the higher pattern. To point out the differences to the models mentioned above, it is not assumed that the pattern of one level has an influence on critical parameters for pattern formation such as diffusion. It is also not assumed that more complex patterns result from a shrinkage of the diffusion range of the molecules.

(ii) Segmentation is assumed to result from the serial repetition of cell states. These cell states activate and stabilize each other on long range but exclude each other locally. No positional, saw-tooth like positional information with a repeat lenght of a segment is to be expected. A segment never exists as such and becomes later on subdivided, but it is created by the periodic repetition of localized gene activities.

I hope that these model will help to understand the complex interctions that are going on during early development of an higher organism.

References

Baker, N.E. (1987). Molecular cloning of sequences from wingless a segment polarity gene in *Drosophila* the spatial distribution of a transcript in embryos . Embo J, **6**, 1765-1774.
Baker, N.E. (1988). Localization of transcripts from the wingless gene in whole *Drosophila* embryos. Development **103**, 289-298.
Bohn, H. (1970). Interkalare Regeneration und segmentale Gradienten bei den Extremitäten von Leucophaea-Larven (Blattari). I. Femur und Tibia. Wilhelm Roux' Archiv **165**, 303-341.
Bohn, H. (1974). Extent and properties of the regeneration field in the larval legs of cockroaches (Leucophaea maderae). I. Extirpation experiments. J. Embryol. exp. Morph. Vol.**31**, 3, 557-572.
Bull, A.L. (1966). Bicaudal, a Genetic Factor which Affects the Polarity of the Embryo in *Drosophila melanogaster*. J. Exp. Zool. **161**, 221-242.
Carroll, S.B. and Scott, M.P. (1986). Zygotically active genes that affect the spatial expression of the *fushi tarazu* segmentation gene during early *Drosophila* embryogenesis . Cell, **45**, 113-126.
Driever, W. and Nüsslein-Volhard, C. (1989). The *bicoid* protein is a positive regulator of *hunchback* transcription in the early *Drosophila* embryo. Nature **337**, 138-143.

Edgar, B.A., Schubiger, G. and Odell, G.M. (1989). A genetic switch, based on negative regulation, sharpens stripes in *Drosophila* embryos. Dev. Genetics **10**, 124-142.

Garcia-Bellido, A., Ripoll, P. and Morata, G. (1973). Developmental compartmentalization of the wing disk of *Drosophila*. Nature New Biol. **245**, 251-253.

Gaul, U. and Jäckle, H. (1987). Pole region-dependent repression of the *Drosophila* gap gene *Krüppel* by maternal gene products. Cell **51**, 549-555.

Gierer, A. (1981). Generation of biological patterns and form: Some physical, mathematical, and logical aspects. Prog. Biophys. molec. Biol. **37**, 1-47.

Gierer, A. and Meinhardt, H. (1972). A theory of biological pattern formation. Kybernetik **12**, 30-39.

Gierer, A. and Meinhardt, H. (1974). Biological pattern formation involving lateral inhibition. Lectures on Mathematics in the Life Sciences **7**, 163-183.

Goodwin, B.C. and Kauffman, S.A. (1990). Spatial harmonics and pattern specification in early *Drosophila* development. 1. Bifurcation sequences and gene expression. J. theor. Biol. **144**, 303- 319.

Goto, T., Macdonald, P. and Maniatis, T. (1989). Early and late periodic patterns of even skipped expression are controlled by distinct regulatory elements that respond to different spatial cues. Cell, **57**, 413-422.

Hafen, E., Kuroiwa, A. and Gehring, W.J. (1984). Spatial distribution of transcripts from the segmentation gene *fushi tarazu* during *Drosophila* embryonic development . Cell, **37**, 833-842.

Harding, K., Hoey, T., Warrior, R. and Levine, M. (1989). Autoregulatory and gap gene response elements of the *even-skipped* promoter of *Drosophila* . Embo J, **8**, 1205-1212.

Harding, K., Rushlow, C., Doyle, H.J., Hoey, T. and Levine, M. (1986). Cross-regulatory interactions among pair-rule genes in *Drosophila*. Science **233**, 953-959.

Hiromi, Y. and Gehring, W.J. (1987). Regulation and function of the *Drosophila* segmentation gene *fushi tarazu*. Cell **50**, 963-974.

Hooper, J.E. and Scott, M.P. (1989). The *Drosophila* patched gene encodes a putative membrane protein required for segmental patterning. Cell **59**, 751-765.

Howard, K. and Ingham, P. (1986). Regulatory Interactions between the Segmentation Genes *fushi tarazu*, *hairy*, and *engrailed* in the *Drosophila* Blastoderm. Cell **44**, 949-957.

Howard, K., Ingham, P. and Rushlow, C. (1988). Region-specific alleles of the *Drosophila* segmentation gene *hairy*. Genes & Development2, 1037-1046.

Howard, K., Ingham, P. and Rushlow, C. (1988). Region-specific alleles of the *Drosophila* segmentation gene *hairy* . Genes Dev, **2**, 1037-1046.

Ingham, P. (1988). The molecular genetics of embryonic pattern formation in *Drosophila*. Nature **335**, 25-34.

Kalthoff, K. and Sander, K. (1968). Der Entwicklungsgang der Missbildung "Doppelabdomen" im partiell UV-bestrahlten Ei von Smittia parthenogenetica. Wilhelm Roux' Archiv **161**, 129-146.

Knipple, D.C., Seifert, E., Rosenberg, U.B., Preiss, A. and Jäckle, H. (1985). Spatial and temporal pattern of *Krüppel* gene expression in early *Drosophila* development. Nature **317**, 40- 44.

Kornberg, T.I., Siden, I., O'Farell, P. and Simon, M. (1985). The *engrailed* locus of *Drosophila*: In-situ hybridisation of transcripts reveals compartment-specific expression. Cell40, 45-53.

Lacalli, T.C. (1990). Modeling the *Drosophila* pair-rule pattern by reaction diffusion - gap input and pattern control in a 4- morphogen system. J Theor Biol144, 171-194.

Lacalli, T.C., Wilkinson, D.A. and Harrison, L.G. (1988). Theoretical aspects of stripe formation in relation to *Drosophila* segmentation. Development104, 105-113.

Lohs-Schardin, M., Cremer, C. and Nüsslein-Volhard, C. (1979). A fate map for the larval epidermis of *Drosophila melanogaster*: Localized cuticle defects following irradiation of the blastoderm with an ultraviolet laser microbeam. Dev. Biol. **73**, 239- 255.

Martinez-Arias, A., Baker, N.E. and Ingham, P.W. (1988). Role of segment polarity genes in the definition and maintenance of cell states in the *Drosophila* embryo. Development **103**, 151-170.

Martinez-Arias, A. and Lawrence, P.A. (1985). Parasegments and compartments in the *Drosophila* embryo. Nature **313**, 639-642.

Meinhardt, H. (1976). Morphogenesis of lines and nets. Differentiation **6**, 117-123.

Meinhardt, H. (1977). A model of pattern formation in insect embryogenesis. J. Cell Sci. **23**, 117-139.

Meinhardt, H. (1978). Space-dependent Cell Determination under the control of a morphogen gradient. J. theor. Biol.**74**, 307-321.

Meinhardt, H. (1980). Cooperation of Compartments for the Generation of Positional Information. Z. Naturforsch. **35c**, 1086-1091.

Meinhardt, H. (1982). Models of biological pattern formation. Academic Press, London.

Meinhardt, H. (1982). The Role of Compartmentalization in the activation of particular control genes and in the generation of proximo- distal positional information in appendages. Amer.Zool. **22**, 209-220.

Meinhardt, H. (1983a). A boundary model for pattern formation in vertebrate limbs. J. Embryol exp. Morph. **76**, 115-137.

Meinhardt, H. (1983b). Cell determination boundaries as organizing regions for secondary embryonic fields. Devl. Biol **96**, 375-385.

Meinhardt, H. (1984). Models for positional signalling, the threefold subdivision of segments and the pigmentation pattern of molluscs. J. Embryol. exp. Morph. **83**,(Supplement) 289-311.

Meinhardt, H. (1985). Mechanisms of Pattern Formation During Development of Higher Organisms: A Hierarchial Solution of a Complex Problem. Ber. Bunsenges. Phys. Chem. **89**, 691-699.

Meinhardt, H. (1986). Hierarchical inductions of cell states: a model for segmentation in *Drosophila*. J. Cell Sci. Suppl.4, 357- 381.

Meinhardt, H. (1986). The threefold subdivision of segments and the initiation of legs and wings in insects. Trends Genetics **2**, 36-41.

Meinhardt, H. (1988). Models for maternally supplied positional information and the activation of segmentation genes in *Drosophila* embryogenesis. In: Development **104**, (Supplement), 95-110.

Meinhardt, H. and Gierer, A. (1980). Generation and regeneration of sequences of structures during morphogenesis. J. theor. Biol. **85**, 429-450.

Nagorcka, B.N. (1989). Wavelike isomorphic prepatterns in development. J. Theoretical Biol. **137**, 127-162.

Nakano, Y., Guerrero, I., Hidalgo, A., Taylor, A., Whittle, J.R.S. and Ingham, P.W. (1989). A protein with several possible membrane-spanning domains encoded by the *Drosophila* segment polarity gene patched. Nature **341**, 508-513.

Nüsslein-Volhard, C. (1977). Genetic analysis of pattern formation in the embryo of *Drosophila melanogaster*. Wilhelm Roux's Archives**183**, 249-268.

Nüsslein-Volhard, C., Frohnhöfer, H.G. and Lehmann, R. (1987). Determination of anteroposterior polarity in *Drosophila*. Science **238**, 1675-1681.

Nüsslein-Volhard, C. and Wieschaus, E. (1980). Mutations affecting segment number and polarity in *Drosophila*. Nature **287**, 795- 801.

Nüsslein-Volhard, C., Wieschaus, E. and Kluding, H. (1984). Mutations affecting the pattern of the larval cuticle in *Drosophila melanogaster*. I. Zygotic loci on the second chromosome. Roux's Arch. Dev. Biol. **183**, 267-282.

Pankratz, E., Seitert, E., Gerwin, N., Billi, B., Nauber, N. and Jäckle, H. (1990). Gradients of *Krüppel* and *knirps* gene products direkt pair rule gene stripe patterning in the posterior regions of the *Drosophila* embryo. Cell, **61**, 309-316.

Pankratz, M.J., Jäckle, H., Seifert, E. and Hoch, M. (1989). Kruppel requirement for *knirps* enhancement reflects overlapping gap gene activities in the *Drosophila* embryo. Nature **341**, 337- 340.

Rijsewijk, F., Schuermann, M., Wagenaar, E., Parren, P., Weigel, D. and Nusse, R. (1987). The *Drosophila* homolog of the mouse mammary oncogene int-1 is identical to the segment polarity gene wingless. Cell **50**, 649-657.

Sander, K. (1959). Analyse des ooplasmatischen Reaktionssystems von Euscelis plebejus Fall. (Cicadina) durch Isolieren und Kombinieren von Keimteilen. I. Mitt.: Die Differenzierungsleistungen vorderer und hinterer Eiteile. Wilhelm Roux' Archiv **151**, 430-497.

Sander, K. (1960). Analyse des ooplasmatischen Reaktionssystems von Euscelis Plebejus Fall (Cicadina) durch Isolieren und Kombinieren von Keimteilen. II. Mitteilung: Die Differenzierungsleistungen nach Verlagern von Hinterpolmaterial. Wilhelm Roux' Archiv **151**, 660-707.

Sander, K. (1976). Specification of the basic body pattern in insect embryogenesis. Adv. Ins. Physiol.**12**, 125-238.

Sullivan, W. (1987). Independence of *fushi tarazu* expression with respect to cellular density in *Drosophila* embryos. Nature **327**, 164-167.

Tautz, D., Lehmann, R., Schnuerch, H., Schuh, R., Seifert, E., Kienlin, A., Jones, K. and Jäckle, H. (1987). Finger protein of novel structure encoded by *hunchback*, a second member of the gap class of *Drosophila* segmentation genes. Nature **327**, 383-389.

Tautz, D., Tautz, C., Webb, D. and Dover, G.A. (1987). Evolutionary divergence of promoters and spacers in the rDNA family of four *Drosophila* species. Implications for molecular coevolution in multigene families. J. Mol. Biol. **195**, 525-542.

Treisman, J. and Desplan, C. (1989). The products of the *Drosophila* gap genes hunchback and *Krüppel* bind to the *hunchback* promoters. Nature **341**, 335-337.

Turing, A. (1952). The chemical basis of morphogenesis. Phil. Trans. B. **237**, 37-72.

Wolpert, L. (1969). Positional information and the spatial pattern of cellular differentiation. J. theoret. Biol. **25**, 1-47.

Yajima, H. (1960). Studies on embryonic determination of the harlequin-fly, Chironomous dorsalis. J. Embryol. exp. Morph. **8**, 198- 215.

van der Meer, J.M. and Miyamoto, D.M. (1984). The specification of metameric order in the insect Callosobruchus maculatus Fabr. (Coleoptera). II. The effects of temporary constriction on segment number. Roux's Arch Dev Biol **193**, 326-338.

REACTION-DIFFUSION PREPATTERNS (TURING STRUCTURES): SUPERCOMPUTER SIMULATION OF CYTOKINESIS, MITOSIS AND EARLY *DROSOPHILA* MORPHOGENESIS

Axel Hunding

Chemistry Department C116
H. C. Ørsted Institute
University of Copenhagen
Universitetsparken 5
DK 2100 Copenhagen Ø, Denmark

Introduction

The spontaneous formation of complex patterns and form in biological systems is largely unexplained. Turing(1952) demonstrated however that autocatalytic biochemical reactions coupled to internal diffusion, but without external control, could break up from the original homogeneous state and form stable well defined inhomogeneous concentration gradients and patterns. General reaction-diffusion systems may be described by

$$\partial \mathbf{c}/\partial t = \mathbf{F}(\mathbf{c}) + \mathbf{D}\Delta \mathbf{c} \tag{1}$$

where $\mathbf{c}$ is the concentration vector, $\mathbf{F}$ is the chemical kinetics rate vector and the last term describes Fickian diffusion. If the Jacobian $\mathbf{J} = \partial \mathbf{F}/\partial \mathbf{c}$, evaluated at the homogeneous steady state, is of one of the forms (a) or (b):

$$\mathbf{J} = \begin{pmatrix} + & - \\ + & - \end{pmatrix} \quad (a) \qquad \mathbf{J} = \begin{pmatrix} - & - \\ + & + \end{pmatrix} \quad (b) \tag{2}$$

spontaneous pattern formation may occur if the rates and diffusion constants satisfy certain inequalities. In the case (a) one speaks of an activation-inhibition system, as c_1 activates both its own formation and that of c_2, and c_2 inhibits both rates. The second class (b) was introduced by Sel'kov (1968) and studied by the socalled Brussels group, the leader of which, I. Prigogine, got the Nobel price in 1977. Their work demonstrates that Turing structures are fully compatible with the second law of thermodynamics since living systems are open systems and they showed (1974) with bifurcation theory that the patterns found in computer simulations are genuine solutions to the nonlinear partial differential equations above. References to early work on such *spontaneous symmetry breaking* in biochemical systems are found in (Nicolis and Prigogine, 1977). A particular Sel'kov type system may be written as

$$\frac{\partial c_1}{\partial t} = \nu - \frac{k_1 c_1 c_2^\gamma}{1 + K c_2^\gamma} + D_1 \Delta c_1 \tag{3}$$

$$\frac{\partial c_2}{\partial t} = \frac{k_1 c_1 c_2^\gamma}{1 + K c_2^\gamma} - k_2 c_2 + D_2 \Delta c_1 \tag{4}$$

Complexity, Chaos, and Biological Evolution, Edited by E. Mosekilde and
L. Mosekilde, Plenum Press, New York, 1991

Here component one is fed homogeneously from a source with rate ν and converted into component two with a rate displaying product activation with Hill constant γ greater than one. Component two is in turn consumed by first order kinetics. Thus the system is thermodynamically open and a flow of free energy through the system keeps it far from equilibrium. Both components satisfy a no flux condition $\partial c_i / \partial n = 0$ where $\partial / \partial n$ denotes the gradient along the outward normal at the surface. The emerging patterns are to a first approximation proportional to eigenfunctions ϕ_{nml} to the Laplacian satisfying

$$\Delta \phi_{nml} = -k^2_{nml} \phi_{nml} \tag{5}$$

For one dimensional model systems of length L, these functions are simple cosine functions $cos(n\pi x/L)$. The general 3 dimensional case is studied here as this makes comparisons to actual biological experiments feasible.

Bifurcation theory

Analytical solutions to Eq(1) may be found by expanding $\mathbf{c}$ from the homogeneous stationary solution $\mathbf{c}_0$. When $\mathbf{z} = \mathbf{c} - \mathbf{c}_0$ is inserted in Eq(1) a linear and a nonlinear part will emerge

$$\mathbf{L}(\mathbf{z}) + \mathbf{N}(\mathbf{z}) = 0 \tag{6}$$

where

$$\mathbf{L}(\mathbf{z}) = \mathbf{J}\mathbf{z} + \mathbf{D}\Delta\mathbf{z} \tag{7}$$

and by construction $\mathbf{L}\mathbf{z} = 0$ for $\mathbf{z} = 0$ (the homogeneous case). Nontrivial inhomogeneous solutions to $\mathbf{L}\mathbf{z} = 0$ may be found by substituting $\boldsymbol{\delta}_{nml}\phi_{nml}$ for $\mathbf{z}$ which yields

$$|\mathbf{J} - k^2_{nml}\mathbf{D}| = 0 \tag{8}$$

This equation for the determinand specifies where bifurcation may occur to inhomogeneous solutions of geometry ϕ_{nml}. Since $\mathbf{J}$ depends on rate constants, Eq(8) defines critical values of rate constants where an exchange of stability occurs between $\mathbf{c}_0$ and $\mathbf{z} = \boldsymbol{\delta}_{nml}\phi_{nml}$. Indeed the matrix $\mathbf{J} - k^2_{nml}\mathbf{D}$ has an eigenvalue with positive real part when the bifurcation point is passed and thus any small fluctuation component along ϕ_{nml} is amplified. The corresponding eigenvector $\boldsymbol{\delta}_{nml}$ defines the *combination* of c_1 and c_2 which then grows. The arising of Turing structures is thus a global phenomenon which is heavily dependent on the boundary conditions (as these define ϕ_{nml}) and thus on the shape of the region studied. It is not fully described as a local competition among activators and inhibitors.

The inhomogeneous stationary solution $\mathbf{z}_p$ (primary bifurcation branch) to Eq(6) may be obtained by expansion in a small parameter ϵ related to the chosen bifurcation parameter. Sel'kov's scheme Eqs(3,4) may be written in the form

$$\partial \mathbf{c}/\partial t = \mathbf{F}(\mathbf{c}) + \alpha\mathbf{G}(\mathbf{c}) + \mathbf{D}\Delta\mathbf{c} \tag{9}$$

with $\alpha = k_1$. Eq(8) then defines α_c, and we may expand:

$$\mathbf{z}_p = \sum_{j=1}^{\infty} \epsilon^j \phi_{j-1} \qquad \alpha - \alpha_c = \sum_{j=1}^{\infty} \epsilon^j \tau_j \tag{10}$$

and collect terms of the same order in ϵ. One obtains a hierarchy of *linear* problems

$$\epsilon^1: \quad \mathbf{L}_0\phi_0 = 0 \tag{11}$$

$$\epsilon^2: \quad \mathbf{L}_0\phi_1 + \tau_1\mathbf{L}_1\phi_0 + \mathbf{N}_2(\phi_0) = 0 \tag{12}$$

$$\epsilon^3: \quad \mathbf{L}_0\phi_2 + \tau_1\mathbf{L}_1\phi_1 + \tau_2\mathbf{L}_1\phi_0 + \mathbf{N}_3(\phi_0, \phi_1) = 0 \tag{13}$$

and so on. Here $\mathbf{L}_0$ is $\mathbf{L}$ evaluated at α_c.

The solution to Eq(11) is

$$\phi_o = \delta_{nml} \sum_{I(N)} \nu_{nml} \phi_{nml} \qquad (14)$$

in the general degenerate case where there is more than one solution $\delta_{nml}\phi_{nml}$. The expansion coefficients ν with indices (nml) taken from the degenerate set $I(N)$ may be obtained from Eq(12). Fredholm's solvability condition for a problem $\mathbf{L}_0 \phi_1 = \mathbf{f}$ is

$$[\mathbf{f}, \phi_0^\dagger(m)] = 0, \qquad m \in I(N) \qquad (15)$$

where $\phi_0^\dagger(m), m \in I(N)$ spans the solution space to the adjoint operator of $\mathbf{L}_0$. $[\mathbf{u},\mathbf{v}]$ denotes inner product $[\mathbf{u}, \mathbf{v}] = \int \mathbf{u}^T \cdot \mathbf{v} d\Omega$ where integration is over the region studied. Application of Fredholm's condition to Eq(12) yields

$$\tau_1[\mathbf{L}_1\phi_o, \phi_o{}^\dagger(m)] = -[\mathbf{N}_2(\phi_o), \phi_o{}^\dagger(m)] \qquad m \in I(N) \qquad (16)$$

which together with a normalisation condition of ϕ_o, $[\phi_o, \phi_o] = 1$, yield a set of $I(N) + 1$ *algebraic* equations in the $I(N) + 1$ unknowns ν_m and τ_1. These *algebraic bifurcation equations* turn out to determine the ν_m's *independent* on the kinetics terms, and thus *the geometry of the solution is independent on the particular chemistry details of the model chosen*. The τ_1, which determines the amplitude of the pattern does depend on the kinetic terms. The solution to the algebraic bifurcation equations may yield some ν's equal to zero: thus *nonlinear selection rules* apply and transitions to patterns with certain geometries are forbidden.

Only two classes of Jacobians allow Turing structures to arise, Eq(2). This means that the geometry of the bifurcating solutions is common to large classes of models with different chemistry details. A similar statement holds for secondary bifurcation (branching from one inhomogeneous pattern to another). Finally this generality of Turing patterns has recently been obtained also in the case of Turing systems of the second kind

$$\partial \mathbf{c}/\partial t = \mathbf{F}(\mathbf{k}(\mathbf{r}), \mathbf{c}) + \nabla \cdot \mathbf{D}(\mathbf{r})\nabla \mathbf{c} \qquad (17)$$

in which one Turing system forces rate constants to be position dependent in a second system (Hunding & Brøns, 1990). It is thus meaningful to study pattern formation in a particular model system numerically as the patterns and pattern sequences recorded are common to a broad class of chemical networks. This makes comparisons to biological experiments reasonable.

Numerical study of prepattern formation

Bifurcations to Eqs(1 & 17) have been investigated by numerical solution. The method of lines was used and thus the system of nonlinear partial differential equations was converted to a large system of ordinary differential equations by discretization of the Laplacian in three curvilinear coordinates: For example the Laplacian for the prolate ellipsoid (elongated sphere) is

$$\Delta = \frac{4/d^2}{\xi^2 - cos^2\phi}\left[\frac{\partial}{\partial\xi}(\xi^2 - 1)\frac{\partial}{\partial\xi} + \frac{1}{sin\phi}\frac{\partial}{\partial\phi}sin\phi\frac{\partial}{\partial\phi} + \frac{\xi^2 - cos^2\phi}{(\xi^2 - 1)sin^2\phi}\frac{\partial^2}{\partial\phi^2}\right] \qquad (18)$$

Divided differences were used: $\partial c/\partial x \simeq (c_{i+1} - c_{i-1})/(x_{i+1} - x_{i-1})$. Coordinate systems investigated include the cube, the sphere, the prolate and oblate spheroids (elongated and compressed sphere respectively), circular and elliptic cylinder, and the torus. The resulting system is large and stiff and solved accordingly (modified Gear code). The Jacobian used in the corrector step is a sparse banded matrix which may be rearranged (chessboard numbering of meshpoints) to yield large blocks within which the solution vector elements may be iterated in parallel (RBSOR method). Implementation on vector computers results in a huge speed up: 2000 equations used for solving pattern formation in a sphere take several weeks of CPU time on an ordinary department computer like a VAX 11/750 FPA using conventional scalar non stiff algorithms with error checks on each computed element. A factor of 500 speed up is achieved with the stiff code reducing the CPU time to approx 15 minutes. The parallel code runs efficiently and close to the top speed of machines like the CRAY X-MP (160 MFLOPS) and the Fujitsu/Amdahl VP1100 (210 MFLOPS) reducing the CPU time to a mere 6 seconds. A total speed up of a factor 150000 is thus achieved. This makes the numerical study of three dimensional pattern formation possible and thus direct comparison to biological experiments feasible.

Turing prepatterns in cytokinesis and mitosis

Patterns in single cells may be studied as bifurcations in a spherical region. Simple gradients from one spontaneously generated pole to another arise in such computer simulations. Such patterns may be involved in cap formation or establishment of the animal-vegetal axis. Experimentally it is found that cap formation may be aligned parallel to many kinds of external perturbations like light, gravity, chemical gradients etc. Computer simulations show that Turing structures arise parallel to any such externally induced inhomogeneity.

More complex patterns may arise as well. For a certain range of parameter values a *bipolar* pattern forms spontaneously in a sphere. This pattern has high concentrations at two spontaneously generated opposite poles and a low concentration band in the equator. Such a pattern should be an ideal platform for the processes of cytokinesis (cell division) and mitosis (chromosome transport to the two poles). This *prepattern theory of cytokinesis and mitosis* has been explored by the present author in a series of papers and it is shortly discussed here.

Cytokinesis and mitosis in higher organisms are complex processes. They are likely to have originated from a less precise but simple mechanism. In many primitive organisms the genome is transferred to the daughter cells in a crude manner such as a simple contraction of the cell around the equator region which separates a large number of presumably identical chromosomes into two bulks. The bipolar Turing pattern may provide the spatial organisation for this process which indeed may have formed the original process of cytokinesis later to be exploited to position a contractile band in animal cells and vesicles in plant cells. Evidence for the existence of this Turing pattern in primitive organisms comes from the radiolarian *Aulacantha Scolymantha*. During nuclear division more than 1000 chromosomes carry out autonomous movements as if they were confined to a succession of highly symmetric regions *inside* the nucleus. This phenomenon is not understandable in terms of microtubule traction forces but may be Turing structures on display. First an equatorial plate of chromosomes is formed. This may be described as the neutral (neither high nor low) region of the recorded pattern j_1Y_{10} which defines a simple gradient in the nucleus. This plate is folded up in a highly symmetrical saddle shaped object. It corresponds to the neutral region of $j_1Y_{10} + j_2Y_{22}$ which emerges subsequently in computer simulations. This object develops into two nearly parallel 'horse shoe' regions also observed in the simulations along the route to the final pattern, the bipolar j_2Y_{20} structure. This transition is robust with respect to deformations of the perfect sphere as seen in simulations of prolate and oblate spheroids. The experimentally established appearance of the highly symmetrical saddle shaped dynamic object and the transition involving horse shoe regions strongly suggest that the chromosomes are under spatial control by the prepatterns recorded numerically (including the bipolar pattern) thus forming experimental support for the existence and involvement of Turing patterns during chromosome transport in primitive cells (Hunding, 1981; 1983; 1984).

Experimental support for the involvement of the Turing structure j_2Y_{20} in mitosis and cytokinesis in animal cells exists as well. First we may note that theories of mitosis based on control from the centrioles have been abandoned by experimentalists as the centrioles may be displaced or destroyed by laser irradiation without affecting the poleward transport of chromosomes during anaphase toward opposite poles. Also plant cells are devoid of centrioles. Current models for mitosis are centriolar free and based on forces linked to the microtubules and these predict forces proportional to spindle fiber length. However some experimental data indicate that equator crossing fibers become much too long to be compatible with such models. Spindle forces on stationary trivalent chromosomes with one equator crossing fiber a toward one pole and two short fibers b and c to the opposite pole should balance each other. If proportionality existed we should obtain $a = b+c$. It appears however that a is much too long. If one assumes instead that the equator crossing part of a yields a force directed to the b, c pole a much better force balance obtains. Such a force is easily obtained if the bipolar Turing pattern is present and yields a force on local microtubule elements directed from the equator toward the poles.

Table 1. Fiber lengths a,b and c in μ. a-(b+c) is consistently too long. Prepattern force (last column) yields much better force balance. The detailed calculations are given in (Hunding, 1981).

a	b	c	a-(b+c)	F(a)-F(b)-F(c)
14.2	4.7	7.0	+2.5	-1.5
16.4	3.9	5.9	+6.6	0.3
15.8	4.6	5.9	+5.3	-0.3
12.3	2.8	5.3	+4.2	0.9

Further arguments in favour of the presence of Turing patterns in dividing cells are given by the experimental observation that sea-urchin eggs compressed to an axis ratio of 1:5 may show direct quadripartitioning: four cells arise, two with chromosomes and two without. Computer experiments have been carried out to demonstrate that a sphere, originally containing the bipolar Turing pattern j_2Y_{20} under compression to this axis ratio yields a quadripolar stable Turing pattern which may govern the described experiment (Hunding, 1985).

Finally a peculiar spatial organisation is observed in ordinary cell division of many species: the socalled animal-vegetal axis of the egg seems to control the spindle to orient itself perpendicular to this axis. So far this phenomenon has gone unexplained by existing models based on intrinsic microtubule forces. If the animal-vegetal gradient imposes a variation of rate constants along this axis in the egg, a bipolar Turing pattern parallel to this axis may still form. This pattern however becomes unstable and changes into a Turing pattern which is exactly perpendicular to the animal-vegetal axis. The computer simulation of this transition was given in (Hunding, 1987) and it was argued that precise spatial organisation perpendicular to an established axis in the egg requires active global governing and the observed series of Turing patterns are ideally suited to provide such governing. Indeed the orientation of the first three divion planes relative to the animal-vegetal axis in the early blastula may be accounted for by Turing structures. Recently a complete bifurcation analysis has been given for this transition confirming the numerically recorded patterns (Hunding and Brøns, 1990).

In conclusion a number of experiments are in support of the assumption that Turing prepatterns exist in cells and play a controlling role in both mitosis and cytokinesis. It is also pleasing that a crude spindle free chromosome segregation and cleavage process may be based on such prepatterns. This is a simple mechanism which goes to the root of the

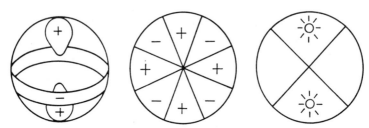

Figure 1 A bipolar pattern $j_2 Y_{20}$ (left) emerges spontaneously from the homogeneous state. This 'mitosis' pattern should be an ideal platform for mitosis and cytokinesis as it contains two spontaneously created poles with high concentration and a belt of low concentration in the equator region. Compression of this pattern yields a quadripolar pattern (middle) which may govern quadripartition to two cells with asters and two without (right) as seen experimentally.

Figure 2 The bipolar pattern (left) may develop into a new bipolar pattern (right) perpendicular to the original axis in systems with gradient dependent rate constants. This may explain spindle orientation and cleavage plane orientation relative to the animal-vegetal axis.

origin of cytokinesis. It has a large potential for being exploited during evolution to yield a basis for precise transport of only a few chromosomes during mitosis in evolutionary advanced cells as those of animals and plants. Thus the interesting possibility exists that a number of seemingly diverse phenomena like animal and plant cytokinesis are different exploitation by nature of a unifying principle for spatial organisation based on Turing patterns.

Turing structures in morphogenesis

Spontaneously created prepatterns in reaction-diffusion systems may act as ideal robust well controllable platforms for spatial organisation in the early embryo. Genes may be activated in regions where the prepattern concentration is high thus generating a reliable system for globally controlled cell differentiation (Wolpert and Stein, 1984), as opposed to local mechanisms relying on cell-cell interactions. Turing structures also readily yield concentration variations between low and high regions of more than a factor 10 which should be enough for reliable activation, resp. inactivation of genes. This makes them favourable compared to models relying on positional information from reading out subtle differences in concentration along a single gradient (Meinhardt, 1988).

When two-dimensional computer simulations appeared in the 1970's it became clear that patterns with wavelength much shorter than the dimensions of the embryo were hard to obtain reproducibly. This occurs because quite a number of different short wave length patterns have overlapping existence regions in parameter space and thus an unpredictable *combination* of such patterns is triggered. Turing patterns also arise in a parameter region which usually only allows D/L^2 to vary some 30 %. Here D is an effective diffusion constant and L is a characteristic length of the embryo. Consequently there may not be much potential for size adaptation: a change of L by a factor of 2 would require D/L^2 to change by a factor $1/(2^2)$. Other models like the mechano-chemical model (Murray, 1989) have an even more severe size regulating problem: they contain the Laplacian acting twice and thus L appears now as L^4.

The problem of size regulation has been solved as D is not the diffusion coefficient in water but rather an effective diffusion constant which describe the combination of pure diffusion and binding of the morphogen to slowly moving proteins. This binding may be influenced by a monitor substance M which in turn is created (or destroyed or simply lost) on the *surface* of the embryo. This would make D a function of L^2 as required and size regulation would easily obtain (Hunding and Sørensen, 1988). The problem of stabilising certain short wave length patterns over other coexisting patterns may be solved by using Turing systems of the second kind Eq(17). The simplest system may be one in which rate constants (enzyme activation) vary with position along a gradient. This was shown to stabilise stripes perpendicular to the gradient (Lacalli et al., 1988).

These recent developments make Turing patterns ideal candidates for spatial governors during early embryogenesis in *Drosophila*. Experiments show the gene *ftz* to be activated in stripes in the middle of the embryo without any prior compartment formation. Supercomputer simulations of three dimensional Turing patterns show that a simple model based on doubling of the number of stripes at each nuclear cycle is not reliable. A haphazard pattern sequence emerges instead.

It is thus necessary to introduce a *hierarchical* model where gradients from maternal genes *bicoid* and *oscar* influence rate constants in the first Turing system. This system then develops a reliable pattern with two periods along the anterior-posterior axis. Thus it has a peak in the middle of the embryo where gap gene *Krüppel* is activated and potential for activation of this gene at both ends as recorded experimentally. Between these peaks gap genes *hunchback* and *knirps* are activated. This level in the hierarchy

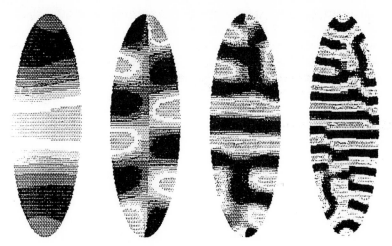

Figure 3   Unreliable haphazard pattern obtains by simple period doubling during each nuclear cycle in *Drosophila*. Thus it is necessary to have a *hierarchical* control of stripe formation

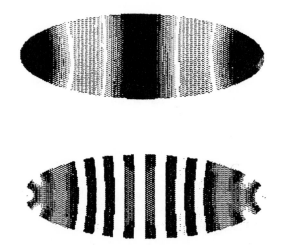

Figure 4   The hierarchical model uses maternal gradients to stabilise a doubly periodic pattern (upper) which may govern gap gene activation. This doubly periodic pattern in turn stabilises short wave length stripes (lower) which may govern pair rule gene activation. No compartments are present, rather a dynamic globally controlled mechanism is suggested

then varies rate constants to be double-periodic along the long axis for the second level in the hierarchy. This second Turing system develops directly into 7 stable stripes in the interior of the embryo with some activation seen near the poles as well. This pattern may then activate the pair rule gene complex *hairy, runt* and *eve*. A number of features of this global model fit experimental data including mutants deficient in one (or all) gap genes (Hunding, Kauffman & Goodwin, 1990).

Acknowledgements: Calculations were performed at Uni*C with support from the Danish Natural Science Research Counsel under grant no. 11-7999.

References

Hunding, A., 1981, Possible prepatterns in mitosis: Mechanism of spindle-free chromosome movement in *Aulacantha Scolymantha, J. theor. Biol.* 89: 353.

Hunding, A., 1983, Bifurcations of nonlinear reaction diffusion systems in prolate spheroids, *J. Math. Biol.* 17: 223.

Hunding, A., 1984, Bifurcations of nonlinear reaction diffusion systems in oblate spheroids, *J. Math. Biol.* 19: 249.

Hunding, A., 1985, Morphogen prepatterns during mitosis and cytokinesis in flattened cells, *J. theor. Biol.* 114: 571.

Hunding, A., 1987, Bifurcations in Turing systems of the second kind may explain blastula cleavage plane orientation, *J. Math. Biol.* 25: 109.

Hunding, A. and Sørensen, P. G., 1988, Size adaptation of Turing prepatterns, *J. Math. Biol.* 26: 27.

Hunding, A. and Brøns, M., 1990, Bifurcation in a spherical reaction diffusion system with imposed gradient, *Physica D* 44: 285.

Hunding, A., Kauffman, S. A. and Goodwin, B. C., 1990, *Drosophila* segmentation: Supercomputer simulation of prepattern hierarchy, *J. theor. Biol.* 145: 369.

Lacalli, T. C., Wilkinson, D. A. and Harrison, L. G., 1988, Theoretical aspects of stripe formation in relation to *Drosophila* segmentation, *Development* 104: 105.

Meinhardt, H., 1988, Models for maternally supplied positional information and the activation of genes in *Drosophila* embryogenesis, *Development* 104 suppl.: 95.

Murray, J. D., 1989, "Mathematical Biology", Springer, Berlin.

Nicolis, G. and Prigogine, I., 1977, " Selforganization in Nonequilibrium Systems", Wiley, New York.

Sel'kov, E. E., 1968, Self-oscillations in glycolysis, *Eur. J. Biochem.*, 4: 79.

Turing, A.M., 1952, The chemical basis for morphogenesis. *Phil. Trans. R. Soc. London,* B237: 37.

Wolpert, L. and Stein, W. D., 1984, Positional information and pattern formation, *in*: "Pattern Formation", G. M. Malacinski and S. V. Bryant, eds., Macmillan, London.

VORTEX FORMATION IN EXCITABLE MEDIA

Stefan C. Müller

Max-Planck-Institut für Ernährungsphysiologie
Rheinlanddamm 201, D-4600 Dortmund 1

TRAVELING WAVES IN CHEMICAL AND BIOLOGICAL SYSTEMS

Excitable systems play an important role in the field of spatio-temporal self-organisation under conditions far from thermodynamic equilibrium [1]. A particular property of these systems is that they can be assumed to exist in one of three different states: as long as a stimulus is absent, they remain in a quiescent state which is excitable; by application of a stimulus, they are excited to an active state; after excitation there follows a refractory period during which the system is not yet excitable, but relaxes to the previous quiescent and newly excitable situation. It is well known that such systems support the formation of traveling waves of excitation with different front geometries, the complexity of which depends on the dimensionality of the system and the influence of internal and external perturbations [2].

Traveling waves and fronts have been studied in biological systems over a long period of time: for instance, in signal propagation in nerves [3,4], in the development of embryos [5], in waves across the chambers of the heart [6,7]. During the last two decades, two laboratory systems have proven to be most suitable for detailed quantitative studies of such propagating waves. These are:
(1) the Belousov-Zhabotinskii (BZ) reaction as a chemical model system, in which malonic acid is catalytically oxidized and brominated by acidic bromate [8], and
(2) the aggregation of the slime mould Dictyostelium discoideum, where a coherent motion of aggregating amoebae cells is stimulated by intercellular wave-like propagation of the signal transmitter molecule cyclic AMP [9].

Many studies have been done on two-dimensional patterns, as obtained in thin excitable layers. In these one of the outstanding features of spatial self-organization is observed, namely the formation of vortices around which spiral-shaped waves of excitation rotate [10,11]. These patterns are induced by appropriate external stimuli or they may form

Complexity, Chaos, and Biological Evolution, Edited by E. Mosekilde and
L. Mosekilde, Plenum Press, New York, 1991

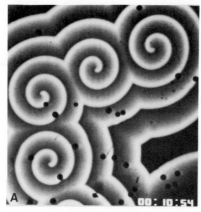

Fig. 1. Rotating spiral waves in a chemical and biological excitable layer. (A) Ferroin-catalyzed BZ reaction, area of image: 15x15 mm^2. Initial concentrations: 0.12 M CH$_2$(COOH)$_2$, 0.33 M NaBrO$_3$, 0.06 M NaBr, 0.37 M H$_2$SO$_4$, and 0.003 M ferroin. (B) Dictyostelium discoideum amoebae, area of image: 19x19 mm^2. For preparation see [12].

spontaneously. Examples are presented in Fig. 1. The snapshot of a chemical spiral pattern in Fig. 1A was taken in transmitted light at 490 nm and shows in its bright parts the spatial distribution of the oxidized state of the catalyst (ferriin). The sharp fronts correspond to the region of maximum production of the activator molecule HBrO$_2$. The spiral pattern of Fig. 1B, taken for the Dictyostelium discoideum system, depicts the actual state of chemotactic cell motion towards an aggregation center along a cyclic AMP gradient. This picture was obtained in dark-field illumination by which the different scattering behaviour of resting and moving cells can be distinguished. Thus, an indirect way is provided to visualize the travelling waves of the activator molecule cAMP.

Beyond the phenomenological similarity between these two examples of spiral waves, we now present various findings that provide quantitative evidence for close analogies between different excitable systems.

PREDICTIONS FROM REACTION-DIFFUSION MODELS

In the recent literature an increasing number of theoretical approaches is presented in order to explain patterns in excitable media, especially the formation of spiral waves. In many cases, model schemes of excitable kinetics are coupled with the diffusion of relevant system variables, e.g. in [13,14]. More abstract models, for instance on the basis of cellular automata [15], are also discussed. There are only a few models which start from the actual reaction kinetics and diffusion properties of the investigated system and thus lead to predictions which can be verified directly by experimentation. For the case of chemical spirals a reaction-diffusion model based on the Oregonator kinetics was proposed in [16]. An analogous model was subsequently introduced for spiral formation in the Dictyostelium discoideum system, in this case based on the Martiel-Goldbeter model for cyclic AMP oscillations [17,18]. Here only the diffusion of one of the variables is taken into account.

These models predict that the shape of spiral wave fronts is close to that of the involute of a circle, which in the outer regions becomes asymptotically identical with an Archimedian spiral. Our evaluation of iso-intensity levels corroborate this prediction, both for the chemical [19] and the biological system [12] under consideration. The relationship between the local curvature of the two-dimensional front, K, and its propagation velocity in normal direction, N, is predicted to follow the linear law

$$N = c - D \cdot K \quad ,$$

where c is the velocity of planar waves and D the diffusion coefficient of the autocatalytic species. The validity of this relationship was proven experimentally both in BZ-reaction [20] and in the slime mould system [12]. Two types of measurements were performed:

(1) Negatively curved fronts were produced by the collision of two circular waves. The collision results in the formation of cusp structures, the temporal evolution of which was followed by extracting iso-intensity lines from a temporal sequence of digital images and fitting hyperbolas to the highly curved portion of these contours. As shown in Fig. 2A and B the predicted proportionality between N and K could be verified both for chemical waves and for aggregation waves in the Dictyo-

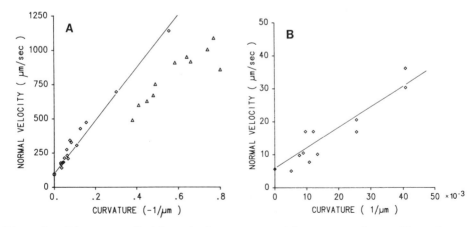

Fig. 2. Linear relation between negative curvature K and normal velocity N determined from cusp structures of colliding waves: (A) Ferroin-catalyzed BZ reaction (from [20]). (B) Dictyostelium discoideum amoebae (from [12]).

stelium discoideum system. The measurements yield a good estimate of the diffusion coefficient of $HBrO_2$ ($2.0 \cdot 10^{-5}$ cm^2/s) and of cAMP ($0.66 \cdot 10^{-5}$ cm^2/s) which fall both in the expected range.

(2) In more difficult experiments involving the triggering of very small circular waves by micro-electrode stimulation, the curvature effects were also investigated for po-sitively curved fronts. The results are consistent with those obtained for negative curvature [12,20].

Since the curvature of a spiral wave increases consi-derably towards its rotation center, the precise determina-tion of the above relationship is of crucial importance for the understanding of vortex formation. A particular conse-quence is the existence of a critical radius

$$R_C = K_C^{-1} = \frac{D}{c} \quad ,$$

below which outward propagation of a circular wave does not take place. In the BZ system this was directly verified by immersing silver-coated capillaries, whereas in the slime mould system a spontaneously triggered circular wave was ana-lyzed. For the investigated situations the critical radius is ≈ 20 μm for the BZ reaction and ≈ 200 μm for the slime mould pattern. In the model given in [16] the critical curvature is assumed to govern the motion of the spiral tip around the singular core region into which it cannot penetrate. Other models may not follow precisely the same picture, but criti-cality of curvature effects also constitutes a major argument [21].

THE STRUCTURE OF THE SPIRAL CORE

According to theoretical considerations there must exist a small area around the center of a vortex - the core area - in which the reaction processes are different from those occurring outside this area. The singular properties of spi-ral cores were characterized experimentally in some detail. In a standard BZ system a sequence of spiral images was ob-tained with a 2D spectrophotometer. Images covering one spi-ral revolution were digitally overlayed [10]. As shown in Fig. 3A, the overlay visualizes a circular area (diameter 0.7 mm) in which the maximum degree of excitation is never reached. The spiral tip rotates on a circle around the center of rotation, which is located in the middle of the black dot. This point is a singular site where the chemical state remains quasi-stationary, while at all points outside the chemical composition varies in time between the two extremes of maximum and minimum excitation. In a three-dimensional perspective representation such a composite spiral picture results in a "tornado"-like structure (Fig. 3B). It clearly shows that inside the core the maximum of excitation de-creases towards the singular site at the center.

In the case of spiral rotation in the slime mould system a first experimental analysis is based on spiral images obtained by the traditional dark-field illumination which provides information about the actual state of cell motion. Fig. 4A shows a spiral-shaped contour map extracted from the

Fig. 3. Spiral core in the BZ reaction. (A) overlay of six
images of a spiral produced in a reaction mixture as
in Fig. 1A, taken at 3s-intervals. (B) The upper
envelope of the concentration variation in an overlay
is shown in three-dimensional perspective (cf. [22]).

outer edge of the dark-field picture in Fig. 1B. An overlay
of 8 contours during one spiral revolution is presented in
Fig. 4B. It visualizes the size of the spiral core, at the
boundary of which the contour maps end. This region is identical with the small dark spot in the center of Fig. 1B
having a diameter of approximately 0.6 mm.

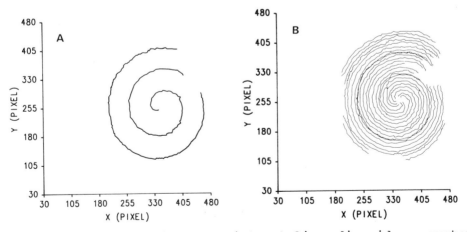

Fig. 4. Spiral core in the Dictyostelium discoideum system
(from [12]). (A) Contour map of spiral pattern
extracted from Fig. 1B. (B) Overlay of 8 spiral
contours, taken at 60s-intervals.

ADDITIONAL TRANSPORT IN CELL AGGREGATION: CHEMOTAXIS

The reaction-diffusion models developed for the explanation of spiral formation in excitable media describe the dynamic behaviour of two variables [2]. In the BZ system these
are the autocatalytic species $HBrO_2$ and the catalyst ferroin
or cerium. In the Dictyostelium discoideum system these variables are the autocatalytically produced signal transmitter
molecule cAMP and the membrane receptor to which the secreted
cAMP is bound, thus stimulating further cAMP production.
While in the BZ system both species are allowed to diffuse,
in the cell aggregation a diffusion term is implemented only
for extra-cellular cAMP. For the BZ system at least one variable can be measured precisely, that is the concentration
of the catalyst, while the $HBrO_2$ is too small for direct detection. For the slime mould patterns, the commonly used
dark-field data supply indirect evidence for the patterns,
because the direct experimental detection of cAMP is very
difficult. Only a small amount of data, obtained with isotope
dilution-fluorgraphy, is available for frozen-in snapshots of
the cAMP structure [11]. In this case the chemotactic motion
of the cells during pattern evolution has to be taken into
account, but the global dynamic features of this motion are
not yet characterized quantitatively.

In order to improve the experimental knowledge about the
dynamic behaviour of cell aggregation patterns, a newly developed quantitative method for measuring the velocity of chemotactic cell motion was used [23]. The method is based on
mutual correlation analysis of temporal sequences of microscopic digital images recorded during the aggregation phases
of the cell population. The main idea is derived from the
fact that the cell motion causes a temporal change of the
grey level at each pixel of a digital image. The local velocity at a certain pixel site is estimated by analyzing the
mutual correlation between the temporal brightness change of
a selected centered pixel and that of neighbouring pixels.
From this, the direction and speed of motion are determined.

The microscopic image shown in Fig. 5A depicts the dense
assembly of a few hundred amoebae which undergo periodic
motion due to the repetitive passage of cAMP pulses. In our
experiments the small areas of the pattern under investigation are oriented such that these pulses propagate in
horizontal direction, i.e. along an x-axis of the recorded
pictures. A result of the velocity analysis is shown in Fig.
6 [24]. The horizontal velocity varies in a periodic fashion,
(period $\approx$ 7 min). The maximum values are 20 to 30 $\mu m/min$ and
the minimum values are close to 0. Within the experimental
error the data suggest that the shape of the velocity
function is non-symmetric.

Whereas this analysis was done in an area of the pattern far away from the aggregation center, we show in Fig. 5B
a microscopic photograph of the cell distribution around a
spiral center. The snapshot shows a rather complex cell
distribution. In the center of this picture there appears to
be a more or less circular region of lower cell density. A
preliminary study of the spatial distribution of cell veloci-
ties reveals that a wave of excitation circulates around this

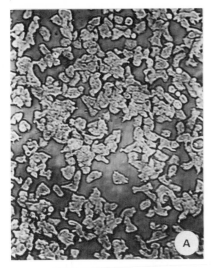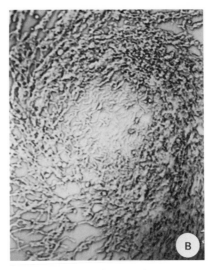

Fig. 5. Microscopic photographs of aggregating Dictyostelium discoideum amoebae in transmitted light. (A) 0.39 x 0.32 mm^2 area outside the aggregation center (from [25]). (B) Area of similar size containing a spiral vortex.

low density region. There is, in fact, a circulating sector of homogeneous cell motion - the tip of the spiral - directed tangentially to the low density region. The scheme presented in Fig. 7 illustrates our finding that the cells move tangentially to the spiral core close to the spiral center, and radially in the periphery of the spiral pattern.

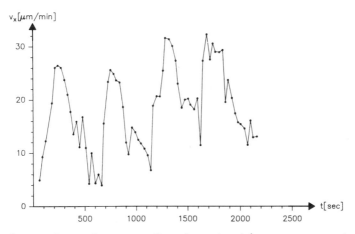

Fig. 6. Time dependence of chemotactic component v_x in direction of the aggregation center, obtained from sequences of images such as shown in Fig. 5A (typical image area: 0.07 x 0.31 mm^2, from [25]).

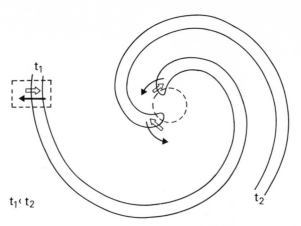

Fig. 7. Schematic illustration of a spiral pattern in the Dictyostelium discoideum system for times t_1 and t_2. Solid arrows: propagating chemical waves; open arrows: direction of chemotactic cell motion; dashed rectangle: typical observation area (compare Fig. 5A); dashed circle: core boundary.

The quantitative results on chemotaxis obtained with this image processing technique indicate that oscillatory behaviour in aggregation patterns can be detected for a long period of time after the macroscopic patterns in dark-field illumination have already disappeared. Thus, a powerful tool is at hand, which probably could be also used to characterize cell motion at very late stages of cell aggregation, when multi-cellular slugs have already grown into the third dimension on top of the spiral center. In fact, circulating motion of individual fluorescence-labelled cells in slugs has been recently reported [25].

Chemotactic cell motion has not yet been implemented in reaction-diffusion models for spiral formation. In order to correlate the observation of cell motion close to the spiral core with the isoconcentration levels of cyclic AMP, appropriate models including chemotaxis have to be developed.

REFERENCES

1. G. Nicolis and I. Prigogine, "Self-Organization in Non-equilibrium Systems", Wiley, New York (1977).
2. J.J. Tyson and J.P. Keener, Singular Perturbation Theory of Traveling Waves in Excitable Media (A Review), Physica D32:327 (1988).
3. R. Fitzhugh, Impulses and Physiological States in Theoretical Models of Nerve Membrane, Biophys. J. 1:445 (1961).
4. G. Matsumoto, H. Shimizu, J. Phys. Soc. Jpn 44:1399 (1978).
5. A. Gierer and H. Meinhardt, "A Theory of Biological Pattern Formation", Kybernetik 12:30 (1972).

6. M.A. Allessie, F.I.M. Bonke and F.J.G. Schopman, Circus Movement in Rabbit Atrical Muscle as a Mechanism of Tachycardia, _Circ. Res._ 41:9 (1977).

7. A.T. Winfree, "When Time Breaks Down", Princeton Univ. Press, Princeton (1987).

8. R.J. Field and M. Burger (eds.), "Oscillations and Travelling Waves in Chemical Systems", John Wiley, New York (1985).

9. G. Gerisch, Chemotaxis in Dictyostelium, _A. Rev. Physiol._ 44:535 (1982)

10. S.C. Müller, Th. Plesser, and B. Hess, The Structure of the Core of the Spiral Wave in the Belousov-Zhabotinskii Reaction, _Science_ 230:661 (1985).

11. K.J. Tomchik and P.N. Devreotes, Adenosine 3',5'-Monophosphate Waves in Dictyostelium discoideum: a Demonstration by Isotope Dilution-Fluorography, _Science_ 212:443 (1981).

12. P. Foerster, S.C. Müller, and B. Hess, Curvature and Spiral Geometry in Aggregation Patterns of Dictyostelium discoideum, _Development_ 109: 11 (1990).

13. E. Meron and P. Pelcé, Model for Spiral Wave Formation in Excitable Media, _Phys. Rev. Lett._ 60:1880 (1988).

14. D. Barkley, M. Kness, and L.S. Tuckerman, Spiral-Wave Dynamics in a Simple Model of Excitable Media, _Phys. Rev. A_ 42:2489 (1990).

15. M. Markus and B. Hess, Isotropic Cellular Automata for Modelling Excitable Media, _Nature_ 347:56 (1990).

16. J.P. Keener and J.J. Tyson, Spiral Waves in the Belousov-Zhabotinskii Reaction, _Physica_ D21:307 (1986).

17. J.-L. Martiel and A. Goldbeter, A Model Based on Receptor Desensitization for Cyclic AMP Signaling in Dictyostelium Cells, _Biophys. J._ 52:807 (1987).

18. J.J. Tyson, K.A. Alexander, V.S. Manoranjan, and J.D. Murray, Spiral Waves of Cyclic AMP in a Model of Slime Mould Aggregation, _Physica_ D34:193 (1989)

19. S.C. Müller, Th. Plesser, and B. Hess, Two-Dimensional Spectrophotometry of Spiral Wave Propagation in the Belousov-Zhabotinskii Reaction. Part II, _Physica_ D24:87 (1987).

20. P. Foerster, S.C. Müller, and B. Hess, Critical Size and Curvature of Wave Formation in an Excitable Chemical Medium, _Proc. Natl. Acad. Sci. USA_ 86:6831 (1989).

21. V.A. Davydov, V.S. Zykov, and A.S. Mikhailov, Kinematical Theory of Autowave Patterns in Excitable Media, _in_: "Nonlinear Waves in Active Media", J. Engelbrecht, ed., Springer, Berlin (1989).

22. J. Ross, S.C. Müller, and C. Vidal, Chemical Waves, _Science_ 240:460 (1988).

23. H. Miike, Y. Kurihara, H. Hashimoto, and K. Koga, Velocity-Field Measurements by Pixel-Based Temporal Mutual-Correlation Analysis of Dynamic Image. _Trans. IEICE Japan E_ 69:877 (1986).

24. O. Steinbock, H. Hashimoto, and S.C. Müller, Quantitative Analysis of Periodic Chemotaxis in Aggregation Patterns of Dictyostelium discoideum, _Physica D_, in press.

25. F. Siegert and C. Weijer, Analysis of Optical Density Wave Propagation and Cell Movement in the Cellular Slime Mould, _Physica D_, in press.

BONE REMODELING

Lis Mosekilde

Department of Connective Tissue Biology
Institute of Anatomy
University of Aarhus
Denmark

INTRODUCTION

All structural materials subjected to frequently repeated cyclical loading are liable to fatigue failure. Bone is no exception, but bone differs from other structural materials in being able to repair itself (Parfitt, 1988). In the adult skeleton of man this bone replacement mechanism - or repair mechanism - is called remodeling.

Bone remodeling occurs in anatomically discrete foci which are active for 4 - 8 months and then "rest" for a period of 2 - 5 years. The bone remodeling process mediates the effect of all agents (hormonal, nutritional or mechanical) - whether beneficial or harmful - that act on the skeleton (Parfitt, 1988). Some facts about the remodeling process of the human skeleton are known from normal bone histological investigations and from bone histomorphometry - but many, and possibly the most important, are still totally unknown.

It is known that in an adult skeleton of normal size an activation of a remodeling process starts every 10 seconds! The process lasts some 4 - 8 months, which means that in total some 1 - 2 million remodeling sites are active at any time! Also, it is known that if the skeleton is in a steady state, remodeling is asynchronous, with different cycles distributed randomly in time.

But the factors which control the location and frequency of activation of the remodeling process are unknown. Whether activation occurs at random or selectively in response to focal, structural or biomechanical requirements is unknown.

It is known that activation is followed by a period of bone resorption, which is then normally coupled to formation of new bone (which fills the resorbed cavity).

But it is not known how control is exerted on the size, shape or depth of a resorption cavity. And it is not known what controls the coupling mechanism - which appears to be intrinsic to bone, to be locally regulated, and automatic.

This presentation covers the following main aspects of bone remodeling: Bone design, bone cells, bone biology, bone remodeling investigated by use of scanning electron microscopy, the cost and the structural and biomechanical consequences of the remodeling process, and final remarks - and the "unknowns".

BONE DESIGN

Gross anatomy

Bone is a tissue formed of two separate components - compact and trabecular bone. Compact bone provides a dense shell around all bones and thereby gives support to the internal trabecular network and at the same time encloses the bone marrow. The cortical bone constitutes approximately 80% of the skeleton. In the central part of long bones, the cortical or compact bone is very thick and gives these bones the extreme strength needed for their support of the body during daily dynamic loading (i.e. acting as a load-bearing frame for the muscular system). At other places - e.g. the vertebral bodies, the femoral neck and the ultra-distal part of the arm - the cortical bone provides a shell only a few hundred μm thick.

Trabecular bone consists of a network of bone plates, columns, and struts with diameters varying from 10 - 400μm. In loadbearing trabecular bone (such as the vertebral body and the femoral neck) the trabecular lattice is obviously designed to withstand dominating vertical loads and strains.

In the vertebral body, the trabecular network has all its thick trabecular columns orientated vertically so as to withstand the dominating compressive forces (Fig. 1A: Vertebral body from a young individual; Fig. 1B: Vertebral body from an elderly individual). Thinner horizontal struts provide support for the vertical columns to prevent their bending when under load. The trabecular network in the vertebral bodies is anisotropic in all directions apart from the horizontal plane. In this plane the network is isotropic.

Ultrastructural and microscopic anatomy

Bone tissue consists of 80% mineral and only 20% organic material. At both microscopic and ultrastructural levels, the relationship between the inorganic and organic components of bone tissue is extremely close and well-organised so as to provide maximum mechanical support with minimum mass.

The inorganic components are mainly calcium-hydroxyapatite crystals and amorph calcium phosphate. The size of the crystals is 4 nm x 60 nm and they are situated in close relationship to the major component of the organic material - the collagen fibers. This special arrangement of organic and inorganic materials gives the bone tissue its characteristic mechanical properties, with the capability of withstanding compressive, bending, and torsional forces.

The ultrastructure's perfect "design", with the strict alignment of crystals and collagen fibers giving maximum mechanical competence, is further confirmed at the microscopic level. In normal lamellar bone, the collagen fibers are arranged in a parallel pattern in a layer of thickness of approximately 3 μm (one lamella), then in the adjacent lamella, the parallel pattern of the collagen fibers has a different angle to the first - and so on throughout the bone structural unit - the osteon. This pattern again provides optimum biomechanical properties for the bone tissue.

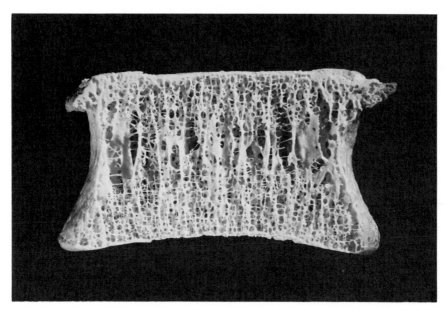

Fig. 1A. Vertebral body from a young individual.

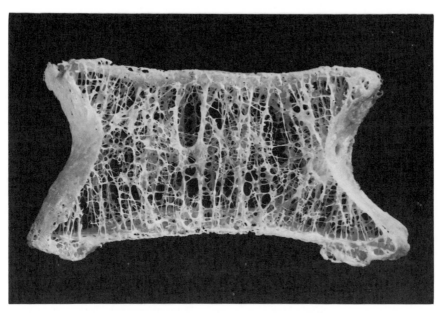

Fig. 1B. Vertebral body from an elderly individual.

In cortical bone, the osteon, identical with the Haversian systems, is a cylindrical structure with a diameter of 150 - 300 μm and a length of 1500 - 4000 μm. The cortical osteons are often aligned parallel to the long axis of the bone and consist of 10 - 20 lamellar rings arranged around a central vascular channel.

The cortical osteons are demarcated by a cement line, which is ground substance without collagen fibers. In between the "perfect" cortical osteons are fragments of previous osteons (interstitial lamellae).

The same units - trabecular osteons or packets - are found in trabecular bone, and the lamellar thickness and arrangement are the same as in cortical bone. However, the trabecular osteons are not cylindrical but are cresent shaped or branching and cover the trabecular structures. The width of the trabecular osteons is fairly constant - approximately 50 - 70 μm. The number of lamellae is, therefore, constant too, and numbers 15 - 25. Trabecular osteons can cover surface areas of more than 1000 x 1000 μm.

BONE CELLS

The four types of bone cells that control the dynamic processes in the bone are: (a) the lining cells (b) the osteoblasts (c) the osteocytes (d) the osteoclasts.

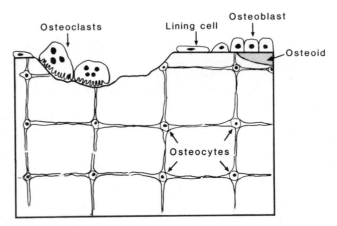

Fig. 2. The four types of bone cells - all involved in the remodeling process.

The resting bone surface, where no dynamic process is ongoing, is covered by a layer of flat, inactive lining cells. These cells lie on an extremely thin layer of non-mineralised matrix (collagen type I). The thickness of this layer is normally only 20 - 100 nm. Despite its thinness, this layer seems extremely important since bone resorption is never initiated on osteoid covered surfaces - this layer has to be removed before any dynamic process can take place (Chambers and Fuller, 1985). This is illustrated in Fig. 2.

The bone matrix is secreted by <u>osteoblasts</u> (activated lining cells) that lie on the surface of existing matrix and deposit new bone onto it. But the osteoblasts are also involved in the bone-resorption process in that they are believed to secrete the enzyme (collagenase) necessary for the removal of the thin non-mineralised surface layer (Chambers et al., 1985).

During active bone formation, the cuboidal osteoblasts lie closely together on the bone matrix and produce osteoid onto it at a generally constant rate. Some of the osteoblasts remain free on the surface, while others gradually become embedded in their own secretion. The newly formed material (osteoid) is rapidly converted into hard bone matrix by the deposition of hydroxyapatite crystals between the closely packed collagen fibers (mineralization).

Once captured in the hard bone matrix, the original bone-forming cells, now called <u>osteocytes</u>, have no opportunity to secrete further matrix, but the osteocytes, each lying in a small lacuna, are not isolated. Tiny channels in the bone tissue enable the cells to form gap junctions with adjacent osteocytes. This network probably plays a major role in controlling both the activity of the osteocytes themselves and the activity of the osteoblasts and lining cells on the surfaces. The osteocyte network seems to be able to sense <u>piezoelectric fields</u> and thereby mechanical stresses and strains applied to the bone. The signals from these cells are transmitted not only to the osteoblasts and lining cells but also to the fourth bone cell type - the osteoclasts.

<u>Osteoclasts</u> are derived from the blood and seem able to erode bone as soon as the thin protective layer of non-mineralised matrix has been removed. Osteoclasts are big (20 - 100 μm), multi-nucleated, and mobile cells with a refined enzymatic system. During bone resorption they are capable of engulfing both collagen, mineral, and osteocytes. The osteoclasts move quickly over the surface as they excavate the cavities - the process normally stops at a depth of 50 -60 μm. The cells leave the surface raw, with collagen layers visible in the bottom of the cavities. This surface is soon covered by ground substance (cement line) and the formative process (coupling) begins.

Like the osteoblasts, the osteoclasts are also controlled from the network of underlying osteocytes - continuously sensing and signalling the mechanical stresses and strains applied to the bone.

BONE BIOLOGY

The "design" and "redesign" of bone involve three different processes: (a) growth, (b) modeling, and (c) remodeling.

During childhood and the early years of adulthood, the skeleton grows in length, and the bones expand in diameter and are modeled.

During bone modeling, osteoblasts and osteoclasts work independently of each other and on different bone surfaces - often over large surface areas. The net balance is positive (i.e. there is increased bone mass) and the bones reach their final external form and high bone density during this period. Again, both the growth and the modeling processes are controlled by mechanical forces - mechanical usage.

Around the age 20 - 25 years, "peak bone mass" is achieved as a result of these processes. The individual "peak bone mass" is dependent on genetic, racial, and hormonal factors and also on external factors, for example, physical activity and nutrition. Establishing a high "peak bone mass" gives the optimum "bone bank" from which to draw during aging. "Peak bone mass" is normally 20 - 25% higher in men than in women.

From the age of 20 - 25 years, bone mass begins to decline. Loss of bone mass with age is unavoidable and is caused by the third process - bone remodeling.

During the remodeling process, osteoclasts and osteoblasts work closely together in time and space (coupling), and they work in units. When a remodeling process has just been completed, a small amount of bone (trabecular or cortical) has been removed and has been replaced by new bone.

The process seems "designed" to renew old bone - which possibly contains microfractures or dead osteocytes - and to reorganise trabecular structures to achieve the maximum strength obtainable with the remaining bone mass.

Trabecular bone is remodeled much more rapidly than cortical bone (5 - 10 times more rapidly). The remodeling process is itself governed by hormonal and mechanical stimuli - as is also seen in the growth and modeling processes.

Each remodeling process (Fig. 3) is initiated by an activation by which osteoblasts start to secrete collagenase which removes the thin layer of unmineralised bone typical of a resting bone surface. This exposes the mineralised bone underneath to the multinucleated, mobile osteoclasts. During osteoclastic bone resorption, lacunae are excavated to a maximum depth of 60 - 70 μm. A short reversal phase, where the cement line is formed, follows, and then bone formation normally begins. If coupling has taken place, osteoblasts produce osteoid (collagen and ground substance) at a rate of 0.5 - 1.0 μm per day. When the osteoid thickness has reached approximately 12 - 15 μm, mineralisation begins from the bottom (mineralisation front). At the termination of each remodeling process, the bone surface is again covered by an extremely thin layer of non-mineralised bone and a layer of flat lining cells. The bone is again converted into a resting surface.

The remodeling process, which is "designed" to renew old bone of inferior quality and to replace bone with microfractures, does so only at a price - the balance is normally negative (i.e., after each remodeling process there is a reduced mass). There is an unavoidable loss of bone mass with age. It also has another cost: it causes disruption of the trabecular network with age. Because the balance is negative, there is a thinning of the trabecular structures in the network, and this makes fortuitous osteoclastic perforations possible. As the normal resorption depth is approximately 60 μm, one resorption site covering more than half of the circumference of a trabecula, or two resorption sites, one on each side of a trabecula, could easily perforate a trabecular structure with a diameter of 100 - 120 μm (100 - 120 μm is the normal thickness of horizontal trabeculae in the vertebral body of elderly individuals).

Perforation and disconnection of the trabecular network with age are, in combination with the age related loss of bone mass, responsible for the extreme loss of bone strength seen with age (Mosekilde et al., 1987).

Furthermore, once lost, connectivity cannot be regained as new lamellar bone cannot refill the gaps (the osteoblasts have no surface on which to work nor any mechanical stimulus to activate them since the structure, once disconnected, is no longer strained)(Parfitt 1984, Parfitt 1987).

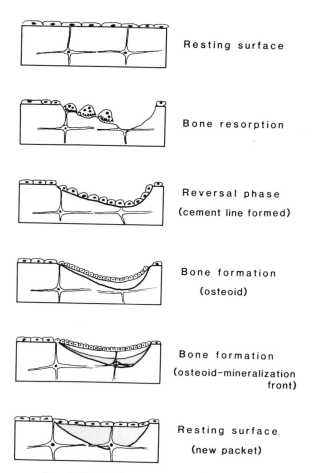

Resting surface

Bone resorption

Reversal phase
(cement line formed)

Bone formation
(osteoid)

Bone formation
(osteoid-mineralization
front)

Resting surface
(new packet)

Fig. 3. The different phases in the remodeling process. Just before bone
resorption begins, activation has
occured.

Bone remodeling in cortical bone involves the same processes as in trabecular bone: Activation, resorption and formation. By the remodeling process, existing Haversian systems are replaced by new ones.

In summary, the three processes - growth, modeling, and remodeling - although different in working methods, are all governed by the same cells and are all dependent on mechanical stimuli. This means that the form, size, density, and strength of the skeleton at any age are dependent on its mechanical usage.

BONE REMODELING INVESTIGATED BY USE OF SCANNING ELECTRON MICROSCOPY (SEM)

Remodeling sites

The remodeling process and the different phases in the remodeling process have traditionally been studied by use of normal light microscopy.

By using SEM it has become possible to describe the remodeling process in the 3-dimensional trabecular network. Additionally, the distribution, shape and size of each remodeling site in the network can be investigated and described.

But it is also possible to elucidate how the remodeling process "destroys" the connectivity of the network and to describe how disconnected and no longer strained or stressed trabeculae are removed.

Perforations of trabeculae

When focussing on a single trabecular structure, it can be seen that the remodeling process "glides" over the surface. As the resorption depth is approximately 50 - 60 μm, only trabeculae with a diameter of less than 100 μm will normally be at risk of being perforated during the resorption process. Fig. 4 shows a trabecula with a diameter of 100 - 120 μm. The deep resorption cavities are clearly visible. It is obvious, too, that this trabecula is not at risk of being perforated during this remodeling process.

In contrast, in Figs. 5A and 5B the thin, horizontal trabecula has been perforated, and resorption pits are clearly demonstrated at the end to the right of the perforation. It is obvious that in this case the thickness of the trabecula (less than 50 μm) before the remodeling process took place was critical.

Remodeling on connected versus disconnected trabeculae

During normal bone remodeling, there is a linkage phase, "coupling", which takes place between bone resorption and bone formation. The essential factor responsible for this coupling is still unknown. Several investigators have, however, pointed out that mechanical stimuli are critical for a successful coupling between the two processes.

Figs. 6A and 6B show the same bone specimen investigated before and after treatment with chlorine to remove organic material. At this surface, coupling has taken place, and new bone is being formed. The surface, therefore, represents early bone formation.

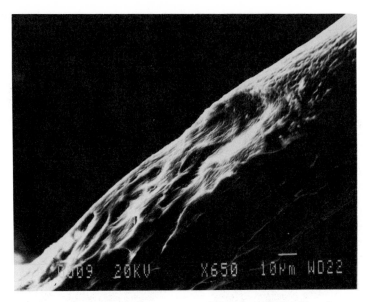

Fig. 4. The remodeling process glides over the trabecular
surface (resting surface is seen at right).

By applying this technique to connected trabeculae, surfaces where coupling
had taken place could be identified at 70 - 80% of the remodeling sites. Figs. 6A
and 6B demonstrate a remodeling site on a connected trabecula. In Fig. 6A a fine
network of collagen fibers can be seen in the middle of the remodeling site. At far
left the bone formation is more advanced, with osteocytes captured in their own
secretion (small pits), and the osteoid is visibly bulging. In Fig. 6B the organic
material has been removed, and it now becomes clear that the bulging appearance
of the osteoid was due to the mineralization front underneath. An osteocyte lacuna
has also been disclosed.

In contrast, on perforated trabeculae which were no longer subjected to stress
or strain, successful coupling between bone resorption and formation was rare (10 -
20% of investigated remodeling sites).

These data firmly stress the importance of mechanical stimulation for
maintaining bone structure -bone connectivity.

THE COST AND THE STRUCTURAL AND BIOMECHANICAL CONSEQUENCES OF THE REMODELING PROCESS

The <u>cost</u> of the remodeling process - which is intended to renew old bone,
replace dead osteocytes and repair microfractures - is tremendous. There is a negative balance as a result of each cycle - and, even more important - there is a gradual
destruction of the perfect 3-dimensional trabecular network seen in younger individuals.

The <u>structural consequences</u> of the remodeling process: By using a special
technique with thick sections of bone specimen (embedded in plastic) investigated
in polarized light, it is possible to demonstrate the loss with age of both bone mass
and connectivity (Mosekilde, 1988).

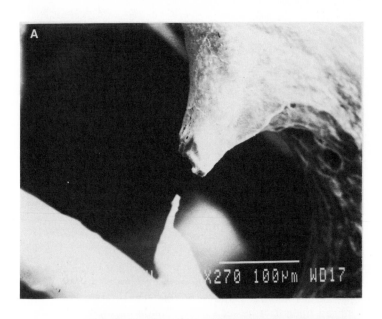

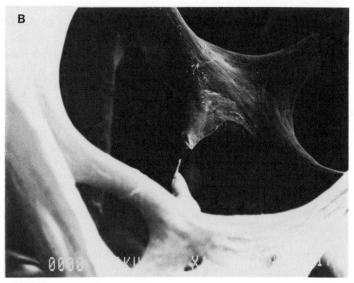

Figs. 5A and 5B. Osteoclastic perforation of a thin horizontal trabecula in the vertebral trabecular network.

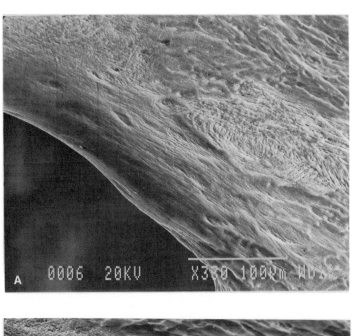

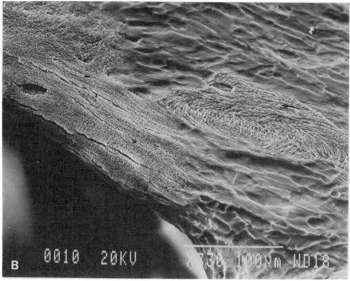

Figs. 6A and 6B. A remodeling site on a connected trabecula before (A) and after (B) treatment with chlorine.

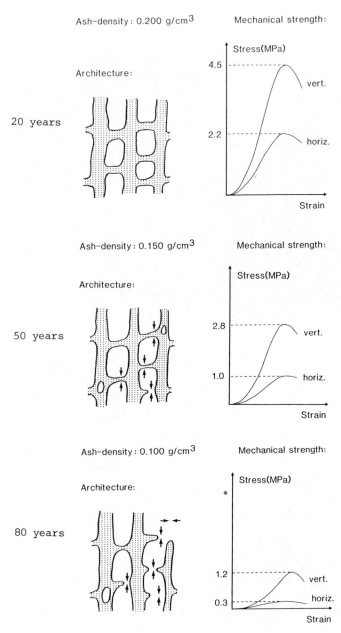

Ash-density: 0.200 g/cm³   Mechanical strength:

Stress(MPa)

Architecture:

20 years

4.5

2.2

vert.

horiz.

Strain

Ash-density: 0.150 g/cm³   Mechanical strength:

Stress(MPa)

Architecture:

50 years

2.8

1.0

vert.

horiz.

Strain

Ash-density: 0.100 g/cm³   Mechanical strength:

Stress(MPa)

Architecture:

80 years

1.2

0.3

vert.

horiz.

Strain

Fig. 7. Changes with age in: ash-density, biomechanical competence, and architecture.

The biomechanical consequences of the remodeling process: Finally, it is demonstrated that the architecture of the trabecular network is extremely important for its biomechanical competence (Fig. 7). The loss of bone biomechanical strength with age is much more pronounced than the loss of bone mass - due to perforations during the remodeling process.

FINAL REMARKS - AND THE "UNKNOWNS"

The skeleton is "designed" in such a way that the bone mass is as small as possible consistent with providing sufficient strength for its mechanical support of the body.

Throughout life, bone mass is lost at an almost constant rate as a consequence of the remodeling process. This means that the repair mechanism, or bone replacement mechanism characteristic for the adult skeleton, is achieved only through great sacrifice.

This presentation has shown how the bone cells work in the 3-dimensional trabecular network. It has demonstrated the size, shape and location of the remodeling sites, and it has shown that the normal sequence: Activation, resorption, coupling and bone formation only takes place when mechanical stimulation is present. It has demonstrated how no longer stressed or strained structures are removed and, finally, the consequences of the remodeling process on the structure and biomechanical competence of trabecular bone.

Many problems are still unsolved: we still do not know why activation occurs at a particular location at a particular time. We still do not know what actually controls the size, shape, and depth of the remodeling cavities. And we still cannot prove that the stimulus for the timely appearance of a sufficient number of new osteoblasts at the base of the resorption cavity is purely mechanical.

Before any medical manipulation of the remodeling process is attempted, all these "unknowns" should be bourne in mind and respected - and, ideally, solved. Perhaps in the meantime we would do better to promote the one factor which is known to initiate a positive balance during bone remodeling - physical exercise.

REFERENCES

1. Chambers T.J., Fuller K. Bone cells predispose bone surfaces to resorption by exposure of mineral to osteoclastic contact. J. Cell Sci. 1985, 76: 155-165.

2. Chambers T.J., Darby J.A., Fuller K. Mammalian collagenase predisposes bone surfaces to osteoclastic resorption. Cell Tissue Res. 1985, 241: 671-675.

3. Mosekilde Lis, Mosekilde Le., Danielsen C.C.. Biomechanical competence of vertebral trabecular bone in relation to ash density and age in normal individuals. Bone 1987, 8: 79-85.

4. Mosekilde Lis: Age-related changes in vertebral trabecular bone architecture - Assessed by a new method. Bone 1988, 9: 247-250.

5. Mosekilde Lis: Consequences of the remodelling process for vertebral trabecular bone structure: a scanning electron microscopy study (uncoupling of unloaded structures). Bone Miner. 1990, 10: 13-35.

6. Parfitt A.M.. Age-related structural changes in trabecular and cortical bone: Cellular mechanisms and biomechanical consequences. Calcif. Tissue Int. 1984, 36: S123-128.

7. Parfitt A.M. Trabecular bone architecture in the pathogenesis and prevention of fracture. Am. J. Med. 1987, 82(1B): 68-72.

8. Parfitt A.M. Bone remodeling: Relationship to the amount and structure of bone, and the pathogenesis and prevention of fractures. In: Osteoporosis: Etiology, Diagnosis and Management. Eds.: B.L. Riggs and L.J. Melton III, Raven Press, New York 1988.

Section VI
Chaos and Hyperchaos

With increasing number of state variables, new types of complex time evolution are realised, including various forms of higher order chaos.

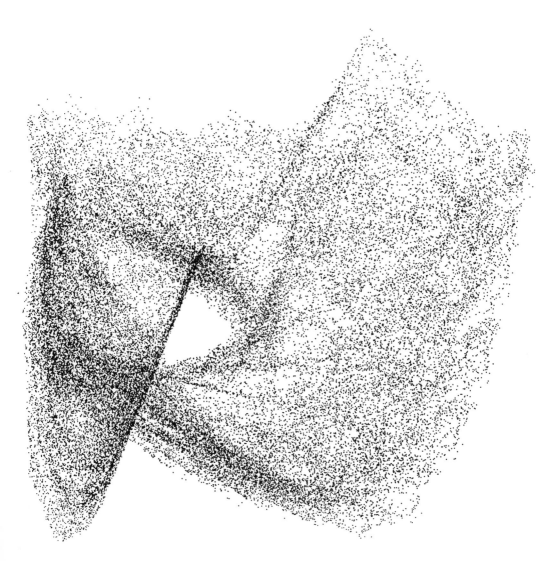

CHILDHOOD INFECTIONS - EXAMPLES OF 'CHAOS IN THE WILD'

L.F. Olsen
Institute of Biochemistry
Odense University
Campusvej 55
DK-5230 Odense M, Denmark

C.G. Steinmetz
Department of Chemistry
Indiana University-Purdue University at Indianapolis
Indianapolis, Indiana 46205, USA

C.W. Tidd and W.M. Schaffer
Department Ecology and Evolutionary Biology
University of Arizona,
Tucson, Arizona 85721, USA

INTRODUCTION

There is a large literature in mathematical biology dealing with the transmission of infectious diseases[1-3]. A significant part of this literature, which dates back to the early 20th century, concerns childhood epidemics - chicken pox, measles, mumps and rubella - because the biology of these diseases is reasonably well understood and because several decades of monthly or weekly notifications - at least in First World countries - have accumulated in public health records. Thus it appears that childhood infections are part of a well-defined epidemiological system for which real-world data and mathematical models may be profitably compared.

The dynamics of childhood infections show different patterns, depending on the disease and on the locality. Sometimes case rates follow an irregular annual cycle, whereas at other times they fluctuate dramatically. The former pattern is typical of chicken pox and the latter is typical of measles. The discovery that fluctuating time patterns in biological systems may be ascribed to deterministic chaos[4-6] has revived the study of childhood infections, and there is mounting evidence that the irregular fluctuations of measles incidences, and possibly also similar fluctuations in incidences of mumps and rubella, are chaotic[7-15].

Complexity, Chaos, and Biological Evolution, Edited by E. Mosekilde and
L. Mosekilde, Plenum Press, New York, 1991

Here we review previous nonlinear analyses of childhood diseases using data obtained from public health records and data simulated by epidemic models. In addition we present some results obtained by applying nonlinear forecasting techniques to such data. On the basis of comparisons of the real data with the data generated by the so-called Susceptible-Exposed-Infective-Recovered (SEIR) model with seasonal variations in contact rate, we conclude that measles dynamics are chaotic whereas the dynamics of chicken pox correspond to those of a noisy periodic oscillation. We also offer an explanation of the dynamics of mumps in terms of noise-induced switchings between coexisting attractors.

CASE REPORTS: PHENOMENOLOGY AND NONLINEAR ANALYSES

Fig. 1 shows the monthly notifications[16,17] of chicken pox, measles and mumps in New York City and in Copenhagen, Denmark over periods of 30-40 years. We note that all three diseases can be characterized by recurrent epidemics. Often there is a yearly cycle suggesting that transmission rates are higher in the winter than in the summer. In addition to this annual cycle there may also be significant year-to-year variations in case rates. The simple annual cycle is observed for chicken pox in Copenhagen and New York and mumps in New York whereas large year-to-year fluctuations are observed for measles in Copenhagen and New York and mumps in Copenhagen. Spectral analyses[9,10,12] of the data show a single peak at 1 cycle per year (cpy) for chicken pox in both Copenhagen and New York, two peaks at 0.4 cpy and 1 cpy respectively for measles in the two cities but only one peak at 1 cpy for mumps in New York as opposed to two peaks at 0.25 cpy and 1 cpy respectively for mumps in Copenhagen. Thus in the case of chicken pox and measles we find the same dynamics in Copenhagen and New York. As for mumps we find different patterns in the two cities: a 'measles pattern' in Copenhagen as opposed to a 'chicken pox pattern' in New York.

The first nonlinear analyses of case reports of childhood diseases were made by Schaffer and Kot[7], who demonstrated that almost unimodal first return maps could be constructed from monthly notifications of measles in Baltimore and New York. The authors computed Lyapunov exponents of 0.5-0.7 bit/year from these maps. The corresponding maps of chicken pox and mumps notifications in New York were randomly scattered points. Schaffer and Kot therefore concluded that the fluctuations in measles incidences in Baltimore and New York were chaotic, whereas those of chicken pox and mumps in New York corresponded to a noisy limit cycle oscillation with a period of one year. Later investigations[9-13] of measles notifications in Aberdeen, Copenhagen, Detroit, St. Louis and Milwaukee revealed similar maps to those in Baltimore and New York, suggesting that measles notifications in First World countries always correspond to chaotic attractors. This suggestion is supported by the computation of maximum Lyapunov exponents from first return maps and from flows constructed by embedding the data in three and higher dimensions using a modification of the method of Wolf et al[18]. Further support comes from computations of correlation dimensions using the Grassberger-Procaccia method[19], from computations of unpredictability profiles[14] and from application of nonlinear forecasting methods to the data[15]. Table 1 summarizes the computations of Lyapunov exponents and correlation dimensions of measles notifications in eight large First World cities. We note that most Lyapunov exponents are around 0.6 bit/year and most correlation dimensions are around 2.5. Positive Lyapunov exponents and finite, non-integer correlation dimensions are often taken as evidence of low dimensional chaos. However, such findings are not always diagnostic. For example, Osborne and Provenzale[20] recently observed that one can compute well-defined correlation dimensions for certain stochastic processes. Consequently, the problem of distinguishing chaos from periodic behavior in the presence of noise for experimental data remains unresolved. The Lyapunov exponents computed for chicken pox notifications in Copenhagen, Milwaukee, New York and St. Louis were usually lower than 0.3 bit/year and the correlation dimensions were usually higher than 3. It was shown earlier that a relatively low Lyapunov exponent and a dimension greater than three comply with

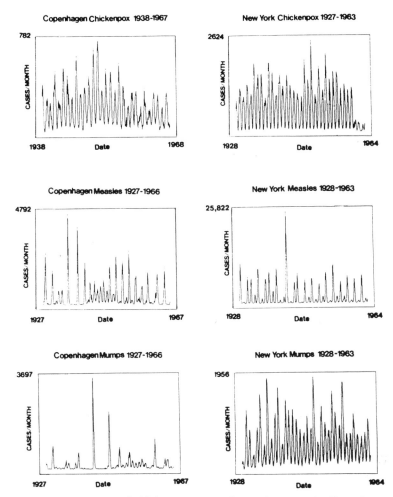

Fig. 1 Monthly notifications of chicken pox, measles and mumps in Copenhagen and New York.

chicken pox dynamics corresponding a noisy limit cycle oscillation[9,10,12]. By contrast Lyapunov exponents and correlation dimensions are more variable for mumps. For Copenhagen mumps the maximum Lyapunov exponent was estimated as 0.52 bit/year and the correlation dimension as 2.96. For mumps in Milwaukee and New York the corresponding numbers were 0.15-0.26 bit/year and 3.2-3.5 respectively. The numbers for Copenhagen therefore suggest chaos, whereas the numbers for Milwaukee and New York suggest a noisy limit cycle oscillation.

Table 1 Maximum Lyapunov exponents and correlation dimensions computed for measles notifications in eight large cities. Data compiled from refs. 7, 9, 10 and 12.

City	Community size	Dates	λ_1 (map) (bit/year)	λ_1 (flow) (bit/year)	D_2
Aberdeen	165,000	1883-1902	-	0.23	2.56
Baltimore	750,000	1900-1927	-	0.40	2.34
Baltimore County	1,285,000	1928-1963	0.60	0.55	2.42
Copenhagen	660,000	1927-1967	0.67	0.60	3.07
Detroit	2,525,000	1920-1962	-	0.59	2.43
Milwaukee	775,000	1916-1965	-	0.71	2.57
New York	9,235,000	1928-1963	0.56	0.44	2.68
St. Louis	1,620,000	1934-1954	-	0.50	2.22

SEIR MODELS

It is useful to compare the nonlinear analyses of case numbers with corresponding data simulated by epidemiological models. Here we consider Susceptible-Exposed-Infective-Recovered (SEIR) models[21]. In these models the population is divided into four categories. Individuals enter the population at birth as susceptibles (S) and leave it by death or by emigration. Susceptibles become exposed (E) by contact with an infective (I). After a latent period - usually one to two weeks - exposed individuals become infectives and later recovered and immune (R). For childhood diseases immunity is assumed to be permanent. The simplest interpretation of the SEIR scheme is a set of four first order nonlinear differential equations:

$$\frac{dS(t)}{dt} = m(N-S(t)) - b(t)S(t)I(t)$$

$$\frac{dE(t)}{dt} = b(t)S(t)I(t) - (m+a)E(t)$$

$$\tag{1}$$

$$\frac{dI(t)}{dt} = aE(t) - (m+g)I(t)$$

$$\frac{dR(t)}{dt} = gI(t) - mR(t)$$

where N is the population size, m is the birth rate, which is assumed to be equal to the sum of death and emigration rates, a and g are the reciprocal latent and infectious periods and b(t) is the contact rate, which is assumed to be seasonally dependent as will be described below. The birth rate, m, can be obtained from census data and the latent and infectious periods from the medical literature. The average contact rate, b_0, has to be estimated indirectly as[21-23]:

Table 2 Parameters used for the SEIR equations. The units are years^{-1}.

Disease	m	a	g	b_0
Chicken pox	0.02	36	34.3	537
Measles	0.02	55	60	1000
Mumps	0.02	30	60	715

$$b_0 = g(1 + L/A)$$

where L is the life expectancy and A is the average age of infection. Table 2 lists the values of m, a, g and b_0 used in the present simulations. In our simulations we assume that the population size $N = S(t) + E(t) + I(t) + R(t)$ is constant and normalize it to 1. The contact rate is not constant, but varies with a minimum in the summer and a maximum in the winter as first demonstrated by London and Yorke[16] and later verified by others[24]. The causes of such seasonal variations may be changes in viability and virulence of the pathogen due to variations of the weather and the assembling of children in schools during terms and their dispersion during holidays. The change in contact rate may be written as:

$$b(t) = b_0(1 + b_1\Phi(t)) \qquad (2)$$

where $\Phi(t)$ is a periodic function with a period of one year. Usually the function $\Phi(t) = \cos(2\pi t)$ is chosen, but here we use:

$$\Phi(t) = 1.5 \frac{(0.68 + \cos(2\pi t))}{(1.5 + \cos(2\pi t))} \qquad (3)$$

because this function more accurately models asymmetries in school year[11]. Fig. 2 shows bifurcation diagrams obtained from eqs. (1)-(3) by slowly varying b_1 and for each b_1 plotting vertically 50 maxima of I(t) following a transient of 200 years. We note that with parameters corresponding to chicken pox we get nothing but a simple annual oscillation for all values of b_1 since each b_1 gives only a single maximum. For parameters corresponding to measles we get a two-year cycle for b_1 around 0.22. When we increase b_1 we observe period doubling bifurcations leading to chaos at b_1 around 0.28. For parameters correspond-ing to mumps we get at least three bifurcation diagrams, depending on the choice of initial conditions. Two of these are shown here. The bifurcation diagram to the left shows more complicated periodic cycles which eventually bifurcate through period doublings to chaos and finally to a simple periodic oscillation as b_1 is increased, wheras that to the right describes a simple annual oscillation for all values of b_1. Thus with mumps parameters there are intervals of b_1 where two attractors coexist. This possibility was already pointed out by Schwartz and Smith[25-27].

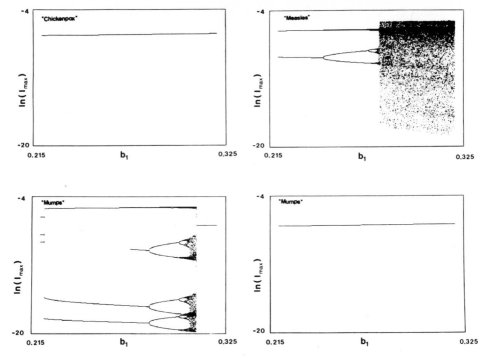

Fig. 2 Bifurcation diagrams constructed for eqs. (1)-(3) with parameters corresponding to chicken pox, measles and mumps as listed in Table 2. The seasonal component, b_1, is increased slowly from 0.22 to 0.32, and for each value of b_1 50 log transformed maxima of I(t) are plotted vertically following a transient of 200 years. The two different diagrams for mumps parameters are obtained at slightly different initial conditions of I(t).

Fig. 3 shows time series of I(t) for simulations of eqs. (1)-(3) with parameters corresponding to chicken pox, measles and mumps and values of b_1 as indicated. The values of b_1 for the measles and mumps simulations were chosen so that they fall in the chaotic regions of Fig. 2. We note that the dynamics of chicken pox and measles appear similar to those shown in Fig. 1. There is also some similarity between the mumps simulation to the right and notifications in New York. However, none of the simulations show any similarity at all to the actual case reports of Copenhagen. Figs. 4 and 5 show time-one maps of chaotic measles and mumps simulations constructed by plotting numbers of infectives against numbers of susceptibles at yearly intervals. These plots have fractal structures, i.e. subsequent magnifications of small subsections continue to reveal new structure. The correlation dimension and maximum Lyapunov exponent of the measles simulation in Fig. 4 was 2.43 and 0.45 bit/year respectively, and those of the mumps simulation in Fig. 5 were 2.20 and 0.15 bit/year. The correlation dimension and the maximum Lyapunov exponent of the chicken pox simulations were 1.00 and 0 bit/year. However, if we perturb these simulations by small amounts of multiplicative noise we obtain a correlation dimension of 3.4 and a maximum Lyapunov exponent of 0.2 bit/year in agreement with the values obtained for real-world chicken pox infections.

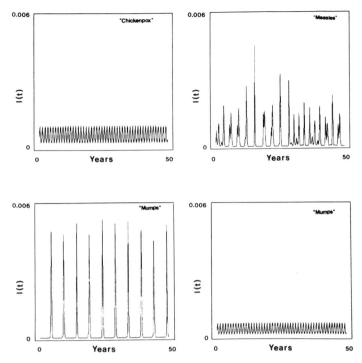

Fig. 3 Time series of I(t) computed from eqs. (1)-(3) with parameters as in Fig. 2 and $b_1 =$ 0.28 (top row) and $b_1 = 0.306$ (bottom row).

It is evident from comparing the simulations of eqs. (1)-(3) of Fig. 3 and the actual case reports of Fig. 1 that the latter have been subjected to noise perturbations. Such perturbations can either be external or internal. The latter type of noise may arise because the parameters of the SEIR model are all average measures and may vary between individuals in the population and because transitions are all discrete events. In order to account for this we have translated the SEIR scheme into a Monte Carlo model. Here we break the scheme into nine individual transformations. To each transition we assign a probability, p_i, instead of a rate as in eq. (1). The transitions and their probabilities are listed below[10]:

i	T_i	p_i
1	----> S	$mN = m(S(t)+E(t)+I(t)+R(t))$
2	S ----> E	$b(t)S(t)I(t)$
3	E ----> I	$aE(t)$
4	I ----> R	$gI(t)$
5	S ---->	$mS(t)$
6	E ---->	$mE(t)$
7	I ---->	$mI(t)$
8	R ---->	$mR(t)$
9	----> I	v

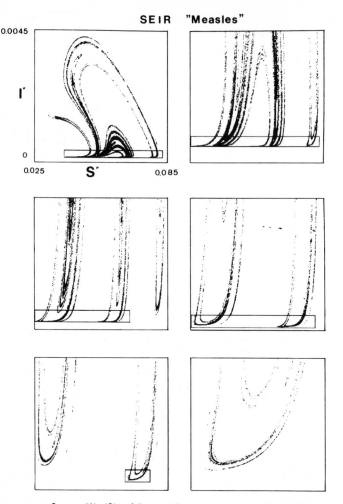

Fig. 4 Time-one map of eqs. (1)-(3) with measles parameters as in Table 2 and $b_1 = 0.28$. The map is constructed by plotting corresponding annual values of S(t) and I(t). The full map is shown in the top left figure. The top right figure is a magnification of the boxed region in the top left figure. The middle left figure is a magnification of the boxed region of the top right figure, etc.

Here v is the probability that an infective can enter the population from the outside, which is assumed to be constant. The parameters m, a, g and b_0 are rescaled such that:

$$\sum_{i=1}^{9} p_i \leq 1$$

Then a random number, r, is generated uniformly in the interval [0,1], and if r satisfies the condition:

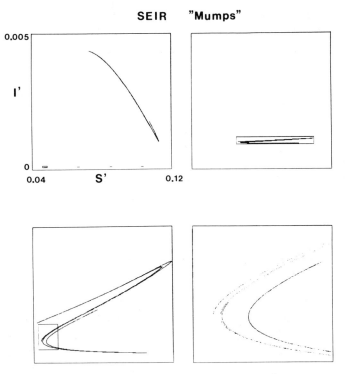

Fig. 5 Time-one map of eqs. (1)-(3) with parameters corresponding to mumps as in Table 2 and $b_1 = 0.306$. Sequence of magnifications as in Fig. 4.

$$\sum_{i=1}^{j-1} p_i \ < \ r < \sum_{i=1}^{j} p_i$$

we perform the transformation T_j. The parameters m, a, g and b_0 used in the Monte Carlo simulations are listed in Table 3. Note that the parameter b_0, unlike the other parameters, depends on the population size, N.

Table 3 Parameters used for the SEIR Monte Carlo simulations. The parameter v is chosen from $v/N = 4 \ 10^{-11} \ min^{-1}$.

Disease	m	a	g	$b_0 N$
	min^{-1}	min^{-1}	min^{-1}	min^{-1}
Chicken pox	$3.858 \ 10^{-8}$	$6.944 \ 10^{-5}$	$6.617 \ 10^{-5}$	$1.036 \ 10^{-3}$
Measles	$3.858 \ 10^{-8}$	$1.061 \ 10^{-4}$	$1.157 \ 10^{-4}$	$1.929 \ 10^{-3}$
Mumps	$3.858 \ 10^{-8}$	$5.787 \ 10^{-5}$	$1.157 \ 10^{-4}$	$1.379 \ 10^{-3}$

Table 4 Correlation dimensions for Monte Carlo simulations in populations with community sizes of 1,000,000 and 5,000,000

	Measles		Mumps	
b_1	$N=10^6$	$N=5 \ 10^6$	$N=10^6$	$N=5 \ 10^6$
0.00	4.56	4.82	4.46	4.61
0.20	3.41	3.38	3.73	3.91
0.28	3.00	2.99	3.34	3.26
0.36	2.86	2.84	3.20	3.08

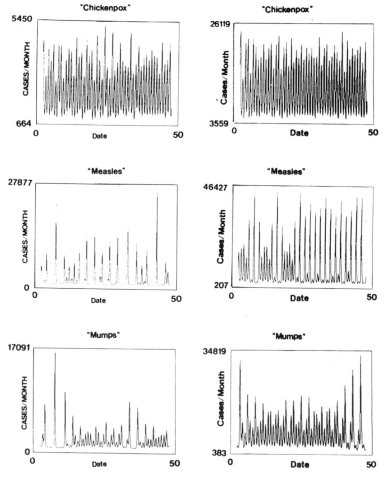

Fig. 6 Monte Carlo simulations of chicken pox, measles and mumps for two populations with community sizes 1,000,000 (left column) and 5,000,000 (right column). Parameters as in Table 3. $b_1 = 0.28$.

Fig. 6 shows simulations of chicken pox, measles and mumps in populations corresponding to 1,000,000 inhabitants (left column) and 5,000,000 inhabitants (right column). We note that the simulations predict a noisy periodic oscillation for chicken pox and aperiodic fluctuations for measles in both populations. For mumps the simulations predict aperiodic fluctuations at $N = 1,000,000$ and a noisy periodic oscillation at $N = 5,000,000$. Thus, the Monte Carlo simulations seem to be able to account for the difference observed in real-world mumps infections between New York ($N > 5,000,000$) and Copenhagen ($N \approx 1,000,000$).

Table 4 lists correlation dimensions computed for the two population sizes with measles and mumps parameters and different magnitudes of b_1. It is evident that the correlation dimensions computed for measles simulations suggest good agreement with the case reports, provided that $b_1 > 0.28$. The correlation dimensions for mumps also suggest $b_1 > 0.28$. However, we note that there is no significant difference between the simulation for $N = 1,000,000$ and that for $N = 5,000,000$ although there is a clear difference between the two time series. Both dimensions are higher than 3, suggesting periodic oscillations perturbed by noise[9,10,12,13]. This could be interpreted in terms of a switching between two coexisting periodic attractors for the simulation with $N = 1,000,000$. A similar explanation may apply to mumps in Copenhagen.

NONLINEAR FORECASTING

Chaotic systems possess a property called 'sensitive dependence on initial conditions[28],' and this property precludes long-term predictions. Nevertheless attempts to predict the future in a chaotic time series can provide useful information about the system generating the time series, and may eventually be used as an alternative procedure for identifying chaotic dynamics. Special routines, called nonlinear forecasting programs, have been developed for such predictions[15,29,30]. Basically these programs finds states in the neighborhood of a point, X(t), whose future is to be predicted. These neighboring points are chosen from previous records of states. Then a local chart that maps points of the neighborhood into their future values is fitted. To make a prediction we evaluate the chart to X(t).

The procedure used here is similar to zero-order local approximation methods described by Farmer and Sidorowitch[29] (see also Casdagli[31]). Sugihara and May[15] applied a slightly different nonlinear forecasting algorithm to the monthly notifications of chicken pox and measles in New York from 1928 to 1972. They used the first half of this data set to generate a library of patterns which then were used to make predictions of each of the points in the second half. For measles they showed that the prediction error increases rapidly with increasing prediction time, as expected if the dynamics were chaotic. For chicken pox the prediction error was essentially unaffected by the length of the prediction period, as expected if the dynamics corresponded to a noisy periodic oscillation.

Here we use nonlinear forecasting as a further test of the validity of the SEIR Monte Carlo simulations. This is done by using 500 years of simulated monthly data as a library for predicting the future of the case reports following normalization of both data sets on the interval [0,1]. As proposed by Sugihara and May[15], the actual time series used were the first difference transformations, $X_{i+1} - X_i$. Otherwise the data were not subjected to any form of smoothing or interpolation. The data were then lagged and embedded in dimension E.

Neighboring points were chosen as those contained in an E-dimensional ball with radius ϵ. Fig. 7 shows a plot of predicted vs. observed for a combination of Monte Carlo measles simulations with different values of b_1 and monthly notifications of measles in Baltimore from 1928 to 1963. We note that the correlation coefficient between predicted and observed values increases with increasing b_1 such that the highest correlation is obtained for $b_1 > 0.28$. Similar good correlations between simulated data and case reports were obtained for measles in Copenhagen and New York also with optimum correlations for $b_1 > 0.28$. These results are therefore in agreement with those in the preceding section.

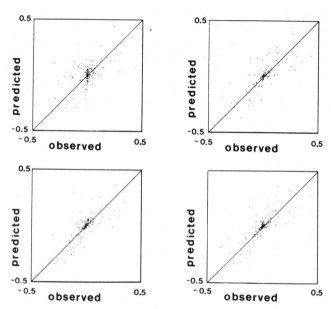

Fig. 7 Plots of predicted vs. observed for nonlinear forecastings of first difference transformed $(X_{i+1} - X_i)$ case reports of measles in Baltimore 1928-1963 using 500 years of Monte Carlo measles simulations with a) $b_1 = 0.0$, b) $b_1 = 0.20$, c) $b_1 = 0.28$ and d) $b_1 = 0.36$ as a library for prediction. Embedding dimension $E=4$, embedding delay $\tau = 1$ month and prediction time $T_p = 1$ month. The squared correlation coefficients r^2 between predicted and observed were a) 0.24, b) 0.51, c) 0.58 and d) 0.61.

NONSTABLE PERIODIC MOTIONS IN SEIR CHAOS

One of the interesting things about chaotic attractors is that they are organized around a countable infinity of nonstable periodic orbits. Sometimes[32] one can show that the periodic orbits are "dense," i.e. that every point on the attractor lies within an arbitrarily small distance of such an orbit. Then by enumerating these orbits one obtains a complete description of the dynamics[33-37]. At a more phenomenological level, nonstable periodic orbits show up as transitory episodes of approximately periodic behavior in otherwise irregular time series. Just such behavior is observed in the SEIR model with parameters appropriate for measles (Fig. 6) as well as in real-world epidemics (Fig. 1). Here, the "periodic" episodes correspond to a rough alternation of high and low years (see, especially, the New York data from 1945 to 1963; also Schaffer[38] and Pool[39]).

Such nearly periodic interludes and their immediate "pre-images" constitute periods of increased predictability for chaotic dynamical systems. We illustrate this finding in Figs. 8-10. Fig 8 shows next-amplitude, i.e. peak-to-peak, maps for the measles attractor given by eqs. (1)-(3) and $b_1 = 0.276$ and $b_1 = 0.28$. These values correspond to chaotic motions just before and after the abrupt transition from small-amplitude to large-amplitude chaos (see Fig. 2, top right). Note that in the case of $b_1 = 0.276$, the chaos is mild, i.e. the system hops around a 4-piece attractor with high and low levels of infectivity alternating from one year to the next. After the transition, the attractor is greatly expanded and the biennial pattern breaks down. Nonetheless, there remains a high concentration of points in the vicinity of the old, low-amplitude attractor.

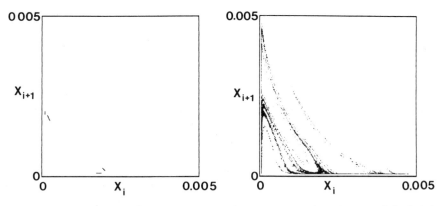

Fig. 8 Next-amplitude maps for the SEIR equations (eqs. (1)-(3)) with $b_1 = 0.276$ (left) and $b_1 = 0.28$ (right). Note the increase in size and complexity of the attractor.

Fig. 9 explores the issue of varying levels of predictability. At the top we compare observed and predicted case rates (right) on up to 14 time units for a non-predictable region of the phase space. Here the initial point of the predicted time series is marked by a "+" at the left. Note that this point is far from the region of biennial oscillations. At the bottom we perform a similar analysis on a point close to the former low-amplitude attractor. Here the system is highly predictable. The improvement in predictability is all the more dramatic if we compare predicted and observed behaviors for the entire phase space (Fig. 10). Here we use a prediction interval of 12 steps of the first difference transformation, $X_{i+1} - X_i$, and we see that the overall predictability is miserable. However, we also note a concentration of points around the 45 degree line. These points correspond to the nonstable 2^n-periodic attractors of Fig. 8. We also emphasize that there are other regions of high predictability, corresponding, we believe, to other nonstable periodic orbits. In sum, detecting nonstable periodic orbits may provide the basis for generating medium-range forecasts for systems that are overall highly non-predictable.

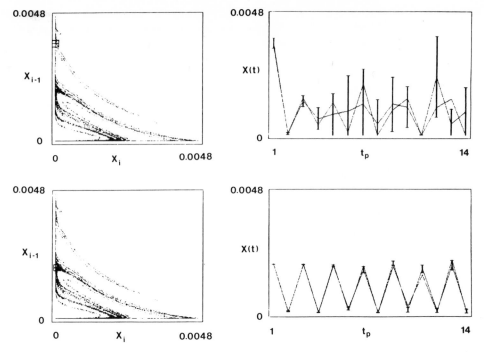

Fig. 9 Predictability in the SEIR equations with $b_1 = 0.28$. (Top) An unpredictable point. The point is marked with a cross at the left. Predicted and observed dynamics are shown at the right. The error bars indicate the confidence intervals of the prediction. The forecasting process breaks down after four time units. (Bottom). A predictable point in the vicinity of what was formerly the low-amplitude "periodic" attractor. Highly accurate forecasts are obtaineable for 14 time steps. A total of 10,000 maxima of $I(t)$ were accumulated. Of these, 2,000 were used to generate predictions for the first point, 2001 for the next point, etc. The embedding dimension was set to 2 and the size of the epsilon ball to 0.002.

CONCLUSION

We have presented nonlinear analyses of chicken pox, measles and mumps notifications in eight North American and European cities and corresponding simulations of the SEIR model. Our analyses suggest that chicken pox notifications correspond to a noisy limit cycle with a period of one year, whereas measles notifications are chaotic. Mumps notifications behave differently: In some cities, e.g., New York and Milwaukee, they correspond to a periodic oscillation, whereas in others, e.g., Copenhagen, they correspond to irregular switchings between coexisting periodic attractors. Our conclusion that measles notifications are chaotic differ from those of others, who propose a periodic oscillation consisting of alternating high and low incidence rates and with a period of two years[16,27,40,41]. Much of the controversy hinges on the estimate of b_1. London and Yorke[16] estimated a value of 0.25 to 0.30. However, others[27] claim a value of around 0.20 or less. Our nonlinear analyses of the SEIR simulations suggest $b_1 > 0.28$, assuming that the simple SEIR scheme is correct, which is hardly to be anticipated. For example it was shown recently that the birth rate, which in the SEIR model is assumed to be constant, shows an annual rhythm[42,43]. It is to be expected that such an additional seasonal component would reduce the value of b_1 giving chaotic fluctuations. Thus, evidence against chaos cannot be given in terms of an estimate of b_1. Finally, we have shown that detecting unstable periodic orbits may facilitate medium-range predictions of chaotic systems. We are currently exploring this possibility for the case reports and Monte Carlo simulations of measles.

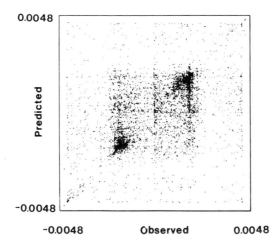

Fig. 10 Predicted vs. observed for the entire SEIR attractor with $b_1 = 0.28$. The data used were the first difference transformations ($X_{i+1} - X_i$). The prediction interval is 12 time units and overall predictability is low ($r_2 = 0.13$). Note the concentrations of points near the 45 degree line which correspond to regions of nearly periodic behavior such as that shown in Fig. 9.

ACKNOWLEDGEMENTS

This research was supported by grants from the Danish Natural Science Research Council to LFO and CGS (grant nos. 11-7421 and 11-8009) and from the National Science Foundation and the National Institutes of Health (USA) to WMS. We thank Dr. Till Roenneberg, Munich University, for sending us preprints of his papers prior to publication.

REFERENCES

1. M.S. Bartlett, "Stochastic Population Models in Ecology and Epidemiology," Methuen, London (1960).
2. N.T.J. Bailey, "The Mathematical Theory of Infectious Diseases and Its Applications," Griffin, London (1975).
3. R.M. Anderson, ed., "Population Dynamics of Infectious Diseases. Theory and Applications," Chapman and Hall, New York (1982).
4. L.F. Olsen and H. Degn, Chaos in biological systems, Q. Rev. Biophys. 18:165 (1985).

5. A.V. Holden, ed., "Chaos," Manchester University Press, Manchester (1986).

6. H. Degn, A.V. Holden and L.F. Olsen, eds., "Chaos in Biological Systems," NATO ASI Series, Series A, Vol. 138, Plenum, New York (1987).

7. W.M. Schaffer and M. Kot, Nearly one-dimensional dynamics in an epidemic, J. Theor. Biol. 112:403 (1985).

8. W.M. Schaffer, Can nonlinear dynamics elucidate mechanisms in ecology and epidemiology ?, IMA J. Math. Appl. Med. Biol. 2:221 (1985).

9. W.M. Schaffer, L.F. Olsen, G.L. Truty, S.L. Fulmer and D.J. Graser, Periodic and chaotic dynamics in childhood infections, in: "From Chemical to Biological Organization," M. Markus, S.C. Müller and G. Nicolis, eds., Springer-Verlag, Berlin (1988).

10. L.F. Olsen, G.L. Truty, and W.M. Schaffer, Oscillations and chaos in epidemics: A nonlinear dynamic study of six childhood diseases in Copenhagen, Denmark, Theor. Pop. Biol. 33:344 (1988).

11. M. Kot, W.M. Schaffer, G.L. Truty, D.J. Graser and L.F. Olsen, Changing criteria for imposing order, Ecol. Modelling 43:75 (1988).

12. W.M. Schaffer, L.F. Olsen, G.L. Truty and S.L. Fulmer, The case for chaos in childhood epidemics, in: "The Ubiquity of Chaos," S. Krasner, ed., American Association for the Advancement of Science, Washington (1990).

13. L.F. Olsen and W.M. Schaffer, Chaos versus noisy periodicity: Alternative hypotheses for childhood epidemics, Science 249:499 (1990).

14. F. Drepper, Unstable determinism in the information production profile of an epidemiological time series, in: "Ecodynamics," W. Wolff, C.-J. Soeder and F.R. Drepper, eds., Springer-Verlag, Berlin (1988).

15. G. Sugihara and R.M. May, Non-linear forecasting as a way of distinguishing chaos from measurement error in time-series, Nature 344:734 (1990).

16. W.P. London and J.A. Yorke, Recurrent outbreaks of measles, chicken pox and mumps. I. Seasonal variations in contact rates, Am. J. Epidemiol. 98:453 (1973).

17. "Medical Report for the Kingdom of Denmark," The National Health Service of Denmark, Copenhagen (1927-1968).

18. A. Wolf, J.B. Swift, H.L. Swinney and J. A. Vastano, Determining Lyapunov exponents from a time series, Physica 16D:285 (1985).

19. P. Grassberger and I. Procaccia, Measuring the strangeness of strange attractors, Physica 9D:189 (1983).

20. A.R. Osborne and A. Provenzale, Finite correlation dimension for stochastic systems with power law spectra, Physica 35D:357 (1989)

21. K. Dietz, The incidence of infectious diseases under the influence of seasonal fluctuations, in: "Lecture Notes in Biomathematics," Vol. 11, Springer-Verlag, Berlin (1976).

22. R.M. Anderson and R.M. May, Directly transmitted infectious diseases: Control by vaccination, Science 215:1053 (1982).

23. R.M. Anderson and R.M. May, Age-related changes in the rate of disease transmission: Implications for the design of vaccination programmes, J. Hyg. 94:365 (1985).

24. P.E.M. Fine and J.A. Clarkson, Measles in England and Wales. I. An analysis of factors underlying seasonal patterns, Int. J. Epidemiol. 11:5 (1982).

25. I.B. Schwartz and H.L. Smith, Multiple recurrent outbreaks and predictability in seasonally forced nonlinear epidemic models, J. Math. Biol. 18:233 (1983).

26. H. Smith, Periodic solutions for a class of epidemic equations, J. Math. Anal. Appl. 64:476 (1983).

27. I.B. Schwartz, Nonlinear dynamics of seasonally driven epidemic models, in: "Biomedical Modeling and Simulation," J. Eisenfeld and D.S. Levin, eds., J.C. Baltzer, Berlin (1989).

28. D. Ruelle, Sensitive dependence on initial conditions and turbulent behavior in dynamical systems, Ann. N.Y. Acad. Sci. 316:408 (1979).

29. J.D. Farmer and J.J. Sidorowich, Predicting chaotic time series, Phys. Rev. Lett. 59:845 (1987).

30. J.D. Farmer and J.J. Sidorowich, Exploiting chaos to predict the future and reduce noise, preprint (1988).

31. M. Casdagli, Nonlinear prediction of chaotic time series, Physica 35D:335 (1989).

32. U. Kirchgraber and D. Stoffer, Chaotic behavior in simple dynamical systems, SIAM Review 32:424 (1990).

33. D. Auerbach, P. Cvitanovic, J.-P. Eckmann, G. Gunaratne and I. Procaccia, Exploring chaotic motion through periodic orbits, Phys. Rev. Lett. 58:2387 (1987).

34. P. Cvitanovic, G.H. Gunaratne and I. Procaccia, Topological and metric properties of Hénon-type strange attractors, Phys. Rev. A 38:1503 (1988).

35. C. Grebogi, E. Ott and J.A. Yorke, Unstable periodic orbits and the dimension of strange attractors, Phys. Rev. A 36:3522 (1988).

36. D.P. Lathrop and E.J. Kostelich, Analyzing periodic saddles in experimental strange attractors, in: "Measures of Complexity and Chaos," N.B. Abraham et al., eds., Plenum, New York (1989).

37. L.A. Smith, Quantifying chaos with predictive flows and maps: Locating unstable periodic orbits, in: "Measures of Complexity and Chaos," N.B. Abraham et al., eds., Plenum, New York (1989).

38. W.M. Schaffer, Chaos in Ecology and epidemiology, in: "Chaos in Biological Systems," H. Degn, A.V. Holden and L.F. Olsen, eds., NATO ASI Series, Series A, Vol. 138, Plenum, New York (1987).

39. R. Pool, Is it chaos or is it just noise?, Science 243:25 (1989).

40. R.M. May and R.M. Anderson, Population biology of infectious diseases: Part II, Nature 280: 455 (1979).

41. J.L. Aron and I.B. Schwartz, Seasonality and period-doubling bifurcations in an epidemic model, J. Theor. Biol. 110:665 (1984).

42. T. Roenneberg and J. Aschoff, Annual rhythm of human reproduction: I. Biology, sociology or both?, preprint (1989).

43. T. Roenneberg and J. Aschoff, Annual rhythm of human reproduction: II. Environmental correlations, preprint (1989).

MULTIFRACTAL ANALYSIS OF MORPHOLOGICAL PATTERNS IN NORMAL AND

MALIGNANT HUMAN TISSUES

Jiri Muller and Jon P. Rambæk

Institute for energiteknikk
Box 40, 2007 Kjeller, Norway

O.P.F. Clausen and Torstein Hovig

Institute of pathology
National Hospital
Pilestredet 32
0027 Oslo 1, Norway

ABSTRACT

Optical microscope pictures of thin histological sections of human tissues from normal organs and from organs with malignant neoplasms have been digitized and numerically studied by multifractal analysis. The results show a marked difference in the behaviour of the multifractal scaling exponents between normal and malignant tissues which could be exploited to characterize various stages of cancer development.

1. Introduction

Interpretation of morphological patterns is of importance in understanding and analysing topology of objects occuring in nature. As many objects in nature obey fractal scaling laws (Mandelbrot 1982), it is therefore natural to use fractal analysis for interpretation and understanding of such patterns.

We have successfully used fractal and multifractal analysis in morphological studies of pore space in oil bearing sedimentary rocks (Muller et al. 1990a). Motivated by these results, we have decided to apply this method to study morphology of cells in malignant tumours and compare it with morphology of normal cells from the same human tissue.

We would like to stress that the present results are only preliminary, and that a more systematic and detailed study will be reported elsewhere.

Complexity, Chaos, and Biological Evolution, Edited by E. Mosekilde and
L. Mosekilde, Plenum Press, New York, 1991

2. Method

Optical microscope pictures of thin histological sections of normal human tissues and malignant neoplasms originating from the same tissue were recorded by video frame grabber with 512 x 512 pixel resolution and stored in a digital form in a computer workstation. This set-up is sketched in Figure 1.

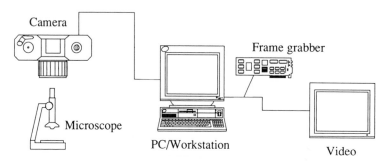

Figure 1. Schematic representation of the experimental set-up for digital image analysis

The digital pictures were then analysed by a multifractal computer programme based on box-counting algorithm described elsewhere (McCauley 1989). The range where the multifractal scaling is valid was found to be in the region from about 16 to 128 pixels. We have found a similar pixel range when analysing multifractal distribution various disordred materials (Muller et al. 1990a,b). The lower limit of 16 pixels is due to the limit of the resolution of our photographs. The details of our multifractal analysis are summarized in the Appendix and are described in further detail by McCauley (1989,1990).

3. Results and Discussion

In the present work, we report the results on the analysis of normal brain tissue and lymph nodes and malignant neoplasms developing in these tissues. The right-hand side of Figure 2a,b shows the microscope pictures of the studied tissues from which we observe a well-known clustering of cells when a normal healthy tissue develops into a malignant tumour.

We have performed a simple one scale clustering analysis on these pictures based on coloured graphics (Muller et al., to be published), which elucidates the clustering effect in clearer details. However, a full multifractal analysis characterized by a series of α and $f(\alpha)$ exponents offers more information than a simple cluster analysis based on only one scaling exponent (e.g. McCauley 1990, Schertzer and Lovejoy 1990).

This can be seen in Figure 2a, b, where we display multifractal $f(\alpha)$
spectra of the four samples. While the tops of the curves
(characterized by D_0 or the so-called Hausdorff Besicovitch dimension) for normal and malignant species do not differ very much, there
is a marked difference in the behaviour of the left-hand sides of
these curves. In particular, the differences in size distribution of
cell clusters in normal and malignant tissues are reflected in the
value of the end points α_{min} and $f(\alpha_{min})$, both being increased as we
go from the normal tissue to the malignant tumour. These findings are
of particular interest because it is well known that left-hand sides
of $f(\alpha)$ curves are reasonably accurate and thus could be used as
numerical indicators of the clustering effect, therefore indicating
various stages of development of malignant tumours.

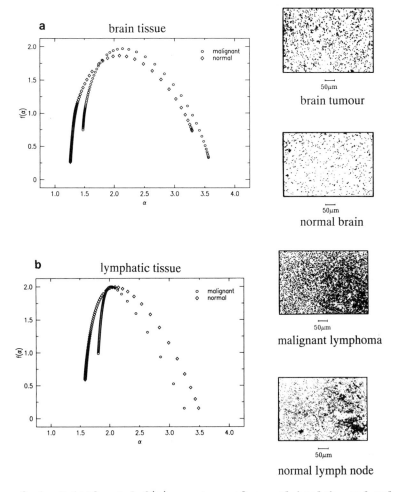

Figure 2a,b. Multifractal $f(\alpha)$ spectrum of normal healthy and malignant tissues of human brain and lymphatic tissue.

In this context it is also of interest to use fractal methods together with image analysis of transmission electron microscope pictures to study differences in nuclear texture of individual malignant and normal cells from the same human tissue. These ideas are presently being investigated.

4. Appendix - Summary of multifractal analysis

Each digitized image is covered by N_n boxes of size ι_n, where the subscript n indicates the nth generation scale. We associate the contents of each box with the fraction P_i of cells that it contains. The generating function X defines the multifractal scaling in terms of scaling exponents α and $f(\alpha)$, and is given by

$$X(q) = \iota_n^{q\alpha - f(\alpha)} \qquad (1)$$

and can be calculated from digitized images by using

$$X(q) = \sum_{i=1}^{N_n} P_i^q \qquad (2)$$

The quantities α and $f(\alpha)$ are given according to the equation

$$\alpha = \tau'(q) = \frac{1}{\ln \iota_n} \left(\sum_{i=1}^{N_n} P_i^q \ln P_i \right) / \sum_{i=1}^{N_n} P_i^q \qquad (3)$$

and the Legendre transformation

$$f(\alpha) = q\alpha(q) - \tau(q) \qquad (4)$$

where

$$\tau(q) = \frac{1}{\ln \iota_n} \ln \left(\sum_{i=1}^{N_n} P_i^q \right). \qquad (5)$$

The procedure for calculating α and $f(\alpha)$ is carried out for n refinements of coarsegraining with q ranging usually from q_{min} (ca -15) to q_{max} (ca 15) in steps of 0.1. Note that a given value of q fixes α through (3), and $f(\alpha)$ through (4) and (5).

The range where the multifractal scaling is valid was found to be in the region from about 16 to 128 pixels with n=4 generation scales. We have observed a similar scaling region with similar distributions of error bars (large error bars for negative q and small error bars for positive q) in our multifractal studies of other disordered materials (Muller et al 1990a,b)

ACKNOWLEDGMENTS

We would like to thank Flavio da Silva, Stuart Kauffman, Arne Skjeltorp, Geir Helgesen and Peter Dvergsdal for valuable comments. This work has been supported in parts by Fina Exploration Norway.

REFERENCES

Mandelbrot B.B, The Fractal Geometry of Nature, Freeman (1982)

Muller J., Hansen J.P., Skjeltorp A.T. and McCauley J.L.,
    Multifractal Phenomena in Porous Rocks, Mat. Res. Symp. Proc.
    vol.176, 719, 1990a Materials Research Society

Muller J. and McCauley J.L., Multifractal Studies of Disordered
    Materials, Mat. Res. Soc. Extended Abstracts (EA-25) 1990b
    Materials Research Society p.147

McCauley J.L., Multifractal Description of the Statistical Equilibrium
    of Chaotic Dynamical Systems, Int. J. Modern Phys. B, 821 (1989).

McCauley J.L., Introduction to Multifractals in Dynamical Systems
    Theory and Fully Developed Turbulence, Physics Reports 189, 5, 225
    (1990)

Schertzer D. and Lovejoy S., Non-linear variability in geophysics:
    analysis and simulations, Fractals, Ed. L. Pietronero, Plenum
    press 49-79 (1990)

METHOD OF COMPATIBLE VECTORFIELD IN STUDYING INTEGRABILITY

OF 3-DIMENSIONAL DYNAMICAL SYSTEMS

Stefan Rauch-Wojciechowski

Department of Mathematics, Linköping University
581 83 Linköping, Sweden

INTRODUCTION

There is an extensive literature about integrable classes of dynamical systems of many different types. However, when it comes to the question of studying integrability of a _given_ system of equations, then there are surprisingly few methods at our disposal.

This paper presents a new method, called here "Method of Compatible Vectorfield" (MCV), which provides a way of finding integrals of motion for a given family of dynamical systems in dimension three. Practical application of this method is discussed through the example of the 3-dimensional Lotka-Volterra system. This presentation is a short description of the ideas and the results of papers [1,2] and claims, essentially, no new results.

Let us consider a 3-dim quadratic autonomous dynamical system of the form

$$\frac{dx_1}{dt} = \sum_{i\,j} a^1{}_{ij} x_i x_j + \sum_i b^1{}_i x_i := P_1(x)$$

$$\frac{dx_2}{dt} = \sum_{i\,j} a^2{}_{ij} x_i x_j + \sum_i b^2{}_i x_i := P_2(x) \qquad (1)$$

$$\frac{dx_3}{dt} = \sum_{i\,j} a^3{}_{ij} x_i x_j + \sum_i b^3{}_i x_i := P_3(x)$$

where summation over repeating indices i,j runs from 1 to 3. This dynamical system is represented by the vectorfield

$$P(x;a,b) = P_1(x)\frac{\partial}{\partial x_1} + P_2(x)\frac{\partial}{\partial x_2} + P_3(x)\frac{\partial}{\partial x_3}$$

For definiteness we shall think of $P_k(x)$ as quadratic functions, but the MCV applies as well to more general vectorfields. However the practical implementation of this method is easier in the case of polynomial and rational $P(x)$.

The process of analytically solving the equations (1)

Complexity, Chaos, and Biological Evolution, Edited by E. Mosekilde and
L. Mosekilde, Plenum Press, New York, 1991

consists of finding time independent integrals of motion. Then every integral of motion reduces the number of independent variables by one. For equations (1) a knowledge of 2 integrals reduces the problem to a first order autonomous ordinary differential equation.

By an integral of motion, we mean here a time independent function $F(x_1, x_2, x_3)$ which is constant on trajectories of P

$$0 = \frac{d}{dt} F(x_1(t), x_2(t), x_3(t)) = (P_1 \frac{\partial}{\partial x_1} + P_2 \frac{\partial}{\partial x_2} + P_3 \frac{\partial}{\partial x_3})F = P(x;a,b)F$$

that is, it vanishes under the action of the vectorfield P. Geometrically, an integral of motion determines a foliation of the phase space, and trajectories of the system (1) stay on the leaves of this foliation. Two functionally independent integrals determine two different foliations of 3-dim phase space, and trajectories of (1) are on 1-dim intersections of their leaves. The first order equation, which remains to be solved, determines the time parametrization of the intersection curves.

The procedure of finding integrals of motion for equations (1) consists of finding integrable values of parameters a^k_{ij}, b^k_i and the corresponding integrals of motion. For arbitrarily chosen a^k_{ij}, b^k_i, it is usually <u>not</u> possible to find an integral which may be expressed by the use of known functions and quadratures (integrations). The best known methods of finding integrals of motion are:

   (i)   ansatz for $F(x;k)$
   (ii)   direct integration of equations (1)
   (iii) Painlevé analysis
   (iv)   Lie symmetry method
   (v)   method of compatible vectorfield

The ansatz method requires a lot of ingenuity, because it is necessary to anticipate the algebraic form of the integral to find the admissible values of the parameters k in the integral. Direct integration of equations (1) is usually possible if there exist some obvious geometrical symmetries. The Painlevé analysis, which consists of expanding the solution x into a Laurent series in complex time, determines only integrable values of parameters a^k_{ij}, b^k_i in equations (1). The integral of motion has to be found by an independent method. The Lie symmetry method [3] requires an ansatz for a symmetry vectorfield which is later used in the calculation of the integral of motion.

Our method of compatible vectorfield(MCV) is a variant of the Lie symmetry method which is based on the Frobenius integrability theorem [4]. In principle the MCV can be reduced to the Lie symmetry method, but the point is <u>not</u> to reduce, since it simplifies considerably all calculations (or even makes them possible).

THE METHOD

The method applies to a given family of vectorfields $P(x;a,b)$ depending on parameters. The main idea is to find a

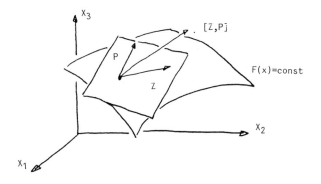

second (compatible) vectorfield Z(x) which satisfies the condition

$$[Z,P] = \alpha(x)P(x) + \beta(x)Z(x) \tag{2}$$

where $[\cdot,\cdot]$ denotes the commutator of vectorfields and $\alpha(x), \beta(x)$ are some functions. Geometrically, the condition (2) means that the vectorfield $[Z,P]$ is in the plane of Z and P. Then for linearly independent Z and P, there is (by the Frobenius theorem) a local foliation of $\mathbf{R}^3$ with leaves tangent to the planes $(Z(x), P(x))$. If the condition (2) is not satisfied, then a continuous distribution of pairs (Z,P) may not have a tangent surface. For instance, $P=\partial/\partial x+z(\partial/\partial y)$ and $Z=\partial/\partial z$ are not compatible, since $[Z,P]=\partial/\partial y$ is linearly independent of P and Z. An equivalent way of writing the condition (2) is

$$\det (Z,P,[Z,P]) = 0 \tag{3}$$

This will be used later for algebraic calculations.
     Notice that for given P the vectorfield Z is not uniquely determined by (2). If Z is a solution of (2), then any vectorfield $Y=f(x)Z+g(x)P$ is also a solution for arbitrary, nontrivial functions $f(x), g(x)$. This nonuniqueness of Z is an advantage, since it leaves us more freedom in choosing an ansatz for $Z(x)$.

     If $Z(x)$ is compatible with $P(x)$, then the integral surfaces of the distribution (Z,P) are level surfaces of the common integral of motion $F(x)$: $PF=ZF=0$. For effective calculation of $F(x)$, we need a vectorfield Z with two explicitly known integrals: $u(x), v(x)$ $(Zu=Zv=0)$. Then the common integral $F(x)=F(u(x),v(x))$ depends on x through u and v and satisfies

$$0 = \frac{dF}{dt} = PF = \frac{\partial F}{\partial u} Pu + \frac{\partial F}{\partial v} Pv$$

This means that $F(u,v)$ is a solution of the first order equation

$$\frac{du}{dv} = G(u,v) \tag{4}$$

385

where $G(u,v)=Pu/Pv$. The quotient Pu/Pv is a function of u and v only since $Z(Pu/Pv)=0$.

Thus the procedure of determining an integral of motion consists of three steps:

(a) finding a compatible vectorfield Z
(b) calculating two integrals u,v of Z
(c) solving the 1st order equation (4)

Finding a compatible vectorfield Z requires an ansatz for Z, and Z has to be such that it is possible to calculate explicitly u and v. For this reason Z is usually assumed to be linear. Solving equation (4) is also a nontrivial step since, (4) is a nonautonomous equation and it may be impossible to express its solution in terms of quadratures. Then the integral of motion is defined implicitly as a solution of (4). Thus the method provides us with a way to handle new kinds of integrals which are not given by a formula.
  The method of compatible vectorfield has clear advantages over the ordinary ansatz method for an integral $F(x;k)$. An ordinary ansatz is like solving a complicated differential equation by a single ingenious guess. Here this guess is replaced by a systematic procedure: the linear ansatz for Z and integration of (4). The linear ansatz looks simple, but the integration of (4) leads to nontrivial integrals of motion. Even the linear ansatz for Z is not a particular limitation, due to the nonuniqueness of the compatible vectorfield Z. The assumption that Z is linear means that in the set of all solutions $Y=f(x)Z+g(x)P$ of (2) there is a linear vectorfield.

The MCV is closely related to the Lie symmetry method which seeks a symmetry vectorfield Y satisfying

$$[Y,P] = \gamma(x)P \tag{5}$$

The vectofield Y is of course compatible with P. On the other hand, the compatibility equation (2) can be reduced to (5) by taking $Y=f(x)Z$ where $f(x)$ satisfies $Pf-\beta f=0$ since

$$[Y,P] = f[Z,P] - (Pf)Z = f\alpha P + (f\beta - Pf)Z = f\alpha P$$

But the function $f(x)$ is not known ($\beta(x)$ as well), and this is the reason why the compatibility condition (2) is more convenient in applications. Moreover, in most of the cases when a nonlinear symmetry vectorfield is known [3], it appears that it has the form $Y=f(x)Z$ with linear Z.

APPLICATION TO THE LOTKA-VOLTERRA EQUATIONS

We consider here the Lotka-Volterra (L-V) equations of the form

$$\frac{dx_1}{dt} = x_1(Cx_2 + x_3 - \lambda)$$
$$\frac{dx_2}{dt} = x_2(x_1 + Ax_3 - \mu) \tag{6}$$
$$\frac{dx_3}{dt} = x_3(Bx_1 + x_2 - \nu)$$

which depends on 6 parameters $A,B,C,$ λ,μ,ν . These equations leave the planes xy, yz, zx invariant and are symmetric with respect to all reflections in these planes. This means that it is sufficient to consider L-V equations only in the first (positive) octant of $\mathbf{R}^3$. The compatibility condition $0=\det(Z,P,[Z,P])$ is a fifth order polynomial in x_1,x_2,x_3 and leads to $21+15+10=46$ algebraic equations on the parameters of the L-V equations. These equations are quadratic and factorize. A complete analysis of all integrable cases has been performed by the use of computer algebra. There are 16 integrable cases, most of which are new. Eleven of these cases have two integrals of motion. The second integral of motion has been determined either from the knowledge of a second independent compatible vectorfield or by the use of the Jacobi last multiplier method.

The Jacobi last multiplier method can be used when a dynamical system has a function $M(x)$ (called an invariant measure) satisfying the condition of vanishing divergence

$$0 = \partial(MP_1)/\partial x_1 + \partial(MP_2)/\partial x_2 + \partial(MP_3)/\partial x_3$$

and when in addition one integral of motion $F(x)$ is known. If this integral $F(x_1,x_2,x_3)=F_0$ can be (locally) resolved with respect to $x_3=\bar{F}(x_1,x_2,F_0)$, then we can calculate a second integral of motion

$$\phi(x_1,x_2,x_3) = \phi(x_1,x_2,\bar{F}(x_1,x_2,F_0)) = \bar{\phi}(x_1,x_2,F_0)$$

as a function of x_1,x_2,F_0. In this case, the expression

$$d\bar{\phi} = \bar{M}(\overline{\partial F/\partial x_3})^{-1}(\bar{P}_1 dx_2 - \bar{P}_2 dx_1)$$

where bar means that everywhere x_3 has been substituted by $\bar{F}(x_1,x_2,F_0)$, appears to be a complete differential in terms of x_1,x_2 only. After integration, F_0 has to be substituted by $F_0=F(x_1,x_2,x_3)$ to obtain the full form of the integral of motion. We shall illustrate the above discussion by the example of the case 4 (from [2]) of integrability of the L-V equations

Example: L-V equations with $ABC+1=0$ and $\nu=\mu B-\lambda AB$

It is easy to check that this L-V vectorfield is compatible with the linear vectorfield $Z=x_2(\partial/\partial x_2)+Bx_3(\partial/\partial x_3)$ which has two integrals $u=x_1$, $v=x_2^B x_3^{-1}$. Further $Pu=u(Cx_2+x_3-\lambda)$, $Pv=v(ABx_3-x_2+\nu-B\mu)$ and by using the conditions $ABC+1=0$, $\nu=\mu B-\lambda AB$ we reduce equation (4) to

$$du/dv = u/ABv$$

Its integral is: $u^{AB}v^{-1}=x_1^{AB}x_2^{-B}x_3=F_0$.

The second integral is simpler to calculate for the sub-case $\lambda=\mu=\nu=0$. Then $x_3=F_0 x_1^{-AB}x_2^B$, $\partial F_0/\partial x_3=x_1^{AB}x_2^{-B}$ and

$$d\bar{\phi} = CF_0^{-1}dx_2 - F_0^{-1}dx_1 + x_1^{-AB}x_2^{B-1}dx_2 + x_1^{-AB-1}x_2^B dx_1$$

yields

$$\phi(x_1,x_2,x_3) = \overline{\phi}(x_1,x_2,F_0) = (Cx_2 - x_1)F_0^{-1} + B^{-1}x_1^{-AB}x_2^{B} =$$
$$= -(ABF_0)^{-1}(ABx_1 + x_2 - Ax_3)$$

So the second functionally independent integral is $F_1 = ABx_1 + x_2 - Ax_3$.

CONCLUDING REMARKS

An analytic approach to solving a given dynamical system consists of two main steps:

(a) transformations of variables in the eqations to reduce them to simpler (or canonical) form in order to recognize the character of the equations
(b) reduction of the number of dependent variables

The reduction (b) is performed by the use of integrals of motion. The method of compatible vectorfield allows for determination of integrable values of parameters in the dynamical system and calculation of an integral of motion. In this method a direct ansatz for an integral of motion is replaced by a two-step procedure of finding linear compatible vectorfield Z and integrating the 1st order ordinary differential equation (4). The MCV is best suited for 3-dimensional dynamical systems of polynomial and rational character. It has been succesfully applied to the L-V equations, the May equations and the Lorentz system [1,2]. It would be worthwhile to use this method to study some other ecological systems and chemical reaction systems.

Acknowledgment: I would like to thank dr.P.Basarab-Horwath for reading the manuscript of this paper.

REFERENCES

[1] J-M.Strelcyn, S.Wojciechowski, Phys.Lett.A 133(1988)207-12
[2] B.Grammaticos, J.Moulin-Ollagnier, A.Ramani,J-M.Strelcyn, S.Wojciechowski, Physica A 163(1990)683-722
[3] L.E.Dickson, Ann.Math.25(1924)287-482
[4] S.Sternberg, Lectures on differential geometry(Chelsea,New York 1983)

DISCRETE STEPS UP THE DYNAMIC HIERARCHY

Gerold Baier

Institute for Chemical Plant Physiology
University of Tübingen
D-7400 Tübingen, FRG

Michael Klein

Institute for Physical and Theoretical Chemistry
University of Tübingen
D-7400 Tübingen, FRG

Abstract

Dissipative continuous and discrete nonlinear systems may show chaotic dynamics. With increasing number of variables new types of complex time evolution and higher chaos are realizable. We give examples of chaotic three– and four–dimensional continuous systems and dynamically equivalent maps. We demonstrate a building block principle for a hierarchy of chaotic attractors in invertible maps.

Introduction

Nonlinear ordinary differential equations (ODEs) are widely used to model the time evolution of experimental systems. In a large number of systems periodic, quasiperiodic, chaotic, and higher chaotic events have been observed. In the following simple three–variable ODE quasiperiodic and chaotic behavior were found for some set of parameters [1]:

$$
\begin{aligned}
\dot{X} &= -Y - Z \\
\dot{Y} &= X \\
\dot{Z} &= a - aY^2 - bZ
\end{aligned}
\qquad (1)
$$

Complexity, Chaos, and Biological Evolution, Edited by E. Mosekilde and
L. Mosekilde, Plenum Press, New York, 1991

Families of two–tori (with two frequencies) can be found in the volume–preserving case $b = 0$. For constant positive dissipation parameter ($b = 0.2$) we study the sequence of attractors as a function of the bifurcation parameter a. For $a < 0.245$, there is a stable limit cycle. This limit cycle undergoes a period doubling sequence to chaos with increasing parameter a. At $a = 0.27675$, there is spiral chaos in the sense of Rössler [1]. Fig. 1a shows the corresponding flow, an expanding spiral with a single fold. The spectrum of Lyapunov characteristic exponents (LCEs) is $[+, 0, -]$. A Poincaré cross–section of the flow consists of a folded one–dimensional line (Fig. 1b). A magnification reveals further foldings and indicates the underlying Cantor set structure of the chaotic flow (Fig. 1c).

Using the technique of Poincaré the invertible two–variable Hénon map can be constructed grasping the dynamic features of the three–variable ODE eq. (1) [2]:

$$x_{n+1} = a - x_n^2 - b\,y_n \qquad (2)$$
$$y_{n+1} = x_n$$

In the volume–preserving case ($b = 1$) this map exhibits a family of closed loops. For $|\,b\,| < 1$ the map is dissipative giving rise to a variety of attractors. E.g. with $a = 2.05$ and $b = 0.2$, a chaotic attractor is found reminiscent of the cross–section in Fig. 1b and 1c (see Fig. 2a and 2b).

A Four–Variable ODE

In higher–dimensional systems we expect new types of dynamics to appear. We now provide an example of a four–variable flow derived from eq. (1). A linear harmonic óscillator (variables Z and W) is added to eq. (1) and the new variable W is linearly coupled back to variable X [3]:

$$\dot{X} = -Y - 0.15 \cdot Z - 0.25 \cdot W$$
$$\dot{Y} = X \qquad (3)$$
$$\dot{Z} = a - a\,Y^2 - 0.1 \cdot W$$
$$\dot{W} = 0.1 \cdot Z - b\,W$$

The dissipative system ($b > 0$) shows dynamical behavior clearly exceeding the complexity of eq. (1). For $a = 0.35$ and $b = 0.05$, we find an attracting two–torus. Increasing a we observe a sequence of torus–doublings, "two–band" chaos, and finally "one–band" chaos at $a = 0.391$. A projection of the "one band" chaotic attractor is displayed in Fig. 3a. Mere inspection of this projection does not reveal the nature of the flow, however. The appropriate tool in such a case is to take a Poincaré cross–section, thus reducing the dimension by one. Fig. 3b presents such a cross–section of the flow in eq. (3) taken at $X = 0$. The cross–section possesses a sheet–like topology equivalent to the spiral chaos in Fig. 1a. The LCE spectrum of the flow is $[+, 0, 0, -]$. We interpret this result as chaos on a distorted (fractalized) three–torus [3].

To model an equivalent dynamical behavior of the cross–section in the four–variable flow (eq. (3)) we designed the following three–dimensional map:

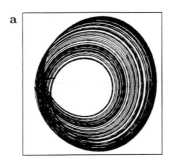

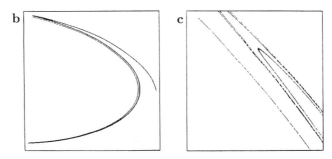

Figure 1. a) Continuous flow: chaotic attractor in eq. 1. $a = 0.27675$, and $b = 0.2$. Projections of three–space with axes: -0.55 ... 0.65 for X, -0.2 ... 0.2 for Y, and -0.5 ... 0.5 for Z. b) Poincaré cross–section at $Y = 0$ in the flow of Fig. 1a. c) magnification of Fig. 1b (upper left corner).

a   b

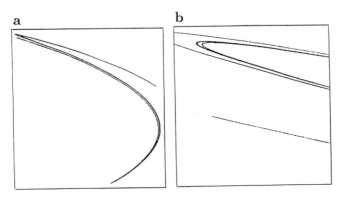

Figure 2. a) Discrete map: chaotic attractor in eq. 2. $a = 2.05$, and $b = 0.2$. State space with axes: -2 ... 2 for x and y. b) Magnification of Fig. 2a (upper left corner). Compare Figs. 1b and 1c.

a   b

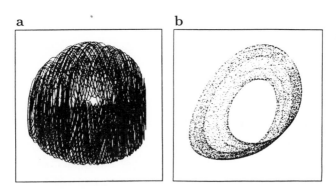

Figure 3. a) Chaos on a distorted three–torus in eq. 3 with $a = 0.391$, and $b = 0.05$ (see [3] for details). Projection of (Y, Z, W)–space with axes: -3 ... 3 for Y, -2 ... 2 for Z, and -3.5 ... 3.5 for W. b) Poincaré cross–section taken at $X = 0$ in the flow of Fig. 3a.

$$x_{n+1} = a - a\,y_n^2 + b\,x_n$$
$$y_{n+1} = y_n - z_n + 0.25 \cdot x_n \tag{4}$$
$$z_{n+1} = y_n$$

There is a two–variable chaotic subsystem in x, y and an additional "harmonic oscillator" in y, z. This construction principle may be compared with the design of eq. (3). The result is as expected: there is an attracting closed loop for small values of parameter a. As a is increased we observe doubling of the loop (corresponding to doubling of tori in the flow of eq. (3)), "two band" chaos and "one band" chaos. The "one band" chaotic attractor (Fig. 4a) is equivalent to the cross–section in Fig. 3b and of spiral chaos topology. A slice of this attractor (Fig. 4b) closely resembles the simple chaos in Fig. 2a.

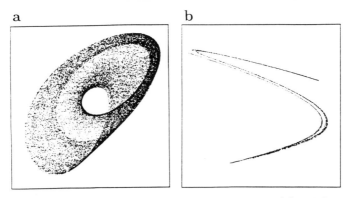

Figure 4. a) Chaotic attractor in the map eq. 4. $a = 1.02$, and $b = 0.8$. y–z plot with axes: -2.5 ... 2 for y and z. b) Slice of the attractor in Fig. 4a taken at $y = 0 \pm 0.007$.

The Discrete Chaotic Hierarchy

We have presented examples of discrete maps which grasp the essentials of the dynamics of chaotic flows. Rössler proposed the idea that with increasing number of variables a hierarchy of chaos (hyperchaos) exist in dissipative systems [4]. Hyperchaos may appropriately be characterized by the existence of more than one positive LCEs. Applying a building–block principle to the Hénon map (eq. (2)) we gain a simple chaos–producing map in arbitrary dimensions [5]:

$$(x_1)_{n+1} = a - (x_{D-1})_n^2 - b\,(x_D)_n \tag{5}$$
$$(x_i)_{n+1} = (x_{i-1})_n$$

for $i = 2, \ldots, D$ with $D \geq 2$; $a, b \in \mathbf{R}$, and $0 < |\,b\,| < 1$. Eq. (5) is a globally invertible D–dimensional discrete map which contracts phase space volume under iteration. The Jacobian equals $b(-1)^D$.

Eq. (5) consists of a single chaos producing quadratic term $(x_{D-1})^2$, $(D-1)$ linear delay variables, a linear coupling term $b \cdot x_D$ (where b is the dissipation parameter), and

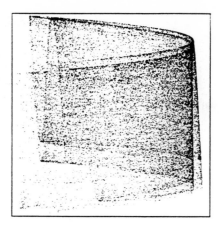

Figure 5. Hyperchaos (folded towel) in eq. (5) with $D = 3$, $a = 1.75$, $b = 0.1$. Projection of three–space with axes: -2 ... 2 for x, y, and z.

the global bifurcation parameter a. By means of the quadratic term the type of chaos generated can be determined. If the square of the delay variable x_{D-1} is iterated the procedure of stretching and folding is successively applied in $(D - 1)$ dimensions due to the coupling of the delay variables. This means that each additional delay creates one more positive LCE. The last variable x_D is coupled back to the first x_1 to yield a globally constant value of the Jacobian and thus a diffeomorphic mapping.

For $D = 2$, eq. (5) yields the Hénon map (eq. (2)). At appropriate values of parameters a and b, it possesses a strange attractor with Cantor set structure in one direction (Fig. 2a and 2b). There is one positive LCE.

For D=3, eq. (5) possesses an attractor of the hyperchaotic type. The attractor shown in Fig. 5 is a folded towel with two positive LCEs [6]. The Hausdorff dimension of the attractor is $2 < D_H < 3$. Slices of the attractor reveal two underlying Hénon–type Cantor sets.

For $D = 4$, the chaotic procedure of stretching and folding may be successively applied in three directions. The fully developed chaotic attractor (Fig. 6a) then possesses three positive LCEs. Its topology might be described as a triply folded hypertowel. The Hausdorff dimension of the attractor is $3 < D_H < 4$. In Fig. 6b a slice of the attractor is presented and it indeed exhibits the expected doubly–folded towel structure.

Summary

Eq. (5) provides a prototypic diffeomorphism for the study of higher chaos in explicit Poincaré maps of flows. A similar equation can be derived for n–tori in volume–preserving maps [7]. Combining these two results on "pure" higher toroidal and higher chaotic dynamics it becomes possible to obtain all types of mixed toroidal and chaotic dynamics. Thus, a set of prototypic maps is available for abstract modelling of low dimensional attractive limit sets, both in theory and for experiments, e.g. hydrodynamic turbulence [8] and brain activity [9].

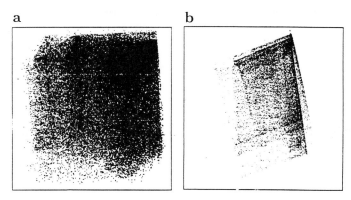

Figure 6. a) Hyperhyperchaos (folded hypertowel) in eq. (5), with $D = 4$, $a = 1.76$, $b = 0.1$. Projection of (x, y, z)–space with Axes: -2 ... 2 for x, y, and z. b) Folded towel structure in a slice of the attractor in Fig. 6a taken at $x_2 = 0 \pm 0.0075$.

References

1. O.E. Rössler, Ann. N.Y. Acad. Sci. **316**, 376 (1979).

2. M. Henon, Commun. Math. Phys. **50**, 69 (1976).

3. G. Baier and M. Klein (eds.): A Chaotic Hierarchy, World Scientific, Singapore, 1990 (to appear).

4. O.E. Rössler, Z. Naturforsch. **38a**, 788 (1983).

5. G. Baier and M. Klein, Phys. Lett. A, in press (1990).

6. O.E. Rössler, Phys. Lett. A, **71**, 155 (1979).

7. G. Baier, M. Klein and O.E. Rössler, Z. Naturforsch. **45**, 664 (1990).

8. A. Brandstatter, J. Swift, H. L. Swinney, A. Wolf, J. D. Farmer, E. Jen, and J. P. Crutchfield, Phys. Rev. Lett. **51**, 1442 (1983).

9. A. Albano et al. and A. Babloyantz, in: Dimensions and Entropies in Chaotic Systems, ed. G. Meyer–Kress (Springer, Berlin, 1986).

HYPERCHAOTIC PHENOMENA IN DYNAMIC DECISION MAKING

Jesper Skovhus Thomsen,[†] Erik Mosekilde[†] and John D. Sterman[‡]

[†]Physics Laboratory III
Technical University of Denmark
DK-2800 Lyngby
Denmark

[‡]Sloan School of Management
Massachusetts Institute of
Technology
Cambridge, MA 02139 USA

ABSTRACT

Most research in psychology and behavioral science focuses on individual choice in static and discrete tasks. In everyday practice, however, we have to deal with much more complicated situations involving time delays, nonlinearities and interactions with other individuals. The purpose of this article is to show how the decision making behavior of real people in simulated corporate environments can lead to chaotic, hyperchaotic and higher-order hyperchaotic phenomena. Characteristic features of these complicated forms of behavior are analyzed with particular emphasis on an interesting form of intermittency in which the trajectory switches apparently at random between two different hyperchaotic solutions.

INTRODUCTION

The discovery of deterministic chaos in many ways represents one of the most significant challenges to our conventional world view since quantum mechanics and the theory of relativity. For 300 years, it has been the underlying presumption of most scientific activity that the behavior of a system could be predicted if the laws of motion and the initial conditions were known with sufficient accuracy, and this presumption has penetrated the philosophical as well as most daily life thinking. We now have to replace the mechanistic world view with a conception in which unpredictability and randomness can be internally generated by the stretching and folding processes of a fully deterministic, nonlinear system. At the same time it has become clear that

Complexity, Chaos, and Biological Evolution, Edited by E. Mosekilde and
L. Mosekilde, Plenum Press, New York, 1991

instability and stochasticity play an essential role in all evolutionary processes, including the formation of form and structure in biological systems.

This may seem like a retreat, a step backwards in understanding. The opposite is true, however. With the development of nonlinear dynamics, we have acquired a variety of new tools which allow us to characterize forms of behavior which previously appeared hopelessly complicated and therefore usually were overlooked or ascribed to exogenous influences. As illustrated by the wealth of detailed examples presented in the preceding chapters of this book, nonlinear dynamics and far-from-equilibrium statistical mechanics are on the way to making revolutionizing contributions to the biological sciences. For the first time we have an idea to work from in our attempts to understand the complexity in structure and behavior that we meet in the living world.

But what about human systems? Are they not complicated as well, and can nonlinear dynamics contribute to a better understanding of their behavior? So far, most research in psychology and behavioral science has focused on individual choice in static and discrete tasks. And microeconomics primarily tells us how to act in particular circumstances, assuming usually complete knowledge of all relevant decision parameters. In the real world, however, human decision makers have to cope with much more complicated situations involving (i) incomplete knowledge about the structure and state of the system in which they operate; (ii) intertemporal couplings such that decisions taken at one instance influence the conditions under which subsequent decisions are to be made; (iii) cooperative, competitive or destructive interactions with other individuals; as well as (iv) mixing and folding phenomena associated with nonlinear constraints.

To explore the nature of human decision making behavior under such conditions, we have performed a series of role playing experiments in simulated microeconomic environments (Sterman 1988, Mosekilde and Larsen 1988, Sterman 1989, Mosekilde et al. 1991). MIT students of management and experienced managers from major U.S. companies were asked to operate a four-stage production-distribution system (the "beer game") consisting of a factory, a distributor, a wholesaler and a retailer. The objective of the individual participant was to minimize costs of maintaining inventories while at the same time avoiding out-of-stock conditions. Performance, however, was systematically suboptimal. By virtue of the built-in delays and nonlinear constraints, many players found that they were unable to secure a stable operation of the system and that, consequently, large-scale fluctuations developed.

Based on post-experiment discussions with the participants and on general concepts from behavioral science (Tversky and Kahneman 1974, Einhorn and Hogarth 1985, Davis et al. 1986), we have proposed a four parameter anchoring and adjustment heuristics for the ordering decisions in the game (Sterman 1988). Econometric estimates have shown that this heuristics can satisfactorily describe the actual decisions (Sterman 1989). Computer simulations of the production-distribution system with the estimated order policies have subsequently revealed a great variety of highly nonlinear dynamic phenomena, including quasiperiodic and chaotic behavior (Mosekilde et al. 1991). For certain regions of parameter space, any neighborhood of a given solution appears to contain a qualitatively different solution, and marginal changes of the ordering policy can completely change the behavior of the system.

The evidence that deterministic chaos can be produced by actual decisions in simplified corporate environments raises a number of issues for economics and social science: How can policy analysis be conducted if changes on the margin produce qualitatively different behavior? To what degree does complexity in behavior slow the discovery of cause and effect relationships by the agents of the economy? How can experience be transferred between apparently similar circumstances? Much further work on nonlinear dynamic phenomena in human systems is clearly required to answer these and related questions. Here, the experimental approach that we have applied in the above studies offers a number of significant advantages (Smith 1982, Roth 1988). First of all, the structure of the system is given. We do not have to discuss the form of the causal relations or the magnitude of the involved time delays. Secondly, the experiments can be repeated under varying circumstances, and we can secure that all data required to interpret the outcome are available.

It turns out, however, that the behavior of our simplified production-distribution system is far more complicated than previously realized. For many parameter combinations, including combinations within the range of realistic policies, the system shows hyperchaotic and higher-order hyperchaotic behaviors. So far, these types of behaviors, which imply that two or more Lyapunov exponents acquire positive values, have only been studied to a very limited extent (Rössler 1979, Rössler and Hudson 1989, Rössler et al. 1990, Kaneko 1990), and in most cases for models which were deliberately constructed to show such behaviors. Developed some 30 years ago at the Sloan School of Management to illustrate how economic structures generate dynamics (Jarmain 1963), on the other hand, the "beer game" has long been used by management schools all over the world, and by now thousands of people have experienced both the aggravation and the fun of trying to control the dynamics of this simple system.

In the present paper we examine the hyperchaotic solutions in greater detail. The largest Lyapunov exponents are calculated as functions of the main parameters of the ordering policy. We also study an interesting form of intermittency where the trajectory apparently at random switches between a hyperchaotic and a higher-order hyperchaotic solution. The paper concludes with a discussion of the significance of our findings for social science.

THE BEER GAME

In order to reach a widespread market, it is customary for breweries (and other industries as well) to utilize a cascaded distribution system with inventory holders at several levels: a distributor receives the beer from the factory and ships it to the main markets; regional wholesalers receive the beer from the distributor and allocate it to local outlets such as liquor stores and bars; and these retailers finally disperse the products for consumption. Besides securing availability of beer at the level of individual customers, the cascaded system is meant to facilitate a swift replacement if a dealer runs low in inventory. The chain should also function as a filter to protect the production line from rapid fluctuations in consumption. Seasonal and other low frequency components in the demand, on the other hand, should propagate towards the factory in a damped fashion.

Figure 1 shows the basic structure of the simplified distribution system considered in our experimental study of human decision making behavior. Orders for beer propagate from right to left, and products are shipped from stage to stage in the opposite direction. To allow hand simulations to be performed, there is only one inventory at each level. The passing of orders, and the production and shipment of beer involve time delays. It is assumed that there is a mailing delay of one week (one time period of the game) from one stage to the next, and in the same way it takes one week to ship beer between two sectors. The production time is taken to be three weeks, and it is assumed that the production capacity of the brewery can be adjusted without limits.

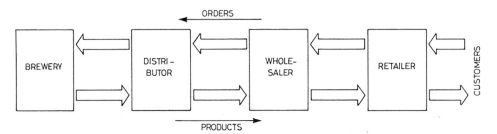

Figure 1. Basic structure of the production-distribution system applied in our experimental study of human decision making behavior. Designed originally as an educational tool, the parameters of the game are chosen to facilitate hand simulation more than to match real-world conditions. All by itself, however, the game constitutes a well-defined managerial system. The detailed procedures for running the game are described by Sterman (1984).

Each week customers order beer from the retailer, who supplies the requested quantity out of inventory. Customer demand is exogenous. In all simulations to be reported here, it consists of a constant demand of four cases of beer per week until week 5, at which time it is stepped up to eight cases of beer per week and maintained at this level for the rest of the game. The game is always initialized in equilibrium with 12 cases of beer in each inventory, and each simulation may be seen as a particular example of the response of the chain to a step increase in demand.

In response to variations in customer demand and to other pressures, the retailer adjusts the orders for beer placed with the wholesaler. As long as the inventory permits, the wholesaler ships the beer requested. Orders which cannot be met are kept in backlog until delivery can take place. Similarly, the wholesaler orders and receives beer from the distributor, who in turn orders and receives from the brewery. According to the rules of the game, orders must always be filled, if they are covered by the available inventory. Orders which have already been placed cannot be canceled, and deliveries cannot be returned.

The objective for the participants is to minimize cumulative sector costs over the length of the game (40 weeks). Because of the costs associated with inventory holding, stocks should be kept as small as possible. On the other hand, failure to deliver on request may force customers to seek alternative suppliers. For this reason, there are also costs associated with having backlogs of unfilled orders. Each stock manager must therefore attempt to keep the inventory at the lowest possible level while at the same time avoiding out-of-stock conditions. If the inventory begins to fall below the desired level, extra beer must be ordered to rebuild inventory. If stock begins to accumulate because of slackening in demand, the order rate must be cut down. Before the hand simulation, the participants are informed that inventory holding costs are $0.50 per case per week, and that the costs of having backlogs are $2.00 per case per week.

For all sectors the decision variable is the amount of beer to be ordered from the immediate supplier in each round. The participants can base their ordering decisions on all information which is locally available to them, i.e., the current value of their inventory or backlog, previous values of these variables, expected orders, anticipated deliveries, etc. In addition, the players can utilize their overall conception of the way in which the distribution chain functions. It is important to realize, however, that we do not assume that the participants moment for moment can maintain a global knowledge about the state of the system. Neither do we assume that they can actually find out how the time delays and nonlinearities of the system influence its dynamics. These assumptions stand in sharp contrast to the assumptions about rational expectations characterizing most economic theory.

To assist the players in developing the necessary information, they are required in each round to plot the values of their effective inventory (inventory minus backlog) and of the orders placed with their supplier. Afterwards, this has allowed us to analyze the course of each game and to estimate the applied decision rules (Sterman 1988, Sterman 1989, Mosekilde et al. 1991). In detail, the simulations, of course, develop differently from trial to trial. Qualitatively, however, the results exhibit a number of significant regularities, which suggests that the participants apply a relatively consistent heuristics to determine their orders.

The hand simulations are characterized by large-scale oscillations which grow in amplitude from retailer to wholesaler and from wholesaler to distributor. Thus, by the time the original stepwise increase in customer orders reaches the factory, it typically leads to an expansion of the production by more than a factor of 6. A second characteristic feature is the wavelike increase in orders which propagates down the chain and depletes the inventories one-by-one to be reflected at the factory, as the large surplus of orders placed during the out-of-stock period is finally produced. These features clearly indicate the existence of an amplification mechanism in the system. At the same time, the behavior is restricted by nonlinearities associated with the non-negativity of orders and shipments.

The amplification phenomenon observed in the hand simulations is connected with the delays built into the distribution system. Assume, for instance, that a particular sector suddenly experiences a significant increase in incoming orders. To find out if this change is of a more permanent character, the player usually hesitates a little before he adjusts his own orders by a similar amount. After all, it is the purpose of the sector inventory to absorb high frequency demand variations. However, because of this hesitation, and by virtue of the built-in mailing and shipping delays, shipments will exceed inventory replacements for several weeks, during which period the inventory will decrease. To bring the inventory back to its desired level, the player must now increase his orders beyond the incoming orders.

Thus, the amplification phenomenon is a natural consequence of the structure of the chain in much the same way as amplification is produced in the real economy by a variety of different accelerator and multiplier mechanisms (Sterman 1985). On the other hand, it is important to realize that the beer distribution chain can be operated in a stable manner. In fact, our experience indicates that many players are capable of doing so, and that only about 25% of the participants produce deterministic chaos or higher-order nonlinear dynamic phenomena (Mosekilde et al. 1991).

THE ORDERING HEURISTICS

The decision task in the beer game is an example of a more general inventory management problem that faces corporate decision makers in any firm. Parts and raw materials must be ordered to maintain inventories sufficient for production to go on, personnel must be hired and laid off to maintain an adequate work force, and units of capital must be ordered or discarded to adjust the production capacity in accordance with present needs. A stock cannot be controlled directly, however, but raw materials and parts must be ordered from the supplier, and the lag in receiving the ordered products is a potential variable. In the beer game, for instance, the retailer will receive the beer requested after 2-3 weeks, only if the wholesaler carries a sufficient inventory. If the wholesaler has run out of stock, the retailer must wait until deliveries have been made from the distributor, etc.

Thus the main question focuses on how participants in the beer game determine their orders. Consistent with behavioral science and with the theory of bounded rationality (Tversky and Kahneman 1974, Einhorn and Hogarth 1985, Davis et al. 1986), we have proposed an anchoring and adjustment heuristics which utilizes information locally available to the individual player only. Surely the participants from time to time watch what happens in the neighboring sectors. However, the spatial arrangement and the pace of the game do not permit them to maintain a complete knowledge about the stock of all sectors. Moreover, in spite of the extreme simplicity of the system compared with real-life managerial systems, the beer distribution chain is much too complicated for the participants to integrate the various pieces of information they have into a globally rational decision rule. Instead, they react to local pressures in terms of growing backlogs, increasing orders, failing supplies, etc.; pressures that they experience in the week-to-week operation of the individual sectors. Based on a variety of practical experiences as well as on theoretical considerations, we believe that this provides a more realistic picture of how human decision makers operate in noisy and complex systems than does the assumption of rational expectations underlying most economic theory.

From discussions with the participants in connection with the debriefing after the game, we have found provision for expected demand to be the main motive (the anchor) in the ordering heuristics. While easy to understand, this motive also agrees well with control theoretical considerations. In equilibrium, where inventory, supply line and other variables are all at their desired values, a stock manager must continue to order enough to cover future demand. Failure to secure sufficient supplies to meet this demand would cause the inventory to fall, reducing the ability of the sector to cope with subsequent orders. Assuming the participants apply adaptive expectations, we express the expected demand as

$$ED_t = \theta \cdot IO_{t-1} + (1-\theta) \cdot ED_{t-1} \tag{1}$$

ED_t and ED_{t-1} are here the expected demand at times t and $t-1$, respectively. IO is incoming orders, and θ $(0 \le \theta \le 1)$ is a parameter that controls the rate at which expectations are updated. $\theta = 0$ corresponds to stationary expectations, and $\theta = 1$ describes a situation in which the immediately preceding value of received orders is used as an estimate of future demand. Econometric analysis of the experimental data for 44 participants shows that θ typically is of the order of 0.25, indicating that expected demand is updated with a smoothing time of about four weeks.

However, provision for expected demand is not sufficient, neither to stabilize the system nor to capture the actually applied ordering policy. Errors in forecasting demand and irregularities in the supply of beer can cause the inventory to wander away from its desired level. Faced with the increasing costs of such a behavior, the stock manager must adjust the orders above or below expected demand so as to bring the inventory back in place. This will introduce a negative feedback that regulates the inventory. For simplicity, stock adjustments are assumed to add a linear correction term

$$AS_t = \alpha_S (DINV - INV_t + BL_t) \tag{2}$$

to the indicated order rate. Here, $DINV$ and INV_t denote desired and actual inventories, respectively, and BL_t is the backlog of orders. The stock adjustment parameter α_S is the fraction of the discrepancy between desired and actual inventory ordered in each round. $DINV$ will, of course, vary from participant to participant. Typically, however, it has a value of 14 cases of beer. A larger desired inventory makes it easier to operate the sector in a stable manner. On the other hand, if stable operation can be achieved, a lower desired inventory tends to give lower costs. Our econometric estimates show that the stock adjustment parameter α_S takes on a spectrum of different values ($0 \leq \alpha_S \leq 1$), depending upon the individual participant. In the following, we shall study the behavior of the distribution chain for different values of α_S.

Adding stock adjustments is still not sufficient for a general ordering policy. In fact, with such a policy, managers would place orders to rectify a lack of inventory, immediately forget that this beer has been ordered, and reorder it in the next round. The problem with such a policy is that it neglects orders which have already been placed, but for which products have not yet been received. This can produce enormous oscillations in the system, as the excess beer requested week after week during an out-of-stock period is finally produced and delivered to the various inventories. Experience shows that many participants take notice of their supply line and try to maintain it at a reasonably stable level. In analogy with the stock adjustments, the supply line adjustments are expressed as

$$ASL_t = \alpha_{SL} (DSL - SL_t) \tag{3}$$

where DSL and SL_t denote the desired and actual supply lines, respectively. α_{SL} is the fractional adjustment rate, i.e., the fraction of the discrepancy between desired and actual supply line ordered in each round.

Defining $\beta = \alpha_{SL}/\alpha_S$ and $Q = DINV + \beta \cdot DSL$, the expression for the indicated order rate becomes

$$IO_t = ED_t + \alpha_S(Q - INV_t + BL_t - \beta \cdot SL_t) \tag{4}$$

Since actual orders must be non-negative, we finally have the order rate

$$O_t = MAX\{0, IO_t\} \tag{5}$$

$DINV$, DSL, and β are all non-negative, implying that $Q \geq 0$. Further, it is unlikely that participants place more emphasis on the supply line than on the inventory itself. The supply line does not directly influence costs, nor is it as salient as the inventory. Consequently, $\alpha_{SL} \leq \alpha_S$, and $\beta \leq 1$. β may be interpreted as the fraction of the supply line taken into account by the participants. If $\beta = 1$, the subjects fully recognize the supply line and do not double order. If $\beta = 0$, orders placed are forgotten until the

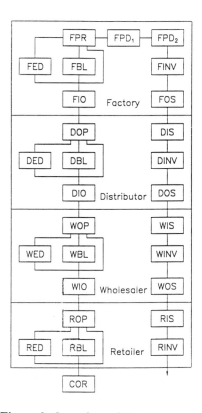

Figure 2. Overview of the structure of the high-dimensional iterated map used to represent the beer model. Altogether the model has 27 state variables. Still it is very simple compared with real-life managerial systems.

beer arrives. In the following we shall see how the behavior of the distribution chain depends on the value of β.

THE SIMULATION MODEL

During a game, the managers of each of the four sectors week by week perform a sequence of operations: receive shipments, advance shipments, receive incoming orders, etc. The corresponding simulation model consists of a high-dimensional iterated map that provides a one-to-one representation of these operations. The structure of the map is illustrated in figure 2, where each box represents a state variable. *COR* depicts the exogenous customer order rate. Other variables are prescribed by a letter indicating the respective sector. Thus R stands for retailer, W for wholesaler, D for distributor and F for factory.

In the wholesale sector, *WINV* is the inventory of beer and *WBL* the backlog of orders. *WIS* and *WOS* represent incoming and outgoing shipments, respectively, and *WIO* is incoming orders. *WED* is the expected demand and *WOP* the orders placed by the wholesaler. One time step later, *WOP* becomes incoming orders to the distributor *DIO*. Similarly, as shipments are advanced, *WOS* becomes incoming shipments to the retailer *RIS*. A similar notation is used in the other sectors, except for the factory, where there is a production rate *FPR* instead of orders placed. The production delay is represented by FPD_1 and FPD_2. In the following, the equations of the model will be presented and explained in terms of the operations conducted in the experiments.

Inventories are updated by adding incoming shipments and subtracting outgoing shipments. To the extent that inventory plus incoming shipments suffice, outgoing shipments are incoming orders plus existing backlog. Otherwise, outgoing shipments are incoming shipments plus inventory, and the new inventory is empty, i.e.

$$WINV_t = WINV_{t-1} + WIS_{t-1} - WBL_{t-1} - WIO_{t-1} \tag{6a}$$

for

$$WINV_{t-1} + WIS_{t-1} \geq WBL_{t-1} + WIO_{t-1}$$

and

$$WINV_t = 0 \quad \text{otherwise.} \tag{6b}$$

Similar expressions hold for *RINV, DINV* and *FINV*. In the same operation, the contents of distributor outgoing shipments are advanced to wholesaler incoming shipments

$$WIS_t = DOS_{t-1} \tag{7}$$

again with similar expressions for *RIS, DIS* and FPD_2.

Backlogs are updated by adding incoming orders and subtracting outgoing shipments. If incoming orders plus backlog are completely covered by incoming shipments plus existing inventory, the new backlog is empty, i.e.

$$WBL_t = WBL_{t-1} + WIO_{t-1} - WINV_{t-1} - WIS_{t-1} \tag{8a}$$

for

$$WBL_{t-1} + WIO_{t-1} \geq WINV_{t-1} + WIS_{t-1}$$

and

$$WBL_t = 0 \quad \text{otherwise.} \tag{8b}$$

Similar expressions hold for *RBL, DBL* and *FBL*. In the same operation, the contents of orders placed by the retail sector are advanced to wholesaler incoming orders

$$WIO_t = ROP_{t-1} \tag{9}$$

with similar expressions for *DIO, FIO* and *FPD*$_1$.

Following the above discussion, outgoing shipments are expressed as

$$WOS_t = MIN\{WINV_{t-1} + WIS_{t-1}, WBL_{t-1} + WIO_{t-1}\} \qquad (10)$$

with similar expressions for *DOS, FOS* and shipments out of retailer's inventory.
Finally expected demand is updated in accordance with Eq. (1).

$$WED_t = \theta \cdot WIO_{t-1} + (1-\theta) \cdot WED_{t-1} \qquad (11)$$

and orders placed are determined in accordance with Eqs. (4) and (5)

$$WOP_t = MAX\{0, WED_t + \alpha_S(Q - WINV_t + WBL_t)$$

$$- \alpha_S\beta(WIS_t + DIO_t + DBL_t + DOS_t)\} \qquad (12)$$

with *WIS + DIO + DBL + DOS* representing the supply line for the wholesaler.
Again, similar expressions hold for the other sectors.

In the experiments (Sterman 1988), each sector manager could apply his own
ordering policy. It appeared, however, that there was little correlation between the
position of a participant in the chain (retailer, wholesaler, distributor or factory) and
the parameters of the applied ordering policy. For simplicity we have therefore
performed the computer simulations of the beer model with the same parameter values
for all four sectors. Moreover, in all simulations to be shown we have taken $\theta = 0.25$
and $Q = 17$ cases of beer.

SIMULATION RESULTS

As described in the previous section, the beer model contains 27 state variables.
Compared with real managerial systems, the model is a strong simplification. It only
considers the flow of a single commodity as opposed to the mutually interacting flows
of materials, personnel and capital units in an actual economy. It supposes that all
sectors operate in synchronism, and it provides a clear and simple purpose for the
individual manager. Thus personal conflicts and opposing goals should not play any
significant role. Compared with most systems investigated in nonlinear dynamics,
however, the model is very complex. In fact, the model produces chaotic behavior for
realistic parameter values with only two sectors. With three sectors, the model
produces hyperchaos, and previous investigations (Mosekilde and Larsen 1988,
Mosekilde et al. 1991) have already revealed a great variety of complex modes in the
four sector model.

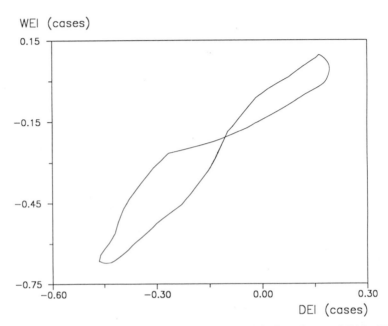

WEI (cases)

DEI (cases)

Figure 3. Quasiperiodic solution obtained for α_S = *0.965* and β = *0.715*. The phase plot shows corresponding values of wholesaler and distributor effective inventories under a very large number of iterations of the 4-sector beer map. For comparison with the following figures, it is worth noticing the very small excursions of the two inventories.

WEI (cases)

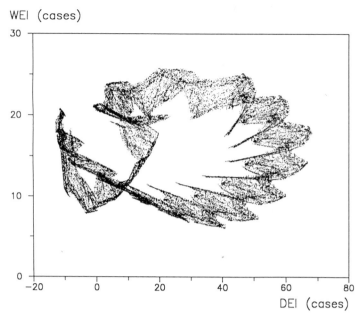

Figure 4. Chaotic attractor observed in the 4-sector beer map for $\alpha_S = 0.45$ and $\beta = 0.075$. More detailed simulations show that the largest Lyapunov is positive $(\lambda_1 = 1.804 \cdot 10^{-2}/weeks)$. The second largest Lyapunov exponent is negative.

As an example of these types of complex behaviors, figure 3 shows the phase plot of the stationary solution obtained for $\alpha_S = 0.965$ and $\beta = 0.715$. Here, the wholesaler effective inventory has been plotted as a function of distributor effective inventory over a long sequence of iterations. As previously noted, effective inventory is inventory minus backlog. Thus, a negative effective inventory corresponds to a backlog of unfilled orders. The fact that the points in the phase plot fall along a one-dimensional curve is indicative of quasiperiodic behavior. For comparison with the subsequent simulations, it is worth noticing the extremely small inventory excursions observed for this solution.

Figure 4 shows an example of a chaotic solution obtained for $\alpha_S = 0.45$ and $\beta = 0.075$. Again, the wholesaler effective inventory has been plotted as a function of distributor effective inventory, but the scale has now been changed by a factor of the order of 100. It is interesting to note the form of this attractor. The points no longer follow a one-dimensional curve, but the iterations produce a folded band somewhat like a nine-fingered flipper. More detailed simulations show that the largest Lyapunov exponent is positive $(\lambda_1 = 1.804 \cdot 10^{-2}/week)$. The second largest Lyapunov exponents is negative, indicating an ordinary form of chaos.

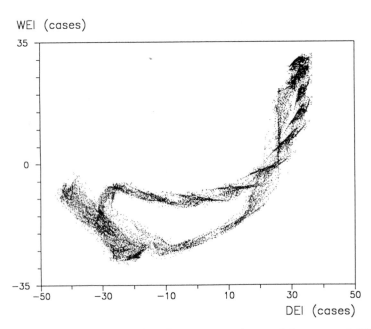

WEI (cases)

DEI (cases)

Figure 5. Phase plot of the hyperchaotic attractor observed for $\alpha_S = 0.655$ and $\beta = 0.54$. Now, the two largest Lyapunov exponents are positive ($\lambda_1 = 7.502 \cdot 10^{-3}$/week and $\lambda_2 = 2.207 \cdot 10^{-3}$/week).

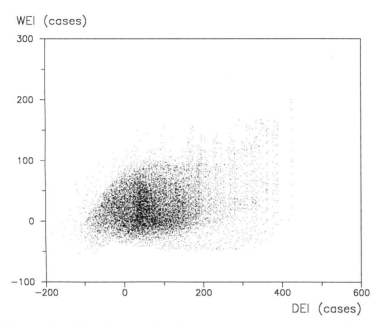

Figure 6. Phase plot of the higher-order hyperchaotic attractor observed for $\alpha_S = 0.75$ and $\beta = 0.175$. This solution has three positive Lyapunov exponents ($\lambda_1 = 4.639 \cdot 10^{-2}/$ week, $\lambda_2 = 2.734 \cdot 10^{-2}/week$ and $\lambda_3 = 1.242 \cdot 10^{-2}/week$). Note the characteristic vertical lines of points.

Figure 5 shows an example of a hyperchaotic attractor obtained for $\alpha_S = 0.655$ and $\beta = 0.54$. Now, the two largest Lyapunov exponents are positive ($\lambda_1 = 7.502 \cdot 10^{-3}$ /week and $\lambda_2 = 2.207 \cdot 10^{-3}/week$) while the third Lyapunov exponent is negative. Compared with figure 4, the phase plot in figure 5 does not appear particularly complicated. Thus, it is not always possible to distinguish hyperchaos from ordinary chaos from the look of a phase plot. Of course, a different projection of the attractor from that presented in figure 5 may reveal more complexity.

Finally, figure 6 shows a higher-order hyperchaotic solution obtained for $\alpha_S = 0.75$ and $\beta = 0.175$. This solution has three positive Lyapunov exponents ($\lambda_1 = 4.639 \cdot 10^{-2}/$ week, $\lambda_2 = 2.734 \cdot 10^{-2}/week$ and $\lambda_3 = 1.242 \cdot 10^{-2}/week$). At the same time, the excursions in phase space are much larger than observed with the previous solutions, and the structure is also very different. The higher order hyperchaotic solution is characterized by vertically running lines of points. Such lines arise when the two inventories are both very much larger than desired. Under such circumstances, the wholesaler does not place any orders with the distributor, and the distributor does not place orders with the factory. As a result, the distributor inventory remains constant while the wholesaler inventory is gradually reduced by shipments to the retailer. Not until the wholesaler inventory is close to its desired value will the wholesaler start to order from the distributor. This type of behavior has also been observed for a time-continuous version of the beer model (Mosekilde and Larsen 1988).

411

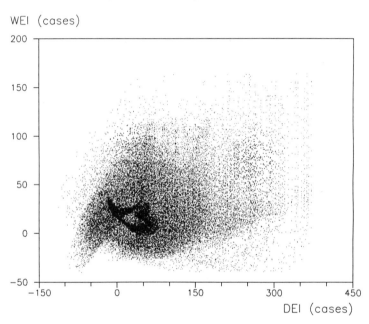

WEI (cases)

DEI (cases)

<u>Figure 7.</u> Phase plot of the stationary solution obtained for $\alpha_S = 0.45$ and $\beta = 0.00$. The system here switches at random between a hyperchaotic and a higher-order hyperchaotic solution.

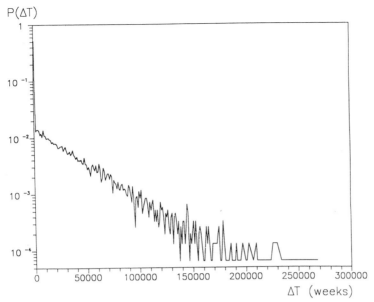

Figure 8. Probability $P(\Delta T)$ that the beer system for $\alpha_S = 0.45$ and $\beta = 0.00$ operates in the hyperchaotic solution for a period ΔT before an explosion into the higher-order hyperchaotic mode takes place. The average lifetime in the hyperchaotic mode is about 33,000 weeks.

As a particularly interesting example of the type of complexity which is met in the beer model, figure 7 shows the stationary solution obtained for $\alpha_S = 0.45$ and $\beta = 0.00$. This pair of parameters is characteristic for subjects which are a little too aggressive in their inventory adjustments and at the same time forget to account for the supply line. The system here switches between a hyperchaotic solution somewhat similar in form to the chaotic solution of figure 4 and a higher-order hyperchaotic solution similar to the solution of figure 6. The model may operate, for instance, for about 47,000 iterations in the hyperchaotic solution to suddenly explode into the higher-order hyperchaotic solution. After another 82,000 iterations, the model may again be caught into the hyperchaotic mode to operate here perhaps for 21,000 iterations before a new explosion occurs. As the system is followed for a longer period of time, it is found to switch apparently at random between the hyperchaotic and the higher-order hyperchaotic modes.

To characterize this intermittency phenomenon in more detail, we have tried to determine the distribution of occupation times in the hyperchaotic solution. With this aim, we have defined an 8-dimensional box in phase space around the hyperchaotic solution. The eight dimensions were chosen to be the backlogs and inventories of the four sectors. By following the system through a couple of billion iterations, we have then determined all intervals in which the model was found to operate within the defined box of phase space for more than 100 iterations. The outcome of this investigation is shown in figure 8, where we have plotted the probability $P(\Delta T)$ that the system operates in the hyperchaotic solution for a period ΔT as a function of ΔT.

WEI (cases)

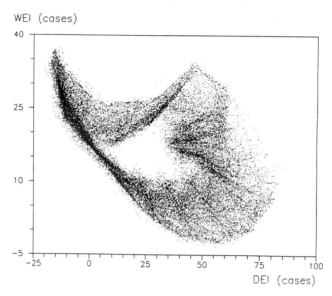

Figure 9. Close-up of the nearly stable hyperchaotic solution in which the map can operate for typically 33,000 iterations before it explodes into the higher-order hyperchaotic solution. The interesting thing is that the system later returns to the much smaller hyperchaotic solution.

It is obvious that $P(\Delta T)$ has a long, nearly exponential tail corresponding to a mean lifetime in the hyperchaotic solution of about 33,000 iterations. For small occupation times there is another and much steeper slope of the $P(\Delta T)$ curve. This could represent the distribution of probabilities that the system operating in the higher-order hyperchaotic mode by chance is confined within the chosen box of phase space for a time ΔT. It is also possible, however, that there is a third nearly stable attractor involved in the intermittency. We are presently investigating this problem in more detail.

To get a better impression of the form of the hyperchaotic solution, we have constructed a phase plot using only those parts of the attractor in which the system is found to operate for 400 or more iterations within the restricted volume of phase space. The result hereof is shown in figure 9.

MODE-DISTRIBUTION IN POLICY SPACE

Figure 10 shows the variation of the largest Lyapunov exponent over the (α_S, β) policy plane for $\theta = 0.25$ and $Q = 17$. Here, the stationary solutions of 200 × 200 simulations have been characterized by means of a grey tone code: light grey indicates stable or periodic behavior for which the largest Lyapunov exponent is negative; dark grey indicates quasiperiodic behavior which is characterized by a vanishing Lyapunov exponent; and black indicates chaos or hyperchaos for which the largest Lyapunov exponent is positive.

414

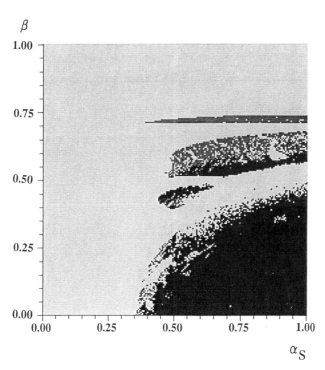

Figure 10. 200 × 200 simulations scan over the (α_S, β) policy plane. Light grey indicates stable or periodic behavior with negative value for the largest Lyapunov exponent; dark grey indicates quasiperiodic behavior; and black indicates chaotic behavior for which the largest Lyapunov exponent is positive.

Inspection of the figure shows that the policy plane contains several regions of aperiodic behavior, separated by fjords of stable or periodic behavior. While the narrow peninsula around $\beta = 0.72$ contains small amplitude quasiperiodic solutions[+] mainly, the other regions of aperiodic behavior are dominated by large amplitude chaotic fluctuations. The occurrence of chaos is most predominant in the lower right

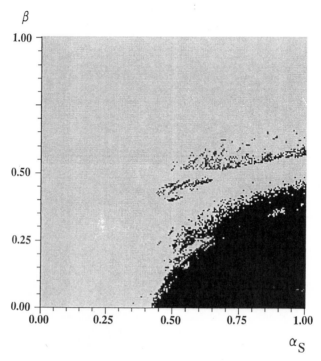

Figure 11. 200×200 simulations scan over the (α_S, β) policy plane for the second largest Lyapunov exponents. Parameter sets which are black yield hyperchaotic solutions.

corner where α_S is large and β relatively small. To stabilize the distribution chain one must apply an ordering policy which accounts for a significant fraction of the supply line and/or which handles inventory discrepancies in a relatively relaxed manner. A particularly interesting feature of the mode distribution in figure 10 is the fractal character of part of the border line between policies that lead to stable or periodic

---

[+] Because of their small amplitude, our automatic mode-recognition procedure has previously mistakenly identified these modes as periodic (Mosekilde et al. 1991).

behavior and policies that lead to chaos. Other parts of this border line are quite regular.

Figures 11 and 12 show the corresponding variations of the second and third largest Lyapunov exponent. As in figure 10, light grey is used to characterize modes with a negative value of the Lyapunov exponent, dark grey modes with vanishing Lyapunov exponent, and black modes with positive exponent. Parameter sets (α_S, β) which in figure 11 are indicated as black thus produce hyperchaotic behavior, while sets

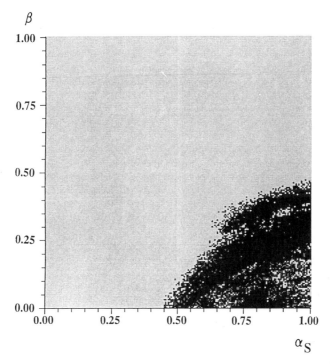

Figure 12. Variation of the third largest Lyapunov exponent over the (α_S, β) plane. Parameter sets that are indicated by a black dot yield higher-order hyperchaotic solutions.

that are indicated as black in figure 12 produce higher-order hyperchaos. It is interesting to note that the higher-order hyperchaotic behavior occurs in bands across the (α_S, β) plane. We have also calculated the next five Lyapunov exponents. In no case have we found more than three positive Lyapunov exponents. We do expect, however, that the number of positive Lyapunov exponents will increase as more sectors are added to the distribution chain.

DISCUSSION

By virtue of their feedback structure and the inherent adjustment delays, corporate systems may oscillate in response to external disturbances (Forrester 1968, Roberts 1978). It is usually assumed, however, that these oscillations are damped, i.e., that managerial systems have stable equilibrium points. The rationale for this is that growing oscillations predetermine a system for collapse and that, consequently, only such systems can have survived which have stable equilibria. This argument is very similar to the argument which has been used to support the idea of homeostasis in biological systems. However, as we know today, the assumption of stable equilibria is unnecessarily restrictive and in many cases even inappropriate for living systems.

The prevalence of deterministic chaos and other forms of complex dynamics in physiological and biological systems naturally motivates a search for similar dynamics in human systems. However, here we are faced with difficulties which do not plague the natural scientist, at least not to the same degree. Human systems are not easily isolated from the environment, they are often unique in structure, and controlled experiments are in most cases impossible. A promising approach is clearly to develop laboratory experiments in which the participants are faced with decisional problems, institutional structures and reward systems similar to those of the real world. This is the approach that we have adopted in our study of human decision making behavior in simulated distribution systems and which has lead us to conclude that deterministic chaos can be produced in simple managerial systems by the locally rational behavior of real people. In the present paper we have demonstrated that hyperchaos and higher-order hyperchaos can also occur. It is particularly interesting to note that the parameter combination $(\alpha_S, \beta) = (0.45, 0.00)$ which leads to the complicated intermittency between two hyperchaotic solutions is quite characteristic for a class of players who, in an attempt to apply a narrow, relatively aggressive inventory control, tend to forget about their supply lines.

These findings are significant in several respects. First of all, our results raise doubts about the conventional wisdom that economic and managerial systems operate at or near equilibrium. Stable equilibria may not exist at all, and non-equilibrium behavior may be as essential to the survival and function of human systems as it is to biological systems. In all circumstances non-equilibrium behavior and randomness are essential in evolutionary processes where qualitative new structures are created. If everyone knew and understood precisely what to do, the world would stop functioning.

Next, realizing that apparently random behavior can be produced by simple and fully deterministic structures must influence the way we distinguish between systematic endogenous processes and random exogenous events. Phenomena which, because of lack of recognizable systematics are considered exogenous, may, in fact, be internally generated. Thus, one of the clearest conclusions that can be drawn from the beer game is that (except for the retailers) all participants believed that the enormous variations in order flows were directly generated by the customer order rate.

Misperception of the dynamics of nonlinear feedback structures lead many to believe that they were following a trajectory in close response to the external signal.

Finally, for nonlinear dynamic systems the assumption of a sort of one-to-one relation between cause and effect implicit in most human logic does not hold. For chaotic systems, a minute perturbation at one instance of time may completely redirect the motion while under other circumstances a substantial blow may be absorbed without significant effect. Clearly, this does not make life easier, neither for the managers of human systems nor for the researchers working with biological and physiological control systems. However, we do believe that it contributes a more truthful perspective of the living world than does the conventional linear paradigm.

ACKNOWLEDGMENTS

We would like to thank Erik Reimer Larsen for a number of stimulating discussions relating to the behavior of the beer distribution system. Janet Sturis is acknowledged for her assistance in preparing the manuscript.

REFERENCES

Davis, H.L., S.J. Hoch, and E.K. Easton Ragsdale (1986): "An Anchoring and Adjustment Model of Spousal Predictions," Journal of Consumer Research 13, pp. 25-37.

Einhorn, H.J. and R.M. Hogarth (1985): "Ambiguity and Uncertainty in Probabilistic Inference," Psychological Review 92, pp. 433-461.

Forrester, J.W. (1968): "Industrial Dynamics," MIT-Press, Cambridge, MA.

Jarmain, W.E. (1963): "Problems in Industrial Dynamics," MIT Press, Cambridge, MA.

Kaneko, K. (1990): "Supertransients, Spatiotemporal Intermittency and Stability of Fully Developed Chaos," Physics Letters 149, pp. 105-112.

Mosekilde, E. and E.R. Larsen (1988): "Deterministic Chaos in the Beer Production-Distribution System," System Dynamics Review 4, pp. 131-147.

Mosekilde, E., E. Larsen, and J. Sterman (1991): "Coping with Complexity: Deterministic Chaos in Human Decisionmaking Behavior" in "Beyond Belief: Randomness, Prediction and Explanation in Science," eds. J.L. Casti and A. Karlqvist, CRC Press, pp. 199-229.

Plott, C.R. (1986): "Rational Choice in Experimental Markets," Journal of Business 59, pp. 301-327.

Roberts, E.B. (1978): "Managerial Applications of System Dynamics," MIT-Press, Cambridge, MA.

Roth, A.E. (1988): "Laboratory Experiments in Economics: A Methodological Overview," Economic Journal 98, pp. 974-1031.

Rössler, O.E. (1979): "An Equation for Hyperchaos," Physics Letters 71A, pp. 155-157.

Rössler, O.E. and J.L. Hudson (1989): "Self-Similarity in Hyperchaotic Data," Springer Series in Brain Dynamics 2, eds. E. Basar and T.H. Bullock, Springer-Verlag, Berlin, pp. 113-121.

Rössler, O.E., J.L. Hudson, M. Klein, and C. Mira (1990): "Self-Similar Basin Boundary in a Continuous System," Nonlinear Dynamics in Engineering Systems, Proc. IUTAM Symposium, Stuttgart, ed. W. Schiehlen, Springer-Verlag, Berlin, pp. 265-273.

Smith, V.L. (1982): "Microeconomic Systems as an Experimental Science," American Economic Review 72, pp. 923-955.

Sterman, J.D. (1984): "Instructions for Running the Beer Distribution Game," Working Paper D-3679, Sloan School of Management, MIT, Cambridge, MA 02139.

Sterman, J.D. (1985): "A Behavioral Model of the Economic Long Wave," Journal of Economic Behavior and Organization 6, pp. 17-53.

Sterman, J.D. (1988): "Deterministic Chaos in Models of Human Behavior: Methodological Issues and Experimental Results," System Dynamics Review 4, pp. 148-178.

Sterman, J.D. (1989): "Modeling Managerial Behavior: Misperceptions of Feedback in a Dynamic Decision Making Experiment," Management Science 35, pp. 321-339.

Tversky, A. and D. Kahneman (1974): "Judgment under Uncertainty: Heuristics and Biases," Science, September 1974, pp. 1124-1131.

PARTICIPANTS

Bjarne Andresen Physics Laboratory, University of Copenhagen, Universitetsparken 5, 2100 Copenhagen Ø, Denmark

Agnessa Babloyantz Service de Chimie Physique, Faculté des Sciences, Campus Plaine, CP 231, U.L.B., B-1050 Bruxelles, Belgium

Gerold Baier Inst. für Chemische Pflanzenphysiologie, Universität Tübingen, Corrensstrasse 41, D-7400 Tübingen, Germany

Clas Blomberg Department of Theoretical Physics, Royal Institute of Technology, S-10044 Stockholm, Sweden

Maarten Boerlyst Bioinformatics Group, University of Utrecht, Padualn 8, 3582 XA Utrecht, Netherlands

Erich Bohl Universität Konstanz, Fakultät für Mathematik, Postfach 5560, D-7750 Konstanz, Germany

H.-G. Busse Universität Kiel, Biochemische Institut, Chr.-Albrechts-Universität zu Kiel, Olshausenstrasse 40 N11, D-2300 Kiel, Germany

Morten Colding-Jørgensen Department of General Physiology and Biophysics, Panum Institute, University of Copenhagen, Blegdamsvej 3c, 2200 Copenahgen, Denmark

Catherine Crawford Cornell University, N 132, MVR Hall, Department of Human Service Studies, Ithaca, NY 14853-4401, USA

Joachim Das Universität Kiel, Biochemische Institut 5, Chr.-Albrechts-Universität zu Kiel, Olshausenstrasse 40-60, D-2300 Kiel, Germany

Hans Degn Biochemical Institute, University of Odense, Campusvej 55, 5230 Odense M, Denmark

Friedhelm Drepper Theoretical Ecology Working Group, Research Center Jülich, P.O. Box 1913, D-5170 Jülich, Germany

Genevieve Dupont Service de Chemie Physique, Faculté des Sciences, Campus Plaine, CP 231, U.L.B., B-1050 Bruxelles, Belgium

Werner Ebeling Sektion Physik, Bereich 04, Humboldt-Universität zu Berlin, Invalidenstrasse 42, 1040 Berlin, Germany

Claus Emmeche Institute of Biological Chemistry B, Sølvgade 83, 1307 Copenhagen K, Denmark

Jacob Engelbrecht Department of Structural Properties of Materials, Bldg. 307, Technical University of Denmark, 2800 Lyngby, Denmark

Leif Eriksson Department of Genetics, Uppsala University, Box 7003, S-75007 Uppsala, Sweden

Doyne Farmer Theoretical Division (T-13), MS B213, Los Alamos National Laboratory, Los Alamos, NM 87545, USA

Brian Goodwin Department of Biology, The Open University, Walton Hall, Milton Keynes, MKT 6AA, U.K.

Björn A. Gottwald Fakultät für Biologie der Universität, Schänzlestrasse 1, D-7800 Freiburg im Breisgau, Germany

Hanspeter Herzel Department of Physics, PB04, Humboldt-Universität zu Berlin, Invalidenstrasse 42, 1040 Berlin, Germany

R.D. Hesch Medizinische Hochschule Hannover, Abteilung Klinische Endokrinologie, Konstanty-Gutschow-Strasse 8, D-3000 Hannover 61, Germany

Niels-Henrik Holstein-Rathlou Department of Physiology and Biophysics, University of Southern California, 1333 San Pablo Street, Los Angeles, CA90033, USA

Axel Hunding Institute of Chemistry C 116, University of Copenhagen, Universitetsparken 5, 2100 Copenhagen Ø, Denmark

Klaus Skovbo Jensen Dansk Bioprotein A/S, Stenhuggervej 7-9, 5230 Odense M, Denmark

Stuart Kauffman Santa Fe Institute, 1120 Canyon Road, Santa Fe, NM 87501, USA

Carsten Knudsen Physics Laboratory III, Bldg. 309, Technical University of Denmark, 2800 Lyngby, Denmark

Heidi Kristensen Physics Laboratory III, Bldg. 309, Technical University of Denmark, 2800 Lyngby, Denmark

Chris Langton Complex Systems Group, Theoretical Division (T-13), MS B213, Los Alamos National Laboratory, Los Alamos, NM 87545, USA

Poul Leyssac Institute of Experimental Medicine, Panum Institute, Bldg. 10.5, University of Copenhagen, Blegdamsvej 3, 2200 Copenhagen Ø

Larry Liebovitch Columbia University, Eye Research, 630 West, 168th Street, New York, NY 10032, USA

Kristian Lindgren NORDITA, Blegdamsvej 17, 2100 Copenhagen Ø, Denmark

David Lloyd University of Wales, College of Cardiff, School of Pure and Applied Biology, Microbiology Group, P.O. Box 915, Cardiff, Wales, U.K.

Henrik Østergaard Madsen Physics Laboratory III, Bldg. 309, Technical University of Denmark, 2800 Lyngby, Denmark

Donald J. Marsh Department of Physics and Biophysics, University of Southern

California, School   of Medicine, 1333 San Pablo Street, Los Angeles, CA 90033, USA

Hans Meinhardt Max-Planck-Institut für Entwicklungsbiologie, Speemannstrasse 35-IV, D-7400 Tübingen, Germany

Federico Moran Dipartimento Bioquimica/Fac. Quimicas, Universidad Complutense de Madrid, E-28040 Madrid, Spain

Erik Mosekilde Physics Laboratory III, Bldg. 309, Technical University of Denmark, 2800 Lyngby, Denmark

Lis Mosekilde Anatomic Institute, University of Aarhus, 8000 Aarhus C, Denmark

Jiri Muller Institute for Energy Technology, Box 40, N-200 7 Kjeller, Norway

Stephan C. Müller Max-Planck-Institut für Ernährungsphysiologie, Rheinlanddamm 201, D-4600 Dortmund I, Germany

Mats Nordahl NORDITA, Blegdamsvej 17, 2100 Copenhagen Ø, Denmark

Tor Nørretranders Onsgaardsvej 19, 2900 Hellerup, Denmark

Lars Folke Olsen Biochemical Institute, University of Odense, Campusvej 55, 5230 Odense M, Denmark

Kenneth S. Polonsky Department of Medicine, University of Chicago, Box 435, 5841 South Maryland Ave., Chicago. IL 60637, USA

Klaus Prank Medical School Hannover, Abteilung Klinische Endokrinologie, Konstanty-Gutschow-Strasse 8, D-8000 Hannover 1, Germany

Steen Rasmussen CNLS and T-13, MC B258, Los Alamos National Laboratory, Los Alamos, NM 87545, USA

Stefan Rauch-Wojciechowski Department of Mathematics, Linköping University, 581 83 Linköping, Sweden

Otto R. Rössler University of Tübingen, Institute for Physical and Theoretical Chemistry, D-7400 Tübingen, Germany

Andreas Schierwagen Karl-Marx-Universität, FG Neuroinformatik, Sektion Informatik, Karl-Marx-Plaz 10-11, Leipzig 7010, Germany

Jeppe Sturis Physics Laboratory III, Bldg. 309, Technical University of Denmark, 2800 Lyngby, Denmark

Alexander Tenenbaum University of Rome I, Dipartimento di Fisica, Piazzale A,\Moro 2, I-00185 Roma, Italy

Jussi Timonen Physics Department, University of Jyväskylä, Seminaarinkatu 15, SF-40100 Jyväskylä, Finland

Andrus Viidik Institute of Anatomy, University of Aarhus, Universitetsparken, 8000 Aarhus C, Denmark

Crayton Walker University of Connecticut, 65 Riverside Road, Mansfield Center, CT 06250, USA

INDEX